D0965893

THE ARTILLERY OF GETTYSBURG

Other Books by Bradley M. Gottfried

Brigades of Gettysburg: The Union and Confederate Brigades at the Battle of Gettysburg

Kearny's Own: The History of the First New Jersey Brigade in the Civil War

The Maps of Gettysburg: An Atlas of the Gettysburg Campaign, June 3–July 13, 1863

Roads to Gettysburg: Lee's Invasion of the North, 1863

Stopping Pickett: The History of the Philadelphia Brigade

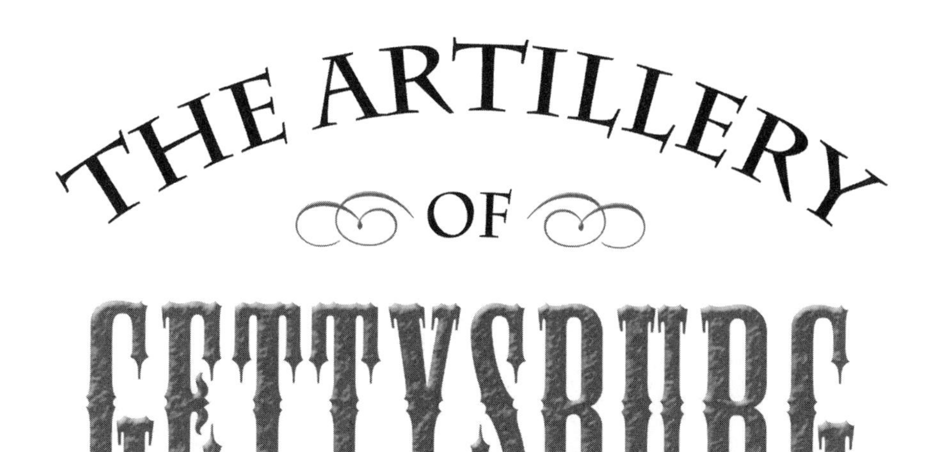

BRADLEY M. GOTTFRIED, PH.D.

CUMBERLAND HOUSE
NASHVILLE, TENNESSEE

The Artillery of Gettysburg
Published by Cumberland House Publishing, Inc.
431 Harding Industrial Drive
Nashville, Tennessee 37211

Cover design by Gore Studio Inc., Nashville, Tennessee

Library of Congress Cataloging-in-Publication Data

Gottfried, Bradley M.
The Artillery of Gettysburg / Bradley M. Gottfried.
p. cm.
Includes bibliographical references and index.
ISBN-13: 978-1-58182-623-4 (hardcover : alk. paper)
ISBN-10: 1-58182-623-0 (hardcover : alk. paper)
1. Gettysburg, Battle of, Gettysburg, Pa., 1863—Artillery operations. 2. United States—History—Civil War, 1861–1865—Artillery operations. I. Title.
E475.53.G66 2008
973.7'349—dc22 2007033198

Printed in the United States of America

1 2 3 4 5 6 7 8 9 10—10 09 08

To my daughters,
Mara and Emily

CONTENTS

MAPS

FOREWORD

ARTILLERY PLAYED A MAJOR role in the battle of Gettysburg, and some historians go so far as to claim that the Federal artillery's superiority in number, placement, mobility, handling, and better ammunition made victory possible. From the very beginning of the battle, when Brig. Gen. John Buford dispersed Lt. John Calef's horse battery to deceive Maj. Gen. Henry Heth's division into believing that the Confederates faced an overwhelming number of Federal troops, to its end, when massed Union artillery slaughtered thousands in the Pickett-Pettigrew-Trimble Charge, the Northern "long arm" played an indispensable role.

Artillery had already gained the respect of both armies by the summer of 1863. Used defensively, it could break up attack formations and change the face of battle. On the offensive, it could soften up enemy positions prior to an attack. Even without tangible results, the psychological effects of strong artillery support could both bolster and discourage infantry, depending on the situation. The need was mutual, for the gunners needed foot soldiers to support them, lest the guns be captured or decimated by enemy fire.

Because of the importance of artillery, the guns were the subjects of a tug of war over who controlled them in battle. This fight was waged throughout the war in varying intensity and involved senior infantry and artillery officers. It was most intense during the early years of the war, when brigade and divisional infantry commanders exercised control over artillery units in battle, and this often led to conflict with high-ranking artillery officers who, at least on paper, commanded the batteries. Artillery attracted the attention of infantry commanders whenever the enemy was beyond the range of their men's rifles. At such times, artillery could help to blunt an attack. As the distance closed, a commander's attention usually shifted to his infantry, sometimes to the complete exclusion of the artillery that he had so keenly watched only moments before.

Some historians have hypothesized that Union Maj. Gen. Joseph Hooker's decentralization of his artillery upon assuming command of the Army of the Potomac contributed to his grinding defeat at the May 1863 battle of Chancellorsville. Realizing the error of his ways after the battle, Hooker permitted Brig. Gen. Henry Hunt, who nominally commanded the artillery, to reorganize his batteries into a brigade system. Five batteries, usually numbering thirty guns, formed a brigade. Each infantry corps was assigned an artillery brigade. Two other artillery brigades reported to the cavalry, and five artillery brigades formed the reserve. While artillery brigade commanders continued to report to the commanders of infantry corps, Hunt maintained direct control over the artillery reserve, which contained 21 batteries, boasting 118 guns, or about one-third of the army's cannon. These could be deployed on battlefields as Hunt directed, and this command system contributed significantly to the Union victory at Gettysburg. Eleventh Corps artillery commander Maj. Thomas W. Osborn confirmed this after the battle, observing, "The artillery reserve proved all that could be expected or even asked of it; without their assistance I do not conceive how I could have maintained the position we held [on Cemetery Hill]."[1]

Robert E. Lee modified the Confederate Army of Northern Virginia's command system on February 15, 1863. Prior to this date, batteries were allocated to infantry brigades. Now he created artillery battalions and assigned one to each infantry division. Two additional artillery battalions formed the reserve in each corps. Each battalion usually contained four batteries of four guns each. The battalion commanders in turn reported to a chief of artillery assigned to the corps. The biggest difference from the Federal system was that reserve battalions were assigned to each corps, not to Lee's artillery chief, Brig. Gen. William Pendleton. Thus Pendleton was less powerful than Hunt, for the Confederate artillery chief had no direct control over any batteries and was forced to consult with divisional infantry and corps commanders prior to moving guns to needed areas elsewhere on the battlefield. This was certainly inferior to the Federal approach, but Lee probably adopted it to compensate for Pendleton's shortcomings as his artillery chief.[2]

During the Gettysburg campaign, the Army of the Potomac's superior artillery organization complemented its superiority in heavy metal, ensuring a mismatch between the two forces. The Federal artillery boasted 372 cannon in 67 batteries—far more than the Confederates' armament of 283 cannon in 70 batteries. The similarities in the number of batteries (but not the total number of guns) resulted from the majority of Federal batteries sporting six guns to the Confederates' four. While all agreed that six-gun batteries were more efficient, the Confederates had to make do with four because of scarcities in both horses and ordnance equipment.[3]

Not only did the Federals have more guns, they also had superior ammunition. According to one Confederate artillery officer, only 20 percent of their fuses functioned properly. While this was probably an exaggerated figure, there is little doubt that the inferior quality of fuses greatly hindered the effectiveness of Confederate artillery. Shells exploded too early, too late, or not at all.[4]

While the effectiveness of Confederate infantry was unquestioned, historian George Stewart disparaged Confederate artillerymen, writing that they lacked "professional skill." He attributed this to a chronic shortage of ammunition, which precluded target practice. Then there was the "disdain for mechanical details." Stewart wrote, "The Confederate ideal was to rush the guns as far forward as possible, and then go *bang-bang.*" That worked well in the wooded terrain of Virginia, but not in the open fields of Gettysburg.[5]

Civil War cannon fell into two categories. Smoothbores, as their name suggests, had smooth interior barrels, while rifled guns had grooved barrels. The latter was the more advanced technology and had the advantage of throwing a projectile farther and with greater accuracy. The 3-inch ordnance rifle contained a rifled barrel that could accurately throw a shell more than eighteen hundred yards. Great for artillery duels or firing at distant targets, it was much less effective at closer ranges, when canister (tin cans filled with metal balls packed in sawdust) was the ammunition of preference. The Napoleon had a smoothbore barrel that was perfect for close-in fighting. While its effective range of less than twelve hundred yards was considerably less than rifled cannon, the Napoleon was deadly against charging infantry because its larger bore dispersed its canister shot over a wider area, with an effect similar to a shotgun blast. The 10- and 20-pounder rifled Parrotts were also found in some batteries. Because of their propensity to burst, a reinforcing band was attached to the breech, giving these guns a distinctive appearance. While versatile and easy to manufacture, many gunners disliked these cannon because their barrels burst more frequently than those of other types of guns and there was also an added unpredictability in the reliability of their projectiles. The Army of Northern Virginia had too many obsolete howitzers—almost thirty. Although lightweight and easy to maneuver, they were limited in range and by the type of ammunition they could use. By this time in the war, the Union embargo had reduced the flow of cannon and ammunition into the Confederacy, causing captures on the battlefield to be the South's major source of new heavy metal. Such was the case earlier in the campaign, when the Confederate 2nd Corps captured twenty-three cannon at the battle of Second Winchester. A shortcoming of this approach was that units could retain the batteries they captured. There was some production of armaments in the South, however. For instance, Lee's army received forty-nine Napoleon-type cannon from Richmond's Tredegar foundry that were cast from obsolete 6-pounder guns.[6]

Because each type of cannon had different uses, artillery commanders had to ensure flexibility in handling different situations. For example, massing enemy troops a mile away called for rifled cannon, but as the enemy approached, the cannon of preference shifted to the smooth-bored Napoleon. The two armies handled this flexibility differently. Confederate batteries were composed of two or three different types of cannon, which permitted each to respond to rapidly changing situations, particularly when some were rifled and others were smoothbores. The negative side of this approach was the distinct possibility that a gun could be loaded with the wrong caliber of ammunition. This was not an unusual occurrence during the heat of battle and often led to a gun's being put out of service when it was most needed. It also complicated the resupply of these batteries. Chief of Artillery Pendleton had recommended a move to homogenous batteries prior to the battle of Gettysburg, but little had been accomplished when the armies were engaged on July 1. Federal artillery was organized differently, with each battery composed of only one type of cannon. Such an arrangement resulted in enhanced firepower and continuity of function, but it also reduced each battery's ability to handle rapidly changing situations and forced artillery brigade commanders to use more than one battery under these conditions.[7]

The cannon used at Gettysburg commonly fired four types of projectiles. Solid shot, or cannonballs, were usually fired by smoothbore guns and caused destruction when they bounded along the ground, smashing into men and equipment. The Federal army apparently had none of this type of ammunition on hand at Gettysburg for its rifled cannon. Instead, they used explosive ammunition, such as common shell, which they did not detonate. The effect was similar to solid shot and precluded the need to stock yet another type of ammunition in the ordnance wagons. This may explain why so many unexploded shells were found on the battlefield after the armies had left the area. Common shell, which could be spherical or conical, was hollow and filled with black powder. A fuse exploded the shell, causing deadly chunks of metal to fly in all directions. Canister charges consisted of large numbers of one-inch iron balls encased in an iron cylinder. When fired, canister acted like a shotgun blast, usually clearing the area immediately in front of the cannon. It was most effective at ranges of up to two hundred yards. As the enemy infantry closed on the artillery's position, gunners often fired double and even triple charges of canister to break the charge. Case shot, or shrapnel, which could also be round or conical, was loaded with small balls that scattered when the shell's fuse caused it to explode. Shrapnel was used instead of canister when firing at more distant targets, and the gunners tried to explode shells above or in front of the enemy's lines, particularly those that were stationary.[8]

Under the limber lid, gunners could find a printed "Table of Fire" that provided useful information. The following was placed under the limber lid of a 12-pounder gun:

> Use SHOT at masses of troops, and to batter, from 600 up to 2,000 yards. Use SHELL for firing buildings, at troops posted in woods, in pursuit, and to produce a moral rather than physical effect; greatest effective range 1,500 yards. Use SPHERICAL CASE SHOT at masses of troops, at not less than 500 yards; generally up to 1,500 yards. CANISTER is not effective at 600 yards; it should not be used beyond 500 yards, and but very seldom and over the most favorable ground at that distance; at short ranges (less than 200 yards) in emergency, use double canister, with a single charge. Do not employ RICOCHET at less distance than 1,000 to 1,100 yards.[9]

Between its limber (which transported the cannon) and caisson (which resupplied the limber from the rear), each gun took four chests of ammunition into the battle. For a Napoleon, a standard chest consisted of twelve shot, twelve spherical case, four shells, and four canister rounds for a total of 112 rounds of long-range ammunition. Caissons were oftentimes kept in the rear to avoid their destruction. They were used to replenish the limbers and periodically returned to the ammunition wagons for resupply.[10]

Not only did the Confederates suffer from inadequate fuses, fewer horses, and poor ordnance, they entered the Gettysburg campaign with less ammunition. Federal artillery averaged more than 250 rounds per gun; the

Confederates had fewer than 200. One modern historian estimated that the Army of the Potomac had enough ammunition after the battle to re-fight it two more times, while the Army of Northern Virginia had enough for only a day of heavy fighting.[11]

The seven cannoneers per gun played several roles during the loading and firing process. Number 1 sponged out the cannon's tube while Number 3 "thumbed the vent." Barrels became very hot during the firing process, and the vent emitted hot gases, so the latter gunner wore a piece of leather called a thumbstall on his left thumb. By closing the vent, no air could enter and potentially discharge the unignited power before being sponged out. During this process, Number 6 removed a round of ammunition and prepared the fuse. He then handed it to Number 7, who carried it to Number 5, who finally gave it to Number 2, who slid it down the muzzle. Number 1 then rammed the charge down the tube while Number 3 continued thumbing the vent. As Numbers 1 and 2 moved away from the barrel to outside of the wheels, Number 3 grabbed the handspike on the gun's trail and moved it while the gun was aimed. Number 3 pricked the powder charge through the vent with a "vent pick," then he stepped away while Number 4 affixed the friction primer to the vent. As Number 3 covered the vent again, Number 4 moved diagonally to the rear. When ready, Number 3 moved clear of the barrel, and at the command, Number 4 pulled the lanyard, discharging the gun. Numbers 1, 2, 3, and 4 then moved the gun back into position, and the loading process was repeated.[12]

When deploying their guns, artillery commanders looked for elevated, level positions with unobstructed fields of fire. The clear, level platforms allowed the gunners to easily maneuver their guns, bring up ammunition, and more easily reposition the guns after recoil. Cannoneers especially liked slightly uphill slopes (to the rear), which reduced the amount of recoil and hence the effort needed to roll the gun back into firing position. A level platform also reduced "cant," or the sidewise tilt of the barrels and axles that affected trajectories and accuracy. Since the guns were almost always vulnerable to infantry attack, it was imperative that the terrain be conducive to a quick getaway.[13]

Experienced gunners had become proficient in maximizing the number of enemy they killed or maimed by this time in the war. Cannoneers manning smoothbored guns fired round shot at the front of approaching infantry columns. The balls bounced through battle lines, bowling over and killing or wounding many infantrymen. Cannoneers often waited until the last moment to pull the lanyards of guns loaded with canister in order to maximize the destructiveness of the charge. Men manning rifled cannon tried to explode their shells directly over advancing columns. Given a choice, all cannoneers preferred to fire into the flank of an attacking column, rather than its front, which maximized the damage inflicted on the infantrymen.[14]

Many people helped make this book possible. Ron Pitman, president of Cumberland House, believed in the project from the beginning and was always accessible for questions. Ed Curtis did a great job of editing the volume. The staff of the Gettysburg National Military Park, particularly John Heiser, was most helpful in providing critical resources. The same is true of Dr. Richard Sommers and his staff at the U.S. Army's Military History Institute. I personally prepared the maps.

THE ARTILLERY OF GETTYSBURG

1

JULY 1

THE FIGHT ALONG CHAMBERSBURG PIKE

Both Sides Take Position

Confederate Maj. Gen. Henry Heth was in for a rude awakening on July 1, 1863. As a new division commander with limited combat experience, Heth believed that only Pennsylvania militiamen defended the town of Gettysburg. His men were fairly well rested after a leisurely march to Gettysburg from the Fredericksburg area. Most of the Army of Northern Virginia had begun the northward march on June 3, but Heth's division, along with the rest of Lt. Gen. A. P. Hill's 3rd Corps, remained near Fredericksburg until June 15, keeping watch on the Federals on the opposite side of the Rappahannock River. In Pennsylvania, Robert E. Lee's army had been fairly well dispersed until the Confederate commander learned that the Army of the Potomac, now under Maj. Gen. George Meade, was rapidly moving north through Maryland. This forced Lee to reassemble his troops, including Lt. Gen. Richard Ewell's 2nd Corps, which had been poised to capture Harrisburg, the Pennsylvania capital.[1]

Marching along Chambersburg Pike, Heth hoped to acquire needed supplies in Gettysburg. Maj. Gen. Dorsey Pender's division followed Heth's, along with Maj. David G. McIntosh's artillery battalion. Lacking cavalry to lead the column, Heth placed Maj. William J. Pegram's artillery battalion in this position, an act historian David Martin called "awkward and even dangerous." The artillerymen were not happy with this turn of events.[2] According to C. R.

Fleet of Capt. Edward A. Marye's Fredericksburg Battery, the battalion "marching behind our whole corps and were considered as in reserve but when the time came for fighting, we opening it." Fleet acknowledged that his battery was in front of the corps that day—"such chances as a batt'n *happening* to be in front after marching some two or three hundred miles in the extreme rear is rather extraordinary are they not?"[3]

No one expected a fight. Lt. John L. Marye of the Fredericksburg Battery recalled that the "morning was lovely. A soft, fresh breeze rippled over ripe wheat fields stretching away on either side of us. We moved forward leisurely smoking and chatting as we rode along, not dreaming of the proximity of the enemy."[4] Heth also discounted the idea of encountering anything but irregulars, so he left behind his own artillery battalion, commanded by Lt. Col. John J. Garnett, on the opposite side of Cashtown. Both Willie Pegram's and David McIntosh's artillery battalions were actually part of the 3rd Corps Reserve.[5]

Up against these fourteen thousand infantry and two battalions of artillery were about twenty-seven hundred cavalrymen from Brig. Gen. John Buford's division and six ordnance rifles of Lt. John Calef's Battery A, 2nd U.S. Maine. These lightweight guns (their wrought-iron barrels weighed only eight hundred pounds) were perfect for service with fast-riding cavalry. Although termed a "horse battery," it was popularly called "flying artillery." These guns differed from the artillery serving with the infantry in that the gunners rode horses rather than the limber and caisson chests, and none ever walked.[6]

Buford knew that two Federal infantry corps were approaching Gettysburg, so his goal on the morning of July 1 was to delay the Confederate advance for as long as possible. Buford was one of the new breed of cavalry officers who fought his men as infantry; horses were used only for rapid transport. When they were close to the enemy, his men dismounted, and three gave their reins to a holder. Similarly, Calef was expected to bring his guns to the battle line and fight close to the enemy. As a result, his caissons carried plenty of canister.[7]

It was appropriate that Calef's battery had the distinction of being the first Federal artillery on the field, for this battery had a long and distinguished history. Disbanded after seeing continual action in the Mexican War, the battery was one of the first to ride into Washington in 1861. As the war progressed, the battery saw action in every battle in the eastern theater, including First Bull Run. It also had the distinction of being the first to play "Taps" at a military funeral.[8]

Buford liked the looks of the high ground west of Gettysburg, so he deployed his cavalrymen there. After throwing out a skirmish line beyond Herr Ridge, he deployed Col. William Gamble's brigade along McPherson's Ridge, straddling both sides of Chambersburg Pike. Col. Thomas Devin's brigade formed on his right and extended the line to the north. Buford's third brigade, commanded by Brig. Gen. Wesley Merritt, was still in Maryland and of no use to him on this field. Knowing that he was up against a powerful Confederate force, Buford ordered Calef's battery spread out over an area more than six hundred yards wide. This deployment not only provided greater coverage of the field, Calef also saw that Buford was trying to deceive the enemy into believing he had more troops and guns than he actually had. The placement was actually prescribed in the artilleryman's "bible," John Gibbon's *Artillerist's Manual*: "A rifled battery when in support of a battle line, opposed by superior numbers of infantry (and artillery), will be split up and placed by the governing general to strengthen the line. These pieces will concentrate their fire toward the enemy's strongest point and display as much fire power as to convince the enemy there are more guns."[9]

Five officers should have been present to oversee the battery's actions, but battle losses reduced the number to just two. Calef's brother officer, Lt. John Roder, commanded a two-gun section that occupied the right of Chambersburg Pike. Sgt. Joseph Newman's section was on the left of the road. About six hundred yards farther to the left (south), just beyond McPherson's Woods and not far from where Union Gen. John F. Reynolds would fall later that morning, Calef posted Sgt. Charles Pergel's section between the 8th New York Cavalry and the 8th Illinois Cavalry.[10]

The first artillery rounds were not fired by Calef's guns, but by Capt. Edward A. Marye's Fredericksburg Battery (two ordnance rifles and two Napoleons), which led Pegram's artillery battalion toward Gettysburg. As the Confederates approached Marsh Creek on Chambersburg Pike (Marye says it was between 7:00 and 8:00 a.m.; C. R. Fleet recalled it was about 9:00 a.m.), Pegram spied Federal troops in the distance. The cannoneers initially believed the Federals were Confederates from Lt. Gen. James Longstreet's 1st Corps, but a newly arrived sergeant dispelled this idea; he had passed those infantrymen many miles back. Unable to ascertain the identity of the unknown troops, Pegram ordered Marye to unlimber a section of ordnance

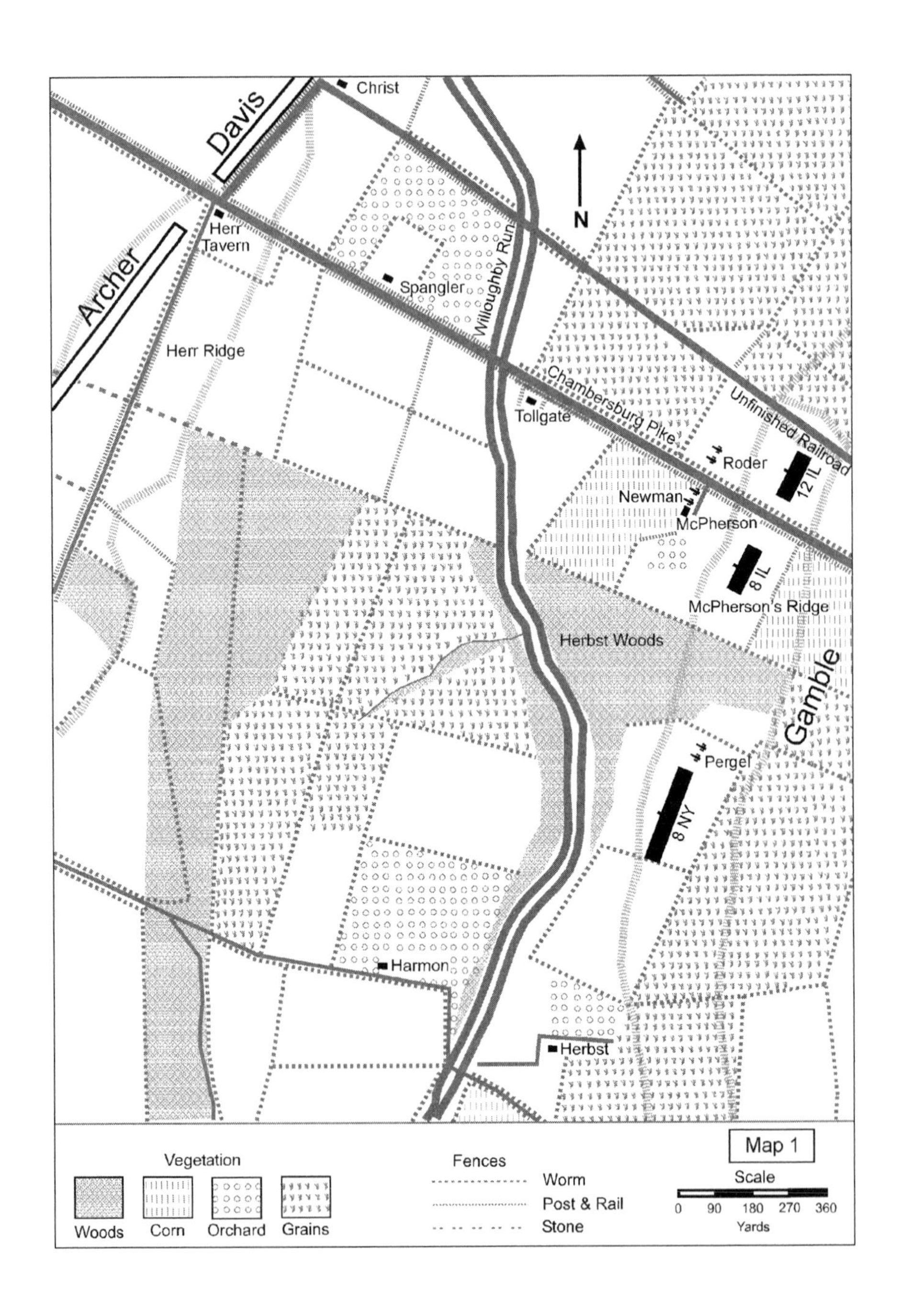

INITIAL CONTACT

rifles along the road.[11] According to the battalion's official report of the battle, a two-gun section of rifled guns was "unlimbered in the road, and opened on a piece of woods to the left of the pike, where was stationed a reconnoitering party of the enemy."[12] Fleet recalled that the two guns were unlimbered in front of a "brick building which looked like an old Virginia county court-house tavern."[13] Just as the order "Load with shrapnel shell" was given, a man rushed out of a nearby house and yelled, "My God, you are not going to fire here, are you?" When the civilian realized that he was having no impact, he threw up his hands and rushed back to his house just as the order to fire was given.[14]

The only response to the shrapnel shell was some Federal cavalry scurrying away in the distance. After firing eight to ten rounds, Archer's and Davis's infantry brigades (Heth's division, 3rd Corps) passed the battery and formed a line of battle farther down the road. The battery limbered up and followed with the rest of Pegram's battalion.[15] According to John Goolsby of Crenshaw's battery, "We sped over the hard, smooth road, the horses in a gallop and just before we reached the field a wheel of one of the guns rolled off right in the main road." Several cannoneers fell off their perch and went barreling along the side of the road, but no one was hurt.[16] There was good reason for the rush: Calef's artillery had opened fire up ahead.

The Confederates halted on Herr Ridge, where Pegram deployed his ten Napoleons and seven rifled guns in his five batteries. Pegram actually had twenty guns in his battalion, but two 12-pound howitzers were not used, and an ordnance rifle belonging to Lt. William E. Zimmerman's battery broke down while galloping into position. The actual positions of Pegram's batteries on Herr Ridge are not clear. It appears that Capt. Joseph McGraw's Purcell Artillery (four Napoleons) was to the left of Chambersburg Pike and Zimmerman's Pee Dee Artillery (four ordnance rifles) was to its left. Marye's Fredericksburg Battery (two Napoleons and two ordnance rifles) dropped trail near Herr Tavern. Lt. Andrew Johnston's Crenshaw Battery (two ordnance rifles and two 12-pounder howitzers) was on its right, and for at least a short time (see below), Capt. Thomas Brander's Letcher Artillery (two Napoleons and two 10-pounder Parrotts) was on his right. The battalion almost immediately drew fire from Calef's guns.[17]

The uniforms of the men serving in Crenshaw's and Marye's batteries could not have been more different. Like most other Confederates, the uniforms of Marye's men were well worn and tattered. Shoes were their greatest concern. Fleet described the "ancient and holey foot-gear . . . kept together by diligent care and sundry strings all through that tedious and muddy march."[18] In contrast, Crenshaw's gunners may have been the best-dressed artillerymen in either army. William G. Crenshaw, who formed the battery in 1861, had left his command to become a purchasing agent in Europe. He never forgot his men, sending a full uniform to each prior to the start of the campaign.[19]

Twenty-two-year-old Maj. William J. Pegram looked anything but a toughened artillery officer. In an army where facial hair predominated, Pegram's face was smooth. This, together with his thick glasses, made him look more like a student than a soldier. Yet both friend and foe respected him on the battlefield. It was said that he would open fire only when he could see the enemy. Since his eyesight was so bad, it meant that he fought his pieces close to the enemy. Confederate foot soldiers knew a fight was in the offing when they saw him. Many would yell, "There's going to be a fight. Here comes that damn little man with the 'specs'!" Perhaps Pegram's greatest fight was at Cedar Mountain, where his four-gun battery tangled with eighteen enemy guns for two hours. Only when high casualties prevented the guns from operating effectively did Pegram finally pull his battery back to safety. Pegram looked pale on the morning of July 1 while he surveyed the Federal positions in front of Gettysburg. His name was on the sick list; he had ridden the last ninety miles in an ambulance rather than miss this important campaign.[20]

Back on McPherson's Ridge, Lieutenant Roder's section of Calef's battery on the right (north side) of Chambersburg Pike spied a group of Confederate horsemen about three-quarters of a mile away and opened fire—the first fire from Federal artillery that morning. Sergeant Newman's section entered the fray soon after, opening fire on Pegram's battalion. Now deployed, Pergel's section on the left also opened fire on Pegram's guns. On paper, it was a mismatch: seventeen Confederate guns against four Federal pieces. But Calef's men held their own. Seeing his men so heavily outnumbered, Calef ordered the four guns to fire slowly and deliberately. He later described the sounds of the battle as the "demoniac 'whir-r-r' of the rifled shot, the 'ping' of the bursting shell and the wicked 'zip' of the bullet." Shells flew everywhere and losses mounted. Concerned about being overpowered, Calef rode over to calm, pipe-smoking Buford and was told, "Our men are in a pretty hot pocket, but my boy, we must hold this position until the infantry come up; then you withdraw your guns."[21]

Support was on the way. The first elements of Maj. Gen. John Reynolds's 1st Corps, Gen. Lysander Cutler's brigade (Wadsworth's division), began arriving after about ninety minutes, and most of the units took position on the right side (north) of Chambersburg Pike. Two Confederate brigades approached from the west—Brig. Gen. Joseph Davis's brigade on the north side of the pike, and Brig. Gen. James Archer's brigade on the south.[22]

Earlier that morning, Capt. Thomas Brander's battery (Letcher Virginia Battery consisting of two Napoleons and two 10-pounder Parrotts) reinforced Davis's brigade while it was about to launch an attack toward Gettysburg. The battery had the distinction of being composed of men "raised from Castle Thunder," Richmond's notorious wartime prison. Davis positioned the battery on a hill just to the left (north) of the Chambersburg Pike, immediately behind his skirmishers, about five hundred yards from Calef's battery. After dropping trail, the four guns immediately opened fire on Cutler's Federal infantry. Pvt. Samuel W. Hankins of the 2nd Mississippi watched the first shell arch over Cutler's line before exploding. "They were in the wheat, however," Hankins recalled, "lying down, though plainly seen, while their officers rode up and down their lines." Calef's battery responded in kind, sending a shell shrieking harmlessly over the gunners' heads.[23]

Calef's gunners soon found the range of Brander's cannoneers, whose exposed position on the top of the hill left them vulnerable. The Federal guns shifted to canister, and Brander's casualties quickly mounted, forcing the battery to seek shelter with the rest of Pegram's battalion on Herr Ridge, which was pounding Calef's gunners. A new threat materialized to Calef's gunners as Davis's Mississippi and North Carolina infantrymen continued their advance. Buford knew a desperate situation when he saw one and ordered the pieces withdrawn. Realizing that the enemy was too close for the guns to limber up and gallop to the rear together, Buford ordered the cannon removed "in each section by piece," which bought time for each gun's safe removal. Utter chaos swirled as enemy shells filled the air, unsettling the Federal infantry. A Confederate shell exploded under the horses of one of Newman's center section's teams, killing or disabling four out of the six animals. It looked as though this gun would be lost, but Newman's men quickly cut the harnesses and dragged the piece to safety with the remaining two horses and hard work. Newman won the rich accolades of the leaders on the field, including several generals from the 1st Corps, which was just entering the field.[24]

Riding back to his left section in McPherson's Woods, Calef saw a terrible sight—Archer's brigade had broken out of the woods and was advancing toward his two guns in a double line of battle. "Their battle flags looked redder and bloodier in the strong July sun than I had ever seen them before," he recalled. Calef ordered Pergel's gunners to aim at the battle flags. They quickly obeyed, throwing the infantry lines into disorder. Reforming, Archer's infantry continued its advance, overlapping the two guns by a considerable distance. Realizing the futility of waiting until the enemy came within canister range, Calef ordered the two guns to the rear to join the rest of the battery. Just then, the men could hear the strains of "The Campbells Are Coming," and looking behind them, they saw the black-hatted Iron Brigade swing into view. The fabled Union brigade quickly deployed on McPherson's Ridge and immediately took on Archer's Alabamians and Tennesseans as they crossed Willoughby Run. Calef never saw what happened next—how the left regiments of the Iron Brigade overlapped the right of Archer's, gained their flank and rear, and forced the Southerners to the rear with the loss of scores of men to capture, including Archer himself. Calef was too busy getting his battery to safety to see these events. The entire battery halted near the Lutheran Theological Seminary on the ridge that bears its name. Here the gunners replenished their ammunition chests.[25]

Buford's report glowingly described Calef's actions during this part of the battle: "Lieutenant Calef, Second U.S. Artillery, fought on this occasion as is seldom witnessed. At one time the enemy had a concentric fire upon this battery from twelve guns, all at short range. Calef held his own gloriously, worked his guns deliberately with great judgment and skill, and with wonderful effect upon the enemy."[26]

Hall's Battery Arrives

With Calef's battery gone, Pegram's gunners turned their full attention to the two Federal brigades attempting to hold the ridges west of Gettysburg. However, Capt. James A. Hall's 2nd Maine Battery (six ordnance rifles) galloped toward the action. It had camped the night before near Marsh Creek, about four miles from Gettysburg. Attached to Brig. Gen. James Wadsworth's 1st Division of the 1st Corps, the battery broke camp at 9:00 a.m. and dashed toward Gettysburg. So desperate was the situation that Hall moved his battery "through fields, gardens, and yards" toward McPherson's Ridge.[27] Capt.

J. V. Pierce of the 147th New York of Cutler's brigade noted, "Hall's splendid battery dashed past us. Horses with distended nostrils, sides white with foam, now wild with excitement, hurried to join the melee."[28] Hall met General Reynolds on Chambersburg Pike near the McPherson barn and was told, "Put your battery on this ridge to engage those guns of the enemy."[29] Reynolds also purportedly added, "I desire you to damage the Artillery to the greatest possible extent, and keep their fire from our infantry until they are deployed, when I will retire you somewhat as you are too far advanced for the general line."[30] Then, according to Hall's postwar account, Reynolds turned to Gen. James Wadsworth and told him that he was positioning the battery to attract the Confederate artillery's attention while his infantry continued deploying. Because of the exposed nature of this position, Wadsworth placed the 147th New York as infantry support on Hall's right.[31]

Although its history was shorter than Calef's, Hall's battery's pre-Gettysburg experiences were almost as illustrious. Formed in 1861, it was a well-trained unit that had shown its mettle on many battlefields. Hall was known for his coolness under fire. Abner Small of the 16th Maine observed this coolness at the battle of Fredericksburg, when a Confederate shell shrieked dangerously close to Hall, who was engaged in conversation with two infantry officers. Hall calmly dismounted, walked over to one of his cannon, and took careful aim at the offending enemy artillery. "The gun was fired and landed a direct hit, dismounting a rebel gun amid a cloud of torn earth and flying splinters. The battery commander walked back to his horse, mounted, and resumed the interrupted conversation as if nothing had happened," reported Small.[32]

As Hall's battery swung into position in the same spot that had been occupied by Roder's two-gun section of Calef's battery on the north side of Chambersburg Pike, his gunners could see a Confederate battery (probably McGraw's) directly in front of them, about thirteen hundred yards away. The enemy battery opened fire on Hall's battery as it unlimbered, but its shells sailed over the Federal gunners. Looking down at his watch, Hall noted that it was 10:45 a.m.[33]

Hall observed that the first six shots his battery fired at Pegram's battalion were exceptionally accurate, causing two of the enemy's pieces to move to the cover of a nearby barn. The battery continued banging away at Pegram's artillery for what Hall thought to be about twenty-five minutes. Suddenly, a line of Confederates appeared within fifty yards of his position. They were from the 42nd Mississippi of Davis's brigade, slowly advancing through the high wheat in front of the battery. Hall ordered the gunners to change to canister, blasting gaping holes in the Mississippians' lines. These green troops, in battle for the first time, were ordered to spread out in a skirmish line to reduce the effectiveness of Hall's guns. "Hall's Battery had been fighting that skimish line in a death grapple," noted Capt. J. V. Pierce of the 147th New York. "Artillery against skirmishers is like shooting mosquitoes with a rifle," he added.[34] Another threat materialized on the right, when additional Mississippians approached under the cover of an unfinished railroad cut to within sixty yards of Hall's right gun. They opened fire, killing and wounding a number of gunners and horses. Hall quickly swung the right and center sections of his battery around and immediately began throwing double-shotted canister loads into the enemy's ranks. The Confederates broke off their attack, but any satisfaction Hall may have felt suddenly turned to dread when he saw his infantry support breaking for the rear. To make matters worse, enemy infantry were taking refuge in the railroad cut and sniping at Hall's men and horses. Hall did not have orders to withdraw, but he reasoned, "If the position was too advanced for infantry it was equally so for artillery." He did not know that the three regiments of Cutler's Federal brigade north of Chambersburg Pike were whipped and heading east toward Gettysburg.[35]

Deciding to withdraw was the easy part—the hard part was deciding how to do it in the face of rapidly advancing Confederate infantry. Hall decided to withdraw the most vulnerable guns first—Lt. William Ulmer's right section. He ordered it to fall back seventy-five yards, drop trail, and cover the withdrawal of the other four guns. Hall did not know it, but other enemy infantry, primarily the 55th North Carolina from Davis's brigade, were far beyond his right. When Ulmer's two guns pulled back, they were hit by small-arms fire that killed or disabled four horses from one gun's team. The gunners managed to drag the gun off by hand, but they were unable to undertake the prescribed assignment to cover the retreat of the four other guns. Hall quickly devised an alternate plan. Not wanting to jeopardize the limbers, he ordered them to reverse, causing them to face the town. The guns were attached and opened fire as the horses slowly pulled the guns to safety. According to Hall, "Under cover of the smoke I had the guns taken down the slope by hand to limber." A stout fence between Chambersburg Pike and the unfinished railroad cut convinced Hall to make his retreat along the strip between these two features. The danger to Hall's guns was

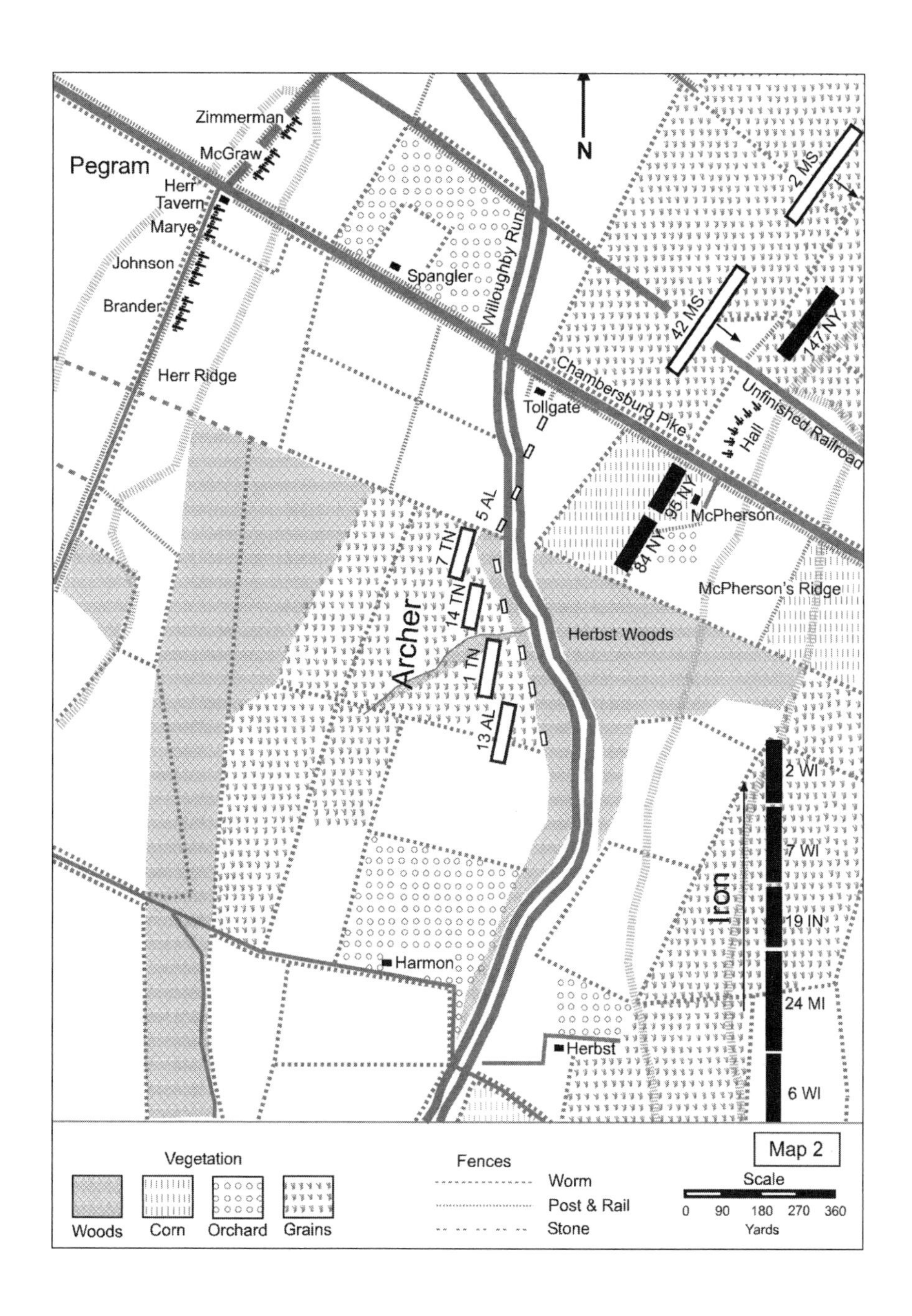

THE MORNING FIGHT FOR McPHERSON'S RIDGE

not yet over. They still had to cross a narrow fence opening on Seminary Ridge that would permit only one gun to pass at a time. It was slow going, permitting Davis's infantrymen to catch up with the gunners. All made it to safety except for the last gun, whose horses went down in a hail of bullets when the Confederates opened fire. With the enemy almost upon them, Hall ordered the gun abandoned.[36]

In the end, Hall was furious. He felt betrayed by Cutler's infantry, who had jeopardized the safety of his battery. With the loss of one cannon and damage to two others, Hall fell into a depression. One of the disabled guns was moving to the rear hitched to a caisson because all of its horses were disabled, and a second bumped along the road with one wheel. This left Hall with only three serviceable guns. Spying Wadsworth, he stormed up to the general and blurted out that it was cowardly for his infantry to abandon him in that situation. Wadsworth merely replied, "Get your guns back to some point to cover the retiring of these troops." Wadsworth meant closer to the town, but Hall responded, "This, General, is that place, right here in the road [Chambersburg Pike on Seminary Ridge]." Wadsworth told him to obey his orders, because the position was untenable, and then yelled, "Loose [*sic*] no time in getting your guns into position to cover the retreat."[37] Before obeying these orders, Hall sent five men back to attempt to retrieve the lost gun. He never saw these men again; all were either wounded or taken prisoner. Hall was actually lucky that he lost only one gun—all could have been lost during their run through the gauntlet of small-arms fire during their journey to safety.[38]

While Hall's men were in action on McPherson's Ridge, Calef's battery was at the Lutheran Theological Seminary, where the men either rested or replenished their ammunition chests. They had not been here long when Buford ordered one gun back to enfilade the unfinished railroad cut now occupied by Davis's brigade. Calef selected John Roder's right piece and ordered it to the designated spot. After driving Cutler's brigade and Hall's battery from the north side of Chambersburg Pike, hundreds of Davis's Mississippians had jumped into the unfinished railroad cut to avoid the devastating volleys coming from three Federal regiments firing into it from the south side of Chambersburg Pike. The railroad cut proved to be a trap: it was too steep for the men to get out, and Federal troops plugged the ends.[39]

As Calef's piece unlimbered just east of the section of the railroad cut that harbored hordes of Davis's men, the Confederates yelled out, "There is a piece—let's take it!" Cpl. Robert Watrous quickly grabbed a round of canister from the chest and ran toward the gun. A bullet slammed into his leg before he reached it, throwing him hard to the ground. Pvt. Thomas Slattery grabbed the round, and within a matter of seconds, the gun fired. "Some of them [Confederates] were so close that when the piece was fired they were literally blown away from the muzzle," Calef reported. He attributed saving the gun from certain capture to Slattery's bravery. The gun fired again and again into the crowded trench, and scores of Confederates fell. Vowing vengeance, a large group of Southerners again approached the lone gun. Just when it appeared that all was lost, three Federal regiments (the 6th Wisconsin and the 84th and 95th New York) launched their attack from Chambersburg Pike, distracting the approaching foe and allowing Roder to hitch his gun and gallop to safety. Although Roder's piece did not spend much time battling the enemy, it helped distract them from capturing portions of Hall's battery.[40]

Midday Artillery Actions

A calm descended on the battlefield. The morning belonged to the Federals, as they had decisively crushed two Confederate brigades (Archer's by the Iron Brigade along the banks of Willoughby Run and Davis's in the unfinished railroad cut). Two Union batteries (Calef's and Hall's) had essentially taken turns battling an entire Confederate battalion. This artillery mismatch was to change when Col. Charles S. Wainwright, commander of the 1st Corps artillery brigade, arrived on Seminary Ridge with four of his batteries at about 11:15 a.m. The fifth battery, Hall's, was already on the battlefield. There wasn't much for Wainwright to do. He later noted, "I put my batteries in a position where they could be got at easily, have them send their battery wagons and forges to the rear, and wait in condition to start at a trot the instant orders came."[41] Wainwright initially massed his batteries at the rear of Seminary Ridge and then rode forward to reconnoiter. Lt. George Breck of Capt. Gilbert H. Reynolds's Battery L, 1st New York, observed that "clouds of cavalry skirmishers, which having been relieved by the infantry, were falling back down the hillsides."[42]

Thirty-seven-year-old Charles S. Wainwright was a wealthy farmer from the Hudson Valley region of New York. With no prior military experience, Wainwright entered the service as a major in the 1st New York Artillery

Regiment and in September 1862 became the 1st Corps artillery chief. A bachelor, Wainwright often had contentious relationships with others, who regarded him as a snob. He had little regard for the infantry and despised the fact that their officers, with little technical knowledge and less concern for the well-being of the guns and cannoneers, could direct the placement of his batteries.[43]

Confederate artillerist Pegram's battalion's seventeen guns also received assistance during the midday lull when Maj. David McIntosh's battalion, accompanying Maj. Gen. Dorsey Pender's division, arrived on Herr Ridge at about 11 a.m. and quickly deployed. Like Pegram's battalion, McIntosh's sixteen guns were part of the 3rd Corps artillery reserve, commanded by Col. R. Lindsay Walker. Capt. R. Sidney Rice's (Danville Artillery) battery of four Napoleons and a section of Whitworth rifles from Capt. William B. Hurt's (Hardaway Artillery) battery joined some of Pegram's batteries on the right (south) side of the road, while Lt. Samuel Wallace's (2nd Rockbridge Artillery) battery was posted just to the left (north) of the road. Farther to the right, on a "commanding hill" near Fairfield Road, McIntosh posted Capt. Marmaduke Johnson's (Richmond) battery of two Napoleons and two ordnance rifles and another section of Hurt's battery, consisting of two ordnance rifles. Historian David Martin noted that the guns were probably posted on the edge of the woods, about one-quarter mile west of the Harman farm.[44] McIntosh noted that the latter six guns "remained during the first day's action without any occasion for an active participation, though frequently under fire."[45]

The two breechloading Whitworth guns in Hurt's battery, imported from England, could throw a "bolt" more than five miles.[46] During this phase of the battle, McIntosh indicated that the Whitworths were used to "shell the woods to the right of town that may have sheltered some men of the newly arriving XI Corps." A South Carolina attorney, McIntosh entered the war in command of a company in the 1st South Carolina infantry. He became an artillerist when his company was transferred to that arm of the service to become the Pee Dee Artillery. McIntosh's abilities were noticed early, and he commanded an artillery battalion in Thomas J. "Stonewall" Jackson's corps. With the reorganization of Lee's army after Chancellorsville, McIntosh was placed in command of a reserve battalion in the 3rd Corps.[47]

Confederate historian Jennings Wise was especially critical of the use of the Southern artillery during the early phases of the battle of Gettysburg. He noted that the thirty-three guns of Pegram's and McIntosh's battalions, "though well up, were unable to gain positions from which to prepare the attack before Heth launched his brigades, and the batteries were left to act as best they could, without any definitive plan or objective." Wise believed that had 3rd Corps commander A. P. Hill placed and utilized his artillery to its best advantage, the guns could have smashed Buford's cavalry before Reynolds's infantry arrived. Both Henry Heth and A. P. Hill were new to divisional and corps level command, respectively, and it showed during the morning hours of July 1. Neither man's performance improved as the battle progressed.[48]

While the batteries of Pegram's and McIntosh's battalions were in action most of the day, Hill permitted John Lane's, William T. Poague's, and Garnett's battalions, totaling forty-eight guns, to rest along Chambersburg Pike far to the rear. They moved somewhat closer at about 11:00 a.m., but only one of Garnett's batteries, Capt. Victor Maurin's (Donaldsville) battery, came into action, when at noon it relieved one of Pegram's batteries, which had expended its ammunition. Maurin's battery was in action through the midafternoon, when the enemy finally vacated McPherson's Ridge. Lt. Col. John Garnett did not believe that the rest of his battalion would be called into the fray, so he permitted the drivers to unhitch their teams so the animals could graze in a nearby clover field. Altogether, it was a miserable performance by Lee's artillery commanders and the generals who commanded them.[49]

A bizarre incident played itself out during the morning of July 1. Upon his arrival in the vicinity, battalion commander Lt. Col. John Garnett was, in his own words, "relieved, and became directly subject to your [3rd Corps artillery chief Col. R. Lindsay Walker's] orders." It was a strange time to remove a battalion commander. A twenty-four-year-old Virginian and West Point graduate, Garnett began the war as a lieutenant in the Washington Artillery. As a major, he commanded the artillery assigned to David R. Jones's division in 1862. Garnett later became inspector of ordnance and artillery in Longstreet's corps, and after Chancellorsville, he jumped to battalion command with the rank of lieutenant colonel. At Gettysburg, Garnett was replaced by Capt. Victor Maurin, although Garnett apparently remained close by, as he was able to file the battalion's report on the campaign. Garnett permanently lost his battalion after the battle because artillery chief William Pendleton believed that Garnett had "proved unsuited to artillery service, despite his training and the high expectations of all."[50]

Pegram's and McIntosh's guns maintained a slow but effective fire through midday, while the infantry of both armies were content to rest. The two battalions' rate of fire increased whenever fresh Union troops approached the defensive line. Such was the case when Col. Roy Stone's brigade (Doubleday's division, 1st Corps) arrived on Seminary Ridge between 11:00 and 11:30 a.m. Gen. Abner Doubleday decided to give the men a speech. They fervently hoped it would be a short one as shells from Pegram's and McIntosh's guns screamed overhead.[51] The brigade was shelled again as it marched to its assigned position on McPherson's Ridge, just to the left (south) of Chambersburg Pike (facing west). Francis Jones of the 149th Pennsylvania observed a "continuous shower of six inch solid shot come over . . . and go on until they struck the ground in our rear and continued their onward rolling and bounding, cutting down men, horses and fences until they passed out of sight, while shells were also bursting all round us."[52]

When Confederate Col. Thomas H. Carter's artillery battalion arrived on Oak Hill to the north (see chapter 2) and opened a deadly enfilading fire on Stone's brigade on McPherson's Ridge, Stone shifted his position from facing west to facing north, along Chambersburg Pike, to respond to this new menace. The men quickly took cover along the south side of the roadway. According to John Bassler of the 149th Pennsylvania, the change in the brigade's orientation was like "jumping out of the frying pan and into the fire for the commander of a rebel battery on the pike [probably McIntosh's battalion], westward, and nearer than the other [on Oak Hill] caught a glimpse of the regiment over the swell of ground before the men lay down, and at once opened a cross-fire upon us."[53] A shell exploded in the midst of Company B of the 149th Pennsylvania, killing three and wounding five. A soldier hopped up to the regiment's commander, Lt. Col. Walton Dwight, and exclaimed, "I am killed, I am killed." Dwight unsympathetically snapped, "The hell you are killed, go back to your place." The soldier dutifully returned to his comrades, stretched out on the ground, and died.[54]

Realizing that there were no Federal batteries in position around midday, Doubleday (now commanding the 1st Corps after Reynolds's death shortly after his arrival on the field that morning) called upon Wainwright to place a battery on the "outer ridge" (McPherson's Ridge). Wainwright initially sent the six ordnance rifles of Capt. Gilbert H. Reynolds's Battery L, 1st New York, trotting toward the action, but he quickly countermanded the order when he realized that its position would be exposed and without proper infantry support. This meant that Wainwright blatantly disregarded the direct orders of an infantry corps commander.[55]

The battle was not yet over for James A. Hall's battery. Just as his three cannon were taking position on Cemetery Hill, Hall received another order from his nemesis, Gen. James Wadsworth. This time, Hall was to take the remnant of his battery and gallop west toward Seminary Ridge and deploy along the unfinished railroad cut where Col. Roy Stone's infantry was making a stand.[56]

Hall's unit dutifully advanced toward the enemy lines for about twelve hundred yards in the railroad cut, all the while being pounded by an intense artillery barrage. There was no turning left or right—his men could only gallop straight ahead. Less-trained cannoneers could not have weathered this storm, but Hall's men did. Finally reaching Seminary Ridge, Hall looked in vain for someone to tell him where to position his battery. He finally decided to park it under the cover of the ridge while he rode ahead. He had not gone far when he encoutered another Wadsworth aide, who told him to continue forward, passing over the crest of the ridge, across a ravine, and into a position near the spot he had occupied earlier in the fight.[57]

Hall could not have felt comfortable with these orders for he knew the enemy was nearby. As soon as the battery ventured forward, another orderly galloped up and screamed that the gunners were approaching the enemy's lines and liable to be captured. As if to confirm this information, Confederate skirmishers, probably from Brig. Gen. Junius Daniel's North Carolina brigade (Rodes's division, Ewell's 2nd Corps), opened fire. Utterly confused, Hall ordered his battery back to the safety of the seminary and again tried to determine where to position his guns. Col. Charles S. Wainwright happened by just then. He agreed with Hall about the danger of his assigned position. "I took the responsibility of forbidding him to put his battery there until I knew there were troops to cover his right flank," Wainwright wrote in his diary.[58] He was still fuming about Hall's earlier loss of a gun and was upset with Wadsworth for sending the guns into further danger. With no sign of his superior officer's wrath subsiding, Hall decided to take a sergeant to try to retrieve the gun he lost earlier.[59] After he found it, he enlisted the aid of members of the 6th Wisconsin, who pulled the cannon closer to the road. An infantryman noted that Hall "thank[ed] us a thousand times."[60] Hall then returned to the rest of the battery and led it to the left of Evergreen Cemetery on Cemetery Hill.

According to historian R. L. Murray, the disagreement between Wainwright and the infantry commanders was due to conflicting priorities. Infantry officers usually wanted to break up the batteries and use the guns to support their troops. Artillery officers, however, were more interested in counterbattery actions and therefore wanted to mass their guns in the most favorable positions. These differing priorities often led to conflict.[61]

His order for artillery support countermanded by Wainwright, Gen. James Wadsworth cast about desperately for assistance. If he couldn't order Wainwright's guns forward, he would again call up John Calef's battery. He told Calef to take two sections of his battery and return to McPherson's Ridge. Calef knew this was suicide, and he strenuously protested the order. Wadsworth was in no mood to put up with another troublesome artillery officer at this point, and he threatened to remove Calef from command if he did not obey the order. A regular army officer, Calef knew he had no choice but to obey. Ordering his men forward, he recalled seeing two abandoned guns as he approached the assigned position, which he claimed were from Hall's battery. Scores of dead and wounded soldiers of both sides were all around the guns. Many wounded men beseeched the gunners not to run over them, so Calef directed his men to drive carefully. Calef quickly realized that the area was devoid of infantry support as he positioned his guns. But orders were orders, and Calef ordered his guns to open fire. His pieces almost immediately became the target of a hail of missiles from three Confederate batteries in front of him and to his left. "In this spot I had most men wounded," Calef later wrote. Infantry support soon arrived in the form of the 6th Wisconsin of the Iron Brigade and the 84th New York of Lysander Cutler's brigade. The only nonchalant beings around the battery were the horses, who were being fed oats left behind by Hall's battery. The horses "ate them with as much relish and little concern as though they were at the picket-rope merely raising their heads if a shell burst near, some of them being killed while munching their grain," noted Calef after the war.[62]

After being pounded for fifteen minutes, Calef saw a Confederate battery to his right (probably Brander's) move farther in that direction to enfilade his guns. Now in a deadly crossfire, Calef moved his guns approximately five hundred yards south, where they dropped trail on the east side of Herbst Woods. This position offered more protection from the Confederate guns on Herr Ridge. Calef miraculously lost no guns during this short but intense encounter; twelve of his men were wounded and thirteen horses were killed.[63] Watching Calef's battery in action, Col. Charles S. Wainwright grudgingly acknowledged that it was superior to any of his 1st Corps batteries. "It could not be expected that the crack Batt'y of the army should be outdone," he noted in his journal.[64]

Watching the enemy's movements through his field glasses, Capt. E. B. Brunson of Pegram's battalion noted that his artillery fire was so effective that it "forced them [the enemy's guns] to limber up and retire their pieces three distinct times." Minimizing the Federals' abilities and bravery, Brunson noted, "They were brought back twice under shelter of the hills, in order to support their advancing infantry, whose lines our guns played upon as they advanced with telling effect." In actuality, Brunson was seeing Calef's guns being replaced by Hall's battery, then by Calef's again, and finally by Reynolds's battery.[65]

Calef never mentioned it, but Capt. Gilbert H. Reynolds's Batteries L & E, 1st New York Light Artillery, with six ordnance rifles, arrived before he left his exposed position north of Chambersburg Pike. Wainwright had sent the battery forward to this hazardous position. Although he continued believing it to be an "ugly place," he realized that Calef's battery was in a dangerous position and needed support. He sent Reynolds to this position on the condition that he would not be required to report to Wadsworth. Even before they could unlimber, Reynolds's guns were hit by enfilading fire from Pegram's and McIntosh's guns on Herr Ridge and from Carter's battalion on Oak Hill. It was only a matter of time before the concentrated enemy crossfire forced the battery back to a more secure location. Reynolds went down with a severe eye injury during the withdrawal, but he refused to leave the field. Lt. George Breck ultimately replaced him.[66] The battery's new position, according to Wainwright, was at "right angles to the ridge [McPherson's Ridge] so that its left was covered by the woods."[67] This was actually next to Calef's position, just east of McPherson's Woods, about five hundred yards to the southeast. The two batteries now opened fire on Carter's battalion on Oak Hill, while their flank was protected from Pegram's and McIntosh's guns on Herr Ridge. The guns remained here, banging away at Carter's, for about an hour. They also fired on the infantry of Rodes's division, who were deploying on and around Oak Hill. Gen. Junius Daniel's brigade on Rodes's right flank received particular attention. One shell killed or wounded nine men of the 2nd North Carolina Battalion.[68]

Roy Stone's brigade, which faced Junius Daniel's brigade, occupied the sector between the unfinished railroad cut and Chambersburg Pike. This site was particularly attractive to several batteries in Pegram's and McIntosh's battalions. Capt. Francis Jones of the 149th Pennsylvania recalled that the sector became a no-man's-land because "the rain of bursting shells and bullets was so thick . . . that the entire hayfield was mown down as if a scythe had cut it off."[69]

Meanwhile, Wainwright's other batteries were arriving. When Capt. James H. Cooper's Battery B, 1st Pennsylvania, of four ordnance rifles arrived at about noon, Wainwright directed it to deploy in an oat field between Seminary and McPherson's ridges, just southwest of the seminary. Wainwright believed this would be the approximate left of Doubleday's 3rd Division, now under Brig. Gen. Thomas Rowley, which was now arriving. Two other batteries, Capt. Greenleaf T. Stevens's Battery E, 5th Maine, and Lt. James Stewart's Battery B, 4th U.S. Artillery, boasting a total of twelve Napoleons, arrived next and were placed on Seminary Ridge as they were not effective for the needed long-range firing against the Confederate positions on Herr Ridge.[70]

Almost as soon as Col. Chapman Biddle's infantry brigade (Rowley's division) swung into position between McPherson's and Seminary ridges, it was hit by Confederate artillery fire. An infantry officer in the 80th New York called it a "warning flight of artillery missiles [that] showed that we were seen by the enemy."[71] Cooper's battery was actually within Biddle's line, with infantry regiments deployed on either side. Cooper did not waste any time, ordering his gunners to immediately fire on the Confederate guns on Herr Ridge. Biddle's infantry were probably most unhappy with this exchange, for they were in the middle of it. Shells burst all around them, and casualties mounted. After Cooper's guns had fired about twenty-five rounds, a battery in Thomas H. Carter's battalion, probably Charles W. Fry's on Oak Hill to the right, took up the firing at about 1:30 p.m. Wainwright, who was nearby, immediately ordered Cooper to "change front to the right" and return fire. One of Cooper's guns snapped an axle during its rapid shift in position, but the other three smartly completed the maneuver. After firing a few minutes, Wainwright ordered the battery back to Seminary Ridge, probably because he misunderstood a conversation. Overhearing a German aide say that Cemetery Hill must be held at all costs, Wainwright confused it with Seminary Ridge and pulled Cooper's guns back to protect it. The guns did not remain idle, however, as they opened fire on Thomas Brander's battery and Junius Daniel's brigade along McPherson's Ridge north of Chambersburg Pike. Some of the cannoneers later claimed that they also saw Alfred Iverson's brigade charge from Oak Hill and lobbed some shells into it. Later, the three guns threw shells at the 45th North Carolina and the 2nd North Carolina Battalion of Daniel's brigade, which were charging toward the unfinished railroad cut in front of Stone's brigade along Chambersburg Pike.[72]

Additional help for the beleaguered Federals was on the way. Gen. John Robinson's division began arriving in the early afternoon, and its first brigade, under Gen. Henry Baxter, was hurried forward to Oak Ridge. Wainwright directed Lt. James Stewart, whose battery had been resting about two hundred yards south of the seminary, to report to Robinson. The battery dropped trail north of Chambersburg Pike on the southern portion of Oak Ridge, which is the name of Seminary Ridge north of the roadway. Stewart divided the battery, placing three guns on either side of the unfinished railroad cut, all facing west. He retained direct control over the guns on the right; Lt. James Davison commanded the guns on the left. The railroad cut caused a hundred-yard gap between the two portions of the battery, which were supported by the remnants of Lysander Cutler's brigade. Capt. Greenleaf T. Stevens's 5th Maine Battery took Stewart's place to the north of the seminary at about 2 p.m.[73]

Stewart's battery initially shelled Daniel's brigade as it swept down from the north to attack Stone's brigade along Chambersburg Pike at about 2:00 p.m. The battery was most effective when it opened fire with canister on the North Carolinians who had taken refuge in the unfinished railroad cut.[74]

The North Carolinians were not the only troops to occupy the railroad cut that afternoon, for the 149th Pennsylvania of Stone's brigade also occupied it during the seesaw fighting in this sector. The movement of the Pennsylvanians into the railroad cut did not go unnoticed. Maj. David G. McIntosh reported that "a fine opportunity was also afforded at this time of enfilading a heavy column of the enemy's infantry, formed in the railroad cut and along a line of fence [along Chambersburg Pike], which was employed to advantage by my batteries in connection with Major Pegram's [probably Thomas Brander's battery], and the enemy, entirely discomforted, disappeared from the field."[75] Watching the battery in operation, C. R. Fleet

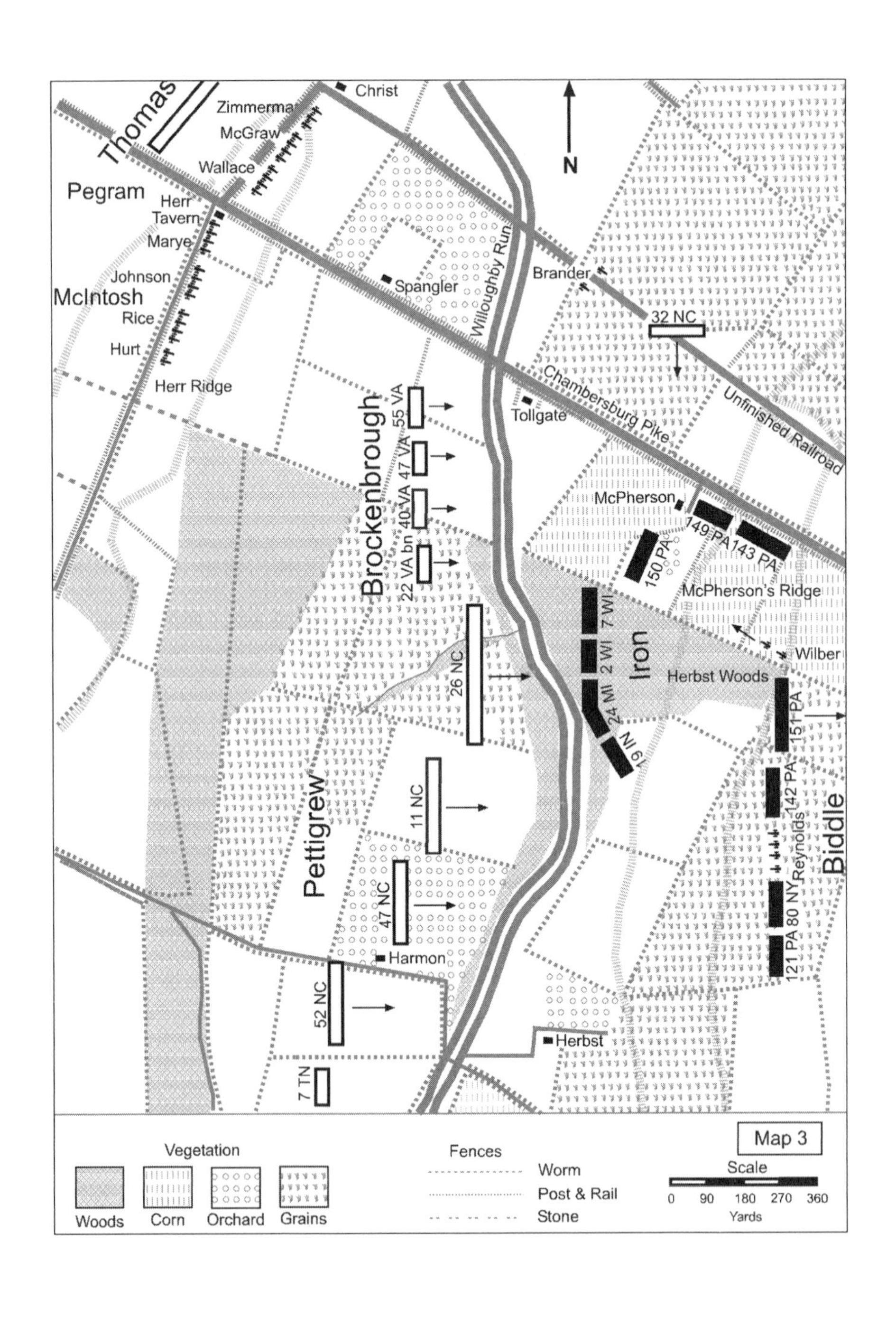

THE AFTERNOON FIGHT FOR McPHERSON'S RIDGE

of Marye's battery wrote, "Two or three shots from Brander's Battery brought the infantry out in 'rough-roll-and-tumble' fashion." Most amusing was Martin Douglas, a large Galway Irishman, who fired gun number four. "Before pulling his lanyard he would, every time, cross himself and mutter, 'Lord, be marsiful to their poor souls,'" noted Fleet.[76]

Wadsworth needed some guns near his troops on McPherson's Ridge, so Wainwright swallowed hard and dispatched Lt. Benjamin Wilber's section of Reynolds's battery to the right of the Iron Brigade, just north of McPherson's Woods. The remaining two sections were sent around the south side of the woods in the open fields between Seminary and McPherson's ridges to take the position vacated by James H. Cooper's battery within Chapman Biddle's brigade. The 142nd Pennsylvania was on its right and the 80th New York on its left. Here it prepared to take on the next Confederate attack.[77]

The Afternoon Fight for McPherson's and Seminary Ridges

Union Col. Charles S. Wainwright closely watched a long line of Confederates advancing against the Federal position on McPherson's Ridge at about 3:00 p.m. This line of battle was composed of Col. John Brockenbrough's and Gen. James Pettigrew's brigades of Heth's division, attempting to break the Union line south of Chambersburg Pike. These brigades had waited in the rear while their brother brigades—Archer's and Davis's—were mauled earlier in the morning. Maj. Gen. Dorsey Pender's division also waited in the rear. According to Wainwright, Pettigrew's men "marched along quietly and with confidence, but swiftly." Most distressing was that they appeared to overlap the Federal left by about half a mile.[78]

A cannoneer in Greenleaf T. Stevens's battery on Seminary Ridge tried to describe the chaos:

> The whole line of battle from right to left was then one continuous blaze of fire. The space between the two ridges was completely filled with the thin blue smoke of the infantry, making it difficult to distinguish friend from foe, while the artillery from their higher position belched forth a tremendous fire of shot and shell, throwing their deadly missiles in rapid succession into the ranks of the enemy advancing on our direct front, covering themselves for the moment in dense clouds of white smoke.[79]

Not many men from Pettigrew's brigade survived the battle of Gettysburg to describe their attack on the Union position on McPherson's Ridge that afternoon. However, a lieutenant of the 47th North Carolina reported that, prior to crossing Willoughby Run, the Federal artillery "at every step rakes through our lines, cutting great gaps, which are quickly filled up by our boys."[80]

It was a mismatch, as the Federal brigades in this sector, the Iron and Biddle's and part of Stone's, faced two powerful Confederate divisions, Heth's and Pender's, approaching from the west. With the Federal line collapsing, this was no place for Lieutenant Wilber's two-gun section, so it hastily withdrew, possibly before even firing a shot. On the opposite side of Willoughby Run, the Confederate batteries on Herr Ridge maintained a slow but steady fire on the Union line. Sergeant Gochenour of Rice's battery wrote that "whilst the infantry advanced in front and flank, we played upon them with fire from our position."[81] The Confederate infantry was advancing so rapidly that it is doubtful whether these thirty-three guns had much of an impact on this initial phase of the attack.[82]

The Federal artillery was poorly managed during this part of the battle. While the Confederate batteries were massed on Herr Ridge, their Federal counterparts were fed into the battle piecemeal and driven off one by one. Wainwright was to blame because he did not trust the abilities of Abner Doubleday and James Wadsworth, and he acted accordingly. Given Wainwright's hesitancy to throw his batteries into the fray, it is not surprising that their losses were fairly small at this time.[83]

When they saw the Iron Brigade's line evaporate, Chapman Biddle's infantry and Gilbert H. Reynolds's artillerymen expected the worst. They did not have long to wait until the Confederate infantry burst into view, and Reynolds's cannoneers nervously watched and waited for orders to open fire. Wainwright was more preoccupied with his own infantry at this time, as they "persisted in getting in front [of the battery]—that being its commander's idea of supporting." Even without this hindrance, Wainwright quickly realized the futility of the situation and noted, "There was not the shadow of a chance of our holding this ridge [McPherson's Ridge] even had our Third Division commanders had any ideas what to do with their men, which they had not." Given the interference from their own infantry and the rapid approach of the enemy, Reynolds's four guns may never have opened fire. Instead, Wainwright pulled them back to their original positions behind a stone wall on Seminary Ridge when the enemy was within two hundred yards.[84] Sgt. William Shelton never forgot this movement. The limbers swung around, and after being hooked to the guns, went off at a "sharp trot" while the cannoneers ran alongside. Enemy shells landed around them, "striking the soft ground and throwing up fountains of black earth."[85]

With the rupture of the defensive line on McPherson's Ridge, Doubleday decided to put up one last fight on Seminary Ridge. James H. Cooper's battery was already in position to the left of Greenleaf T. Stevens's battery, which was to the right of and in front of the seminary. Benjamin Wilber's section of Gilbert H. Reynolds's battery, which had briefly supported the Iron Brigade, dropped trail along Chambersburg Pike, and the rest of the battery took position at the opposite end of the line, about a hundred yards south of the seminary. James Stewart's six guns remained in position astride the unfinished railroad cut north of Chambersburg Pike. Wainwright's collection of eighteen cannon now occupied a line of about a thousand yards, but in some cases they were more concentrated, such as to the right (north) of the seminary, where the guns were less than five yards apart, rather than at the prescribed fifteen-yard interval. Scattered along the ridge were also the remnants of the Federal infantry units that had tried to hold McPherson's Ridge.[86]

They did not have long to wait, for a fresh line of Confederate infantry soon hove into view. Three brigades from Maj. Gen. Dorsey Pender's division slowly but steadily advanced across the open plain between the two ridges. Gen. Alfred Scales's brigade, with Col. Abner Perrin's on his right, advanced directly in front of Wainwright's massed guns and the three infantry brigades. To their right, Gen. James Lane's brigade moved against William Gamble's cavalry brigade.

The Federal batteries massed on Seminary Ridge were forced to hold their fire while retreating Union infantry blocked their fields of fire. Seeing the cannoneers about ready to open fire, some infantrymen simply dashed up and threw themselves at the base of the guns. "When our front was clear and within canister range, using double charges, the guns . . . were turned on the enemy, and when their first line was within one hundred yards of the Seminary it was brought to a halt," noted a cannoneer in Greenleaf T. Stevens's battery.[87] Capt. Robert Beecham of the 2nd Wisconsin (Iron Brigade) recalled, "Almost at the same moment, as if every lanyard was pulled by the same hand, this line of artillery opened, and Seminary Ridge blazed with a solid sheet of flame, and the missiles of death that swept its western slopes no human beings could endure. After a few moments of the belching of the artillery, the blinding smoke shut out the sun and obstructed the view." Stevens's guns expended about fifty-seven rounds of canister in a matter of moments, an amazing feat, considering the number of infantrymen who had replaced many hardy lumbermen killed or wounded three months earlier at the battle of Chancellorsville.[88]

This was not glorious war—it was suicide. Rather than halting his victorious infantry on McPherson's Ridge and bringing up even a few guns from Pegram's, McIntosh's, or Garnett's battalions to blow apart the defenders on what was an almost impregnable position, Confederate 3rd Corps commander A. P. Hill did nothing. The infantry commanders were left to their own devices.

Confederate Gen. Alfred Scales, whose brigade was on the left of the line, may have had something to prove. He had watched with disdain as John Brockenbrough's brigade in front of him made a halfhearted attack on McPherson's Ridge that gained success only because of the heroics of James Pettigrew's brigade to their right and Junius Daniel's brigade on their left. As Scales rode forward at the front of his supporting line of battle, he encountered Brockenbrough's victorious Virginians lounging around the McPherson barn. They claimed their ammunition was exhausted and they could go no farther. Scales disgustedly wrote in his report, "I ordered my men to march over them, they did so." After briefly dressing his lines, Scales ordered the charge on Seminary Ridge. In addition to the desperate infantry in front of him, Scales faced twelve smoothbore Napoleons from Greenleaf T. Stevens's and James Stewart's batteries and five ordnance rifles from James H. Cooper's battery and Benjamin Wilber's section of Gilbert H. Reynolds's battery.[89]

Scales's 38th North Carolina on the left of the line, across Chambersburg Pike, charged into the very teeth of the Lt. James Davison's three Napoleons of Stewart's battery on the left of the unfinished railroad cut, which was supported by the 76th, 84th, and 147th New York of Lysander Cutler's brigade. Cannoneer Augustus Buell reported, "First we could see the tips of their [the North Carolinians'] color-staff coming over the little ridge, then the points of their bayonets, and then the Johnnies themselves, coming on with a steady tramp, tramp, and with loud yells."[90] The artillerymen were cautioned to hold their fire as the enemy troops closed the distance. Closer and closer they came, until Davison screamed, "Load—canister—double!"

and then a short time later, "Ready!—By piece!—At will!—Fire!"[91] The results were as gory as they were predictable. C. V. Tervis of the 84th New York described the scene: "With the rebel yell they rushed up the slope again and again in a splendid series of charges, advancing in line of battle, as if on parade. They were checked again and again with the murderous fire . . . but as one line was wiped out and broken up by grape and canister and musketry another would be reformed at the bottom of the hill."[92] The day was hot, causing many of the cannoneers to shed their jackets and roll up their sleeves, methodically loading and firing their guns under the watchful eye of Davison.[93]

Capt. James Stewart's battery possessed a long and distinguished record, dating back to the Revolutionary War. It saw subsequent action in the War of 1812 and the Mexican War. At the battle of Buena Vista in the latter war, the battery bravely attempted to withstand a Mexican attack, alone and without infantry support. Its losses were horrendous, and ultimately all of its guns were captured, but the gunners bought time enough for the army to regroup, which then beat back the enemy. The battery's exploits continued during the Civil War. Its current commander, Stewart, had the distinction of being promoted from the ranks because of his meritorious service.[94]

Both Stevens's and Stewart's batteries consisted of bronze Napoleons, whose smoothbore barrels could throw a shell less than sixteen hundred yards. These guns were meant for close-in fighting and were usually in the hottest places, in close proximity to the enemy. Because of the probability that they would be engaged in hand-to-hand combat, each cannoneer carried a saber. Because these were large, heavy, and unwieldy, the men quickly abandoned them in favor of revolvers, handspikes, rammer staffs, stones, and even fists.[95]

It was a hot time for Stewart's guns. Augustus Buell reported:

> Up and down the line men reeling and falling; splinters flying from the wheels and axles where bullets hit; in the rear, horses tearing and plunging, mad with wounds or terror; drivers yelling, shells bursting, shot shrieking overhead, howling about our ears or throwing up great clouds of dust where they struck; the musketry crashing on three sides of us; bullets hissing, humming and whistling everywhere; cannon roaring; all crash on crash and peal on peal, smoke, dust, splinters, blood, wreak and carnage

> indescribable; but the brass guns of Old B still bellowed and not a man or body flinched or faltered! . . . Out in front of us an undulating field, filled almost as far as the eye could reach with a long, low, gray line creeping toward us, fairly fringed with flame![96]

The cannoneers resorted to "thumbing their vents" when the guns became hot. According to Sgt. Simon Hubler of the 143rd Pennsylvania of Roy Stone's brigade, "One man would hold a piece of leather over the vent while another would ram home the charge. As soon as he would remove his thumb from the vent the charge would be exploded."[97]

While James Davison's three left guns were wreaking havoc on the 38th North Carolina on the south side of the unfinished railroad cut, the right three cannon on the opposite side of the cut under Capt. James Stewart were battling the North Carolinians of Junius Daniel's brigade. Supported by several regiments from Lysander Cutler's brigade, the battery faced five of Daniel's regiments and the 3rd Alabama of Col. Edward O'Neal's brigade, keeping all at a respectable distance.

After pounding the 38th North Carolina into submission, Davison shifted his guns ninety degrees to the left, to face Chambersburg Pike. Here they fired canister into the flank of the 13th North Carolina and the rest of Scales's line. Buell commented, "From our second round on a gray squirrel could not have crossed the road alive."[98] Hit in front and flank, Scales's ranks were blown apart. Seeing an opportunity for redemption, the remnants of the 38th North Carolina surged forward again and opened fire, felling a number of artillerymen, including Davison, who was hit in two places. A bullet shattered his ankle, but he refused to leave the field. Hobbling about, he yelled, "Feed it to 'em, God damn 'em! Feed it to 'em!" Davison eventually lost so much blood he had to relinquish command of his guns to Sgt. John Mitchell. It was a grim time for these three pieces, but the 6th Wisconsin and the 11th Pennsylvania advanced and helped reduce the North Carolinians' ardor.[99]

Augustus Buell graphically related the situation: "The very guns became things of life—not implements, but comrades. Every man was doing the work of two or three. At our gun at the finish there was only the Corporal, No. 1 and No. 3, with two drivers fetching ammunition. The water in Pat[Wallace]'s bucket was like ink. His face and hands were smeared all over with burnt powder."[100]

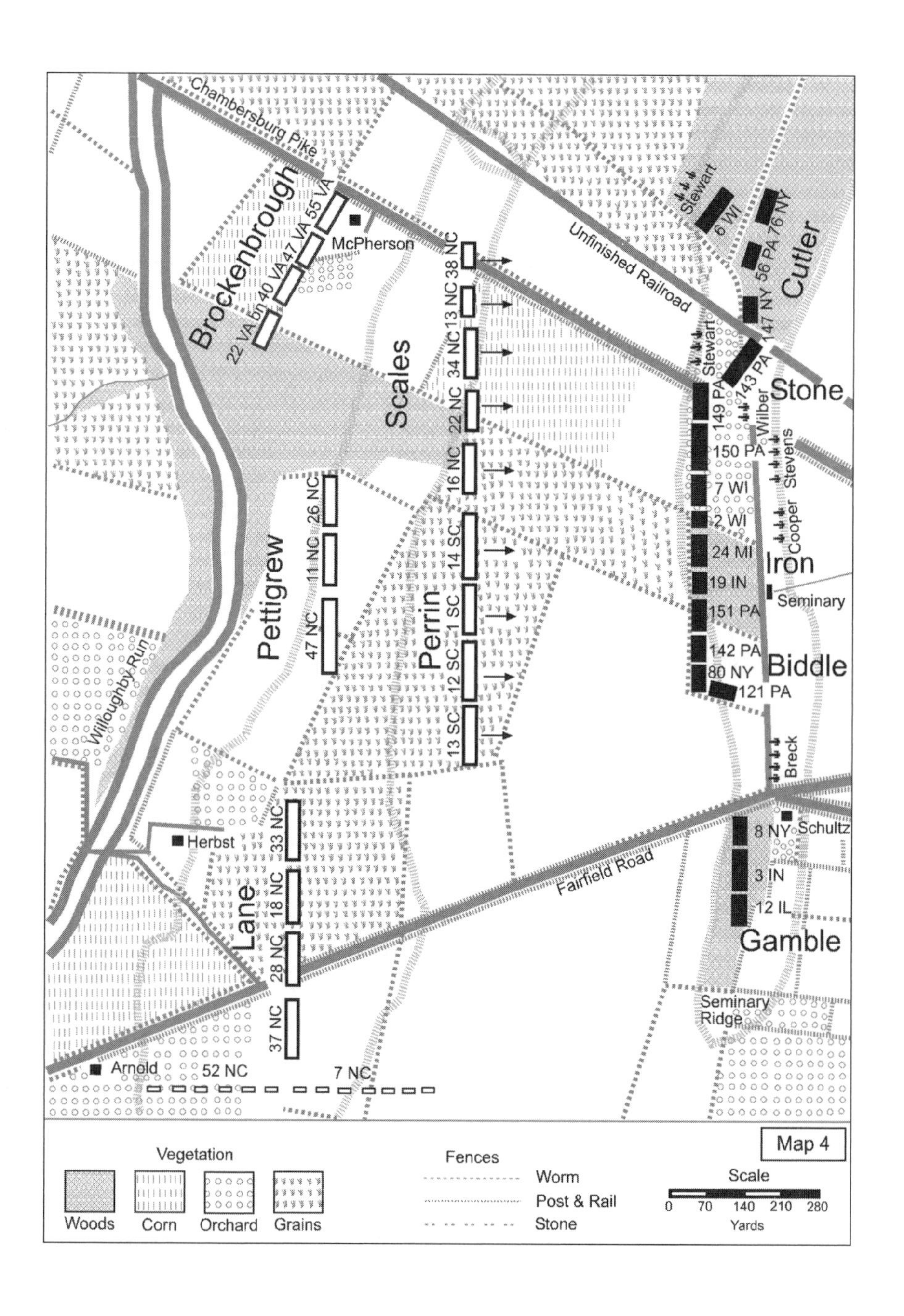

Chambersburg Pike
Brockenbrough
22 VA bn
40 VA
47 VA
55 VA
McPherson
Unfinished Railroad
Scales
38 NC
13 NC
34 NC
22 NC
16 NC
Perrin
14 SC
1 SC
12 SC
13 SC
Pettigrew
26 NC
11 NC
47 NC
Willoughby Run
Herbst
Lane
33 NC
18 NC
28 NC
37 NC
Arnold
52 NC
7 NC
Fairfield Road
Stewart
6 WI
76 NY
56 PA
147 NY
Cutler
Stewart
143 PA
Stone
149 PA
Wilber
150 PA
Stevens
7 WI
2 WI
Cooper
24 MI
Iron
19 IN
Seminary
151 PA
142 PA
Biddle
80 NY
121 PA
Breck
8 NY
Schultz
3 IN
12 IL
Gamble
Seminary Ridge
Map 4
Vegetation
Woods
Corn
Orchard
Grains
Fences
Worm
Post & Rail
Stone
Scale
0 70 140 210 280
Yards

The outlook for the rest of Scales's brigade resolutely attacking Seminary Ridge on the south of Chambersburg Pike was grim. Scales's official report contains the best description:

> We passed . . . up the ascent, crossed the ridge, and commenced the descent just opposite the theological seminary. Here the brigade encountered a most terrific fire of grape and shell on our flank, and grape and musketry in our front. Every discharge made sad havoc in our line, but still we pressed on at a double-quick until we reached the bottom, a distance of about 75 yards from the ridge we had just crossed, and about the same distance from the college, in our front . . . our line had been broken up, and now only a squad here and there marked the place where regiments rested.[101]

With the concentrated fire of Stevens's and Cooper's batteries, along with Wilber's section of Reynolds's battery in front of them and Stewart's three guns on their left, Scales's brigade was almost destroyed. Every field officer, save one, went down, and within a matter of moments the brigade sustained 545 casualties. Col. Alexander Wainwright marveled at the North Carolinians' courage. "Never have I seen such a charge. Not a man seemed to falter. Lee may well be proud of his infantry; I wish ours was equal to it," he wrote in his journal.[102]

The situation was not much better with Abner Perrin's brigade to the right or south of Scales. In front of them were the remnants of the Iron Brigade and Chapman Biddle's brigade and two sections of Gilbert H. Reynolds's battery. The South Carolinians had been ordered not to stop to fire during the charge, but to "give them the bayonet; if they run, then see if they can outrun the bullet."[103] Federal infantry fled before them as the Confederates steadily and methodically marched toward Seminary Ridge. Looking beyond them to the ridge, they saw the makeshift Federal defensive line. J. Caldwell of the 1st South Carolina recalled, "The artillery of the enemy now opened upon us with fatal accuracy. They had a perfectly clear, unobstructed fire upon us. Still we advanced, with regular steps and a well-dressed line. Shell and canister continued to rain upon us." Perrin called this the most destructive fire he had ever experienced. Still the South Carolinians marched on. As one officer astutely noted after the war, "to stop was destruction, to retreat was disaster, to go forward was orders." Meanwhile, the right of Perrin's brigade, composed of the 12th and 13th South Carolina, veered to the right and took on William Gamble's cavalry. John Calef's battery was active in this sector, and in the words of its commander, "made some excellent shots, but my ammunition being nearly exhausted, the firing was very deliberate."[104]

Confederate Col. Abner Perrin did not know what to do. Only temporarily in command of the brigade because its commander had been wounded at Chancellorsville, he could see that Scales's brigade to his left had broken off its attack, and there was no sign of James Lane's brigade on his right. He did not know that the latter had encountered some of Gamble's cavalry along with Calef's guns.[105]

But Perrin's South Carolinians were not to be denied. Finding a gap between Biddle's brigade and Gamble's cavalry brigade, the 1st and 14th South Carolina surged against it and pierced the Union line. Like dominos, the rest of the line collapsed, as both the infantry and artillery were forced to beat a hasty retreat toward Gettysburg. It was very similar to what had occurred a short time before on McPherson's Ridge.[106]

Riding over to Greenleaf T. Stevens's battery, Wainwright was shocked to see it limbering up. Halting the men, he was told that they were simply obeying General Wadsworth's orders. Wainwright was probably seething about what he considered continued incompetence, for he himself had heard from Doubleday that Seminary Ridge must be held at all costs. Just then he realized the enemy was streaming toward his position from the north, south, and west. He could also see the Federal infantry in full flight toward the town. Knowing he could wait no longer, he ordered all of his batteries to limber up and walk to safety. Wainwright was very clear about the latter, because he knew that if the batteries left at a trot, it could create panic among the infantry who now clogged the road. Several cannoneers recalled that the roads were so congested they couldn't have broken into a trot even if they tried.[107]

Wainwright thought it was too late to save his guns, for just as he was ordering James H. Cooper's battery to the rear, he saw a large Confederate soldier plant his regiment's standard on a pile of rails within fifty yards of the battery. To its right, the enemy had reached the works in front of Greenleaf T. Stevens's guns. Across the road, James Stewart's men were also in danger, and according to Augustus Buell, the enemy was within fifty yards when the last cannon galloped away. Wainwright sat on his horse by the side of the road, yelling at his men not to trot. "Each minute," he recalled, "I expected to hear the order to surrender for our infantry had all gone from around me, and there was nothing to stop the advancing line." With a little luck, and much dedication and expertise, the guns were limbered up and began their journey to safety without a loss. They were not safe yet, however.[108]

Not content with capturing Seminary Ridge, Abner Perrin wanted the Federal artillery. Pressing his men forward, they opened fire, hitting a number of artillerymen and horses with their volleys. Perrin later lamented to South Carolina governor Luke Bonham, "If we had any support at all we could have taken every piece of artillery they had and thousands of prisoners."[109]

The onrushing Confederates hurt their chances of capturing Wainwright's guns because their volleys caused the Federal infantry to jump into the unfinished railroad cut and retreat along this protected avenue. The Confederate artillery opened on the cut with surprising accuracy. A Mississippi soldier noted that "every shell seemed to bring down a dozen men," and a North Carolinian added that he could "almost hear their bones crunch under the shot and shell."[110]

Wainwright ordered his teams to gallop to safety along the now clear Chambersburg Pike. The roadbed was wide enough for the guns to retreat three across, and soon all eighteen guns were almost out of the enemy's reach. However, some Confederate artillery now opened on the column, causing three of Stewart's caissons to break down, and a fourth was hit by a shell and destroyed. One cannon in Wilber's section of Reynolds's battery was also lost when the Confederates killed four horses and one of the drivers. Wainwright was devastated when he learned of the piece's loss, but he was consoled with the realization that it truly was a wonder that any of his cannon had reached safety, as the enemy was closing in from three sides.[111]

Another piece was almost lost when one of Stevens's guns suddenly lost a wheel, causing its axle to scrape along the ground. The gunners jumped to the road, raised the gun by hand, and quickly replaced the wheel. Greenleaf T. Stevens, "springing from his horse and seizing the gunner's pinchers, which he inserted handle first as a lynch-pin, and so saved the gun from capture," reported one of the men. The Confederates were closing in on the gun when it finally galloped to safety.[112]

Capt. James Stewart's battery was probably the last "organized" unit to leave Seminary Ridge. The two three-gun detachments pulled out separately. Realizing that their infantry supports had departed, Stewart first ordered his three cannon removed from the north side of the unfinished railroad cut.

Two of the pieces crossed without difficulty, but the third caught some rocks, breaking its pintle hook. Quick thinking prevailed, though, as the men tied the trail to the limber, but not before the enemy had approached to within sixty yards. The other two cannon opened fire in an attempt to dissuade the enemy from trying to take the gun. The repair finally completed, the gun galloped off, but not without the loss of some horses and one man. Stewart turned his horse around as these guns made their way toward Cemetery Hill and rode toward the three he had left under Lt. James Davison on the south side of the cut. His travels took him directly into a group of Confederate infantry, who ordered his surrender. Stewart quickly turned his horse and galloped to safety, but not before a shell fragment plowed into his hip. It was not until later that Stewart learned that these guns had already successfully completed their retreat. When the enemy was within a hundred yards, the limbers were quickly brought up, which caused the Confederates to redouble their rifle fire. A number of horses and men were hit, but the gunners loaded the guns with canister and fired and then worked with precision to hitch them to the limbers. One gun halted in the town, unlimbered, and loaded with canister, prepared to face the approaching foe.[113]

Many soldiers marveled about how few cannon were captured by the Confederates during this chaotic period of the battle. Augustus Buell correctly noted, "I was astonished at the caution of the enemy at this time. He seemed to be utterly paralyzed at the punishment he had received from the First Corps, and was literally 'feeling every inch of his way' in his advance on our front." While few Federal cannon changed hands, seven of the twenty-eight 1st Corps cannon and four of the twenty-six 11th Corps cannon were not serviceable by the end of the day.[114]

With the Federal troops in full retreat, the Confederate batteries now left Herr Ridge and redeployed on Seminary Ridge, where they opened fire on the retreating Union infantry in the town.[115] Many infantrymen avoided being hit by this cannon fire by "hitting the ground and then jumping up and running immediately after the Rebel artillery pieces discharged," wrote C. V. Tervis of Cutler's brigade. Shells striking the masonry buildings showered the men with ragged-edged stones, causing minor wounds among many.[116]

Watching the Federal troops retreating in disarray, Robert E. Lee turned to his artillery chief, Gen. William Pendleton, and asked "whether positions on the right could not be found to enfilade the valley between our position and the town and the enemy's batteries next to the town."[117] Pendleton complied by sending Capt. Marmaduke Johnson's battery and two ordnance rifles from Hardaway's battery, both of David G. McIntosh's battalion, to where Fairfield Road crosses McPherson's Ridge. The guns did not open fire, however, because they lacked infantry support. Historian David Martin was critical of this decision, as it effectively ended all artillery actions in that sector for the remainder of the day.[118]

Confederate artillery chief William Pendleton bore a striking resemblance to Robert E. Lee. Both men were West Point graduates, but Pendleton resigned his commission to enter the ministry. He did not take up arms again until Virginia seceded, when at age fifty-one he became the captain of the Rockbridge Artillery. Despite his position, Pendleton continued preaching to the troops whenever he could. During a battle in Stonewall Jackson's Valley campaign, Pendleton purportedly raised his hand in benediction and yelled, "May the Lord have mercy on their poor souls." With that, his hand swept down as he yelled, "Fire!" Although nominally in charge of Lee's artillery, he had a reputation of not performing well on the battlefield, and most men knew it. In command of the reserve artillery at the battle of Malvern Hill during the Seven Days campaign, Pendleton left twenty batteries idle as their Federal counterparts blew away line after line of attacking Confederate infantry. He again performed badly at the battle of Antietam about two months later. John Chamberlayne described Pendleton as "Lee's weakness" and noted, "Like an elephant, we have him & we don't know what on earth to do with him, and it costs a devil of a sight to feed him." Why Lee retained Pendleton, when he actively extracted countless others from their comands, remains a mystery of the war. Perhaps it was Pendleton's close ties to President Jefferson Davis or Lee's reluctance to hurt an old friend that protected him. Pendleton was, however, a superb organizer and was most effective in camp. Lee attempted to compensate for Pendleton's shortcomings by converting his position into an administrative one with little direct impact on the battlefield, but this would come back to haunt him at Gettysburg, for it prevented the artillery from being deployed as a strong, cohesive force.[119]

Pendleton now ordered up two fresh 3rd Corps artillery battalions—John J. Garnett's and William T. Poague's—that had been in reserve behind Herr Ridge. They deployed for action along Seminary Ridge, south of Pegram's and McIntosh's battalions, on both sides of Fairfield Road. "The position was within range of the hill beyond the town, to which the enemy was retreating, and where he was massing his batteries," Pendleton wrote in his report.[120] Just as these two battalions were about to open fire, Gen. Stephen Ramseur rode up to Pendleton and, in the latter's own words, "requested that our batteries might not open, as they would draw a concentrated fire on his men, much exposed."[121] Pendleton agreed and noted, "Unless as part of a combined assault, I at once saw it would be worse than useless to open fire there."[122] As a result, Garnett's guns were ordered to move closer to the town, but these batteries did not open fire. This was another missed opportunity for Lee's long arm. Battalion commander David G. McIntosh wrote of the exchange after the war, "The suggestion was a[s] untimely and ill-judged as its acceptance was weak and unfortunate."[123]

Among the Federals, Wainwright's and Calef's men had every reason to be satisfied with their exploits during the fight for Seminary Ridge in the late afternoon of July 1. Although ultimately defeated, they poured a continual fire into both the Confederate artillery and infantry, inflicting heavy casualties among the latter. The amount of firepower was impressive. During the latter portions of the fighting for McPherson's and Seminary ridges, James H. Cooper's three cannon (its fourth was disabled early in the fight) fired about 375 rounds. Wainwright's losses were heavy, however—eighty-three officers and men and about eighty horses.[124]

The Confederate guns battling Wainwright's, on the other hand, were poorly utilized during this phase of the battle. Remaining on Herr Ridge, far from the attacking lines, they played only a minor role in the fight for McPherson's Ridge and an even lesser one for possession of Seminary Ridge. The historian of the Confederate artillery noted that the "batteries were left to act as best they could, without a definite plan or objective. . . . Buford's dismounted troopers . . . would have been unable to hold their lines had they first been subjected to a heavy artillery fire. Heth would almost certainly have been able by a proper concert with Pegram and McIntosh to seize Buford's position before the latter was reinforced."[125]

Pegram's losses for the day amounted to two men killed, eight wounded, and six horses lost. McIntosh's battalion lost one killed, one wounded, and several horses disabled. The other battalions—Lane's, Garnett's, and Poague's—totaling twelve batteries, were not engaged at all. Instead, they were permitted to rest well to the rear along Chambersburg Pike.[126]

Although A. P. Hill's infantry ultimately knocked the Federal troops off of two ridges west of the town, both inflicting and sustaining serious losses in the process, the victory could have been much more decisive had Hill used his infantry, and especially his artillery, more effectively.

While the controversy surrounding Gen. Richard S. Ewell's failure to capture Cemetery Hill during the evening of July 1 has endured for well over a century, A. P. Hill's inactivity has not been emphasized. Lt. Col. David G. McIntosh believed that if Hill's batteries had taken proper positions on Seminary Ridge on the evening of July 1, they could have shelled the Federal troops off Cemetery Hill. He wrote after the war, "Had this been done, and the demoralized troops on Cemetery Heights been subjected to an artillery fire, it is certain that the effect must have been disastrous, and might have led to an abandonment of the position."[127]

One of Hill's battalions—John Garnett's—actually had the opportunity to test this idea when Lee ordered it to a position near the seminary where it could fire on the troops moving toward Emmitsburg Road. It did so but was quickly hit by massive counterartillery fire from Cemetery Hill. Any massing of artillery on Seminary Ridge on July 1 would have probably brought an immediate response from the better-placed Federal guns on Cemetery Hill, but the Union guns could have been ineffective, and quickly silenced, had the Confederate battalions ringed the hill and poured an enfilade fire upon it.[128]

2

JULY 1

THE BATTLE NORTH OF GETTYSBURG

CARTER'S BATTALION ARRIVES

The infantry of the Federal 1st Corps deployed west of Gettysburg was grateful for the midday lull in the fighting. Danger, however, approached from the north, as Lt. Gen. Richard S. Ewell's 2nd Corps neared the battlefield. This corps had traveled far to the north—briefly capturing Carlisle and York, Pennsylvania, and threatening Harrisburg—before being recalled by Robert E. Lee. These roads converged on Gettysburg, and Maj. Gen. Robert E. Rodes's division arrived first, deploying on Oak Hill north of Gettysburg. Rodes carefully noted the terrain and the enemy's depositions, quickly ascertaining the Federal infantry's vulnerable position on McPherson's Ridge far in front of him. Slamming into their flank from the north would assure their rout, he reasoned.[1]

As Rodes deployed his division in two lines on Oak Hill, two batteries of Lt. Col. Thomas H. Carter's artillery battalion unlimbered in front of them. Thirty-two-year-old Carter was a Virginia Military Institute graduate and a physician. A cousin of Robert E. Lee, Carter managed his family's large plantation during the war. He entered the conflict as captain of the King William Artillery and was promoted to major and given D. H. Hill's division's artillery during the following winter. A year later, Carter, nicknamed "Old Raw Hide," was again promoted and given command of a battalion attached to Rodes's division. The historian of Lee's artillery called Carter "gentle," adding that he was "noted for the purity and strength of his character, beloved and respected by all."[2]

Carter brought two batteries with him: Capt. William P. Carter's (King William) battery (two Napoleons and two 10-pounder Parrotts) and Capt. Charles W. Fry's (Orange) battery (two 10-pounder Parrotts and two ordnance rifles). They dropped trail on the south slope of Oak Hill, with Fry's guns next to the woods and Carter's on his right. The two Virginia batteries immediately opened fire on the Federal infantry directly to the south on McPherson's Ridge and along Chambersburg Pike. This fire, coupled with that of Pegram's and McIntosh's battalions on Herr Ridge, made these sectors hot places for the Federal troops. Carter's eight active guns' first target was Col. Roy Stone's brigade (Doubleday's division, 1st Corps) in position near Chambersburg Pike on McPherson's Ridge. Carter wrote that his two batteries fired with a "very decided effect."[3] The Federal infantrymen were initially shocked when Carter's artillery opened on them, believing that the batteries were their own. The first shells hit an orchard, showering the men with limbs and splinters. When he realized the brigade's danger, Stone shifted its position to face Chambersburg Pike and Carter's artillery.[4] Most upsetting to the men at this time were the "whistling shrieks" of the Whitworth guns from William B. Hurt's battery on Herr Ridge.[5]

The shelling was so severe that Stone attempted a ploy to save his men. Sending the 149th Pennsylvania's color guard forward about twenty-five yards to the west and forty-five yards to the north to crouch behind a pile of fence rails, Stone hoped to draw the Confederate artillery fire away from his infantry. The ruse succeeded, for Capt. John Bassler could see "how the shells from the enemy's batteries now exploded in the vicinity of the colors which were just visible to them over the wheat, one shell passing through the starry field of the state flag."[6]

Carter's guns also tormented Chapman Biddle's brigade in its exposed position between McPherson's and Seminary ridges. The cannonade was so intense that the brigade was forced to change its position again and again. At one point the brigade was ordered to the side of Fairfield Road, where the men were told to lie down. The characteristic noise told Col. Theodore Gates that his 80th New York was being hit with shells from Whitworth rifles. The reprieve by the side of the road was a short one, for Carter's guns opened fire again within ten minutes. James H. Cooper's battery, which was now positioned within Biddle's brigade, shifted its position to take on Carter's battalion, but it was a mismatch, and the Federal guns soon grew silent.[7]

Several Federal batteries effectively responded to Carter's battalion on Oak Hill, including at various times Reynolds's, Cooper's, and Calef's. Carter noted that the Federal guns "replied slowly." Because Carter's battery was so exposed, it sustained several casualties during this period—four men killed and seven more wounded. Abner Doubleday could easily see the growing danger to his right flank and immediately sent Henry Baxter's brigade (John Robinson's division, 1st Corps) rushing north to take position along Oak Ridge.[8]

Gen. Oliver O. Howard's 11th Corps also began arriving around noon. The corps had spent the night at Emmitsburg, Maryland, and began its final march to Gettysburg that morning. Four companies of the 45th New York (George von Amsberg's brigade) quickstepped to the Hagy farm on Mummasburg Road. They were to connect with the left of Baxter's brigade, facing north, but this never occurred, probably because of the intense fire from Carter's guns on Oak Hill, which created a four-hundred-foot no-man's-land between the two corps.[9]

Concerned about these rapidly arriving Federal reinforcements, Rodes ordered Carter to deploy his remaining two batteries from their resting places along Middletown Road to the left of the batteries already in action. Capt. Richard Page's Morris Battery (four Napoleons) dropped trail at the foot of Oak Ridge and opened fire on the New Yorkers, forcing them to take refuge behind a fence. It didn't take long for these infantrymen to calm themselves and open fire on Page's gunners. Carter's fourth battery, Capt. William Reese's Jeff Davis Artillery (four ordnance rifles), deployed farther to the east.[10]

Capt. Hubert Dilger's Company I Battery, 1st Ohio, was the first 11th Corps battery to arrive, and Brig. Gen. Alfred Schimmelfennig ordered it into position between the Mummasburg and the Carlisle roads. Dilger complied by placing Lt. Clark Scripture's two-gun section on high ground to the southeast of the 45th New York, where it immediately opened fire on Carter's guns on Oak Hill. Dilger's remaining four pieces remained near Carlisle Road. "A four-gun battery of the enemy [Page's] immediately opened fire at about 1,400 yards on this section, and compelled me very soon to bring my whole battery into action," noted Dilger. After completing the deployment of his remaining four guns by 1:00 p.m., Dilger could see additional 11th Corps infantry from Schimmelfennig's division arriving.[11]

Union commander George G. Meade was fortunate that Dilger and his battery were among the earliest arrivals on the battlefield. Known as "Leatherbreeches," Dilger gained artillery experience while in his native Germany, but a revolution made him an immigrant to the United States. The battery and its commander distinguished themselves during Stonewall Jackson's Valley campaign, when Dilger gained command of the unit. His battery performed brilliantly at Chancellorsville, when it used a leapfrog technique to keep the victorious Confederate infantry at bay. He would ultimately receive the Medal of Honor for his actions here. Blunt and a bit arrogant, Dilger was loved by his men but not by his superiors. He brought a well-trained artillery unit with him to Gettysburg, one that he had commanded for more than a year.[12]

Dilger opened an oblique fire on Page's battery, relieving the pressure on the 45th New York, whose remaining companies were now arriving. An infantry officer wrote,

> The first shot from the Ohio Battery flew over the Confederate Battery. At this time the rebels yelled in derision. Capt. Dilger now sighted the gun himself and fired it. The shot dismounted a rebel gun and killed the horses. Capt. Dilger tried it a second time, sighting and firing the gun. No effect being visible with the naked eye, Col. [Philip] Brown who was near, asked, "what effect, Capt. Dilger?" Capt. Dilger, after looking through his glass, replied, "I have spiked a gun for them plugging it at the muzzle."[13]

Seeing the Federal 11th Corps deploying in ever-greater numbers on his left flank, Rodes decided to attack immediately. The plan was for three brigades—Edward O'Neal's, Alfred Iverson's, and Junius Daniel's—to step smartly off Oak Hill and sweep down on the Federal infantry on Oak Ridge. After crushing the position, the attack would roll south to hit the enemy infantry on McPherson's Ridge or turn and hit the 11th Corps in the flank while Jubal Early's division, soon to arrive on the left, took on its front. The plan went awry from the start, as O'Neal launched his attack prematurely, with only three of his five regiments. The Alabamians were easily repulsed. Seeing the attack, Iverson quickly ordered his brigade forward without support on its right or left, and the North Carolinians were blown apart by

Henry Baxter's sheltered infantry on Oak Ridge. Carter's guns apparently did not support these attacks, as none of the myriad Federal accounts mention enemy artillery fire at this time. Instead, Carter was seemingly content to shell the Federal troops and batteries on McPherson's Ridge and, somewhat later, the 11th Corps to his left.[14]

The Federal artillery played a more active role in blunting some of the Confederate attacks during this period. As O'Neal's brigade approached Oak Ridge, Capt. Francis Irsch of the 45th New York requested artillery support to help smash the attack. Dilger was only too happy to comply. The New York infantrymen threw themselves to the ground as Dilger shifted his guns in their direction and then poured canister into O'Neal's left flank. An officer in the 45th New York noted that Dilger's battery was so effective that the "massed enemy began to halt and waver."[15] This brisk fire, together with small-arms fire in their front and flank, caused the Alabamians to break off the attack.

After repulsing Rodes's initial attack, Dilger's battery again changed position and resumed its close-in artillery duel with Richard Page's guns, whose positions were restricted by the slope of the hill. With much less ability to maneuver, the Confederate battery received the worst of it, losing two men killed, two mortally wounded, and twenty-six wounded. Seventeen horses were also killed or disabled. Carter reported that the battery had "caught every shell & was almost torn to pieces." Page quickly sent for Carter, who found the guns deployed one above the other on the hill, "like seats in an amphitheatre." Realizing the battery's terrible predicament, Carter, who had not positioned the guns, galloped over to Rodes, "as mad as a hornet," and asked, "General, what fool put that battery yonder?" An awkward pause accompanied suspicious facial expressions, and Rodes quietly said, "You had better take it away, Carter." As Carter rode away from the group, one of the general's aides quietly told the irate battalion commander that Rodes had posted the battery. The battery finally limbered up and beat a hasty retreat, ending its terrible beating at the hands of Dilger's gunners.[16]

Carter's last reserve battery, William Reese's Jeff Davis Artillery, took position on a rise just south of the Cobean farmhouse and opened fire on Dilger's guns. These rifled cannon were beyond the range of Dilger's smoothbore Napoleons, so the German artillerist requested rifled cannon from the 11th Corps artillery commander, Maj. Thomas W. Osborn. Within a half hour of

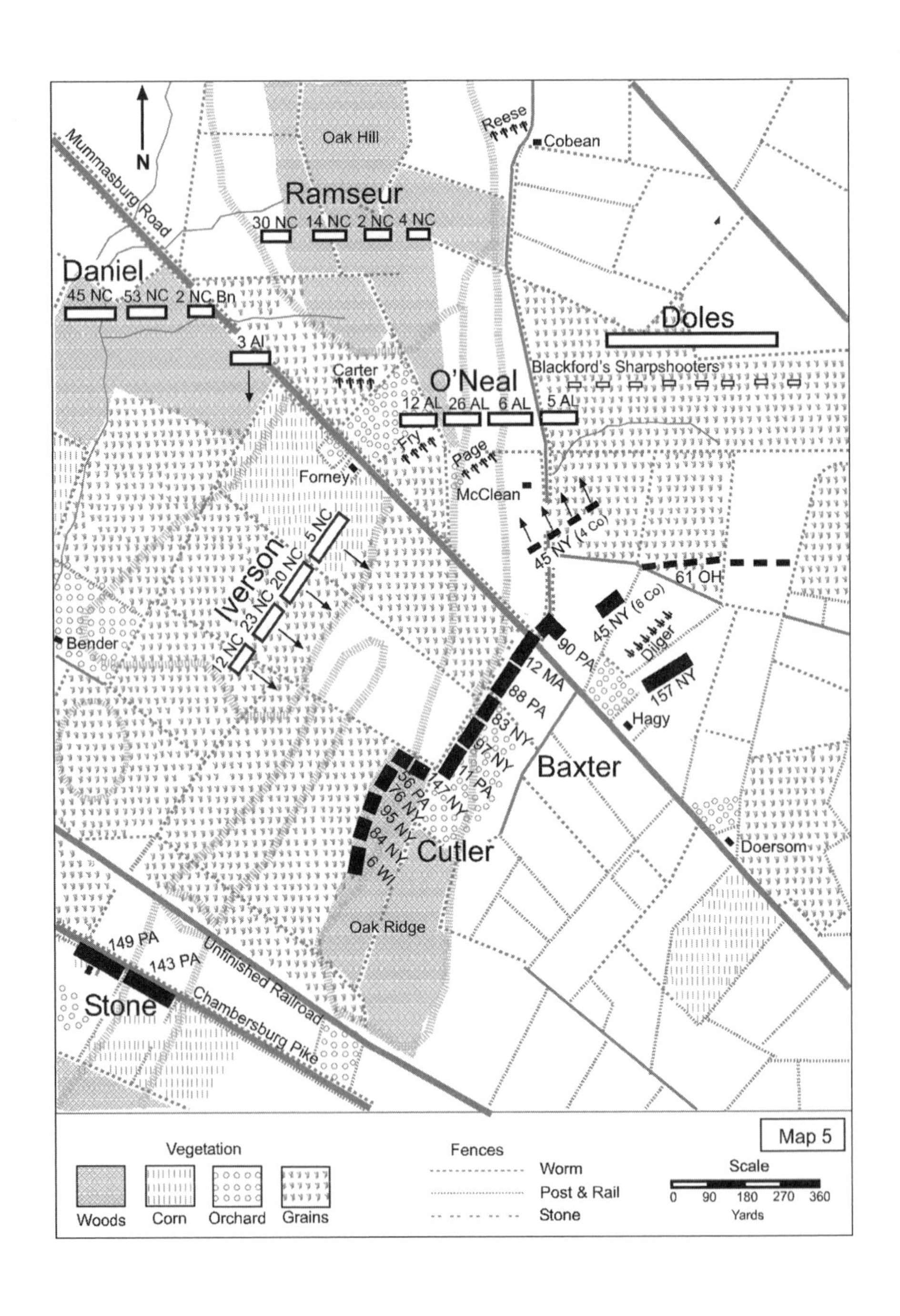

THE INITIAL ATTACKS ON OAK RIDGE

Dilger's battery going into action, Lt. William Wheeler's 13th New York Independent Battery (four ordnance rifles) arrived from its perch on Cemetery Hill and took position on its right. The battery had been ordered to hurry to the battlefield earlier in the day, when it was about four miles away. Wheeler ordered the battery into a trot, but he worried about the effect of the rough roadway on his battery's wheels, which were already in poor condition. As a safeguard, Wheeler had his men dismount from the limber chests and run beside the guns. The halt on Cemetery Hill was a short one, for the guns were almost immediately ordered north of town to assist Dilger. During the gallop through Gettysburg, two of Wheeler's caissons broke down on the rough roads. One was quickly repaired, but the other was too badly shattered and had to be left behind. Dilger was placed in command of all nine guns. The 157th New York and 82nd Illinois of Col. George von Amsberg's brigade arrived and formed a supporting line behind these two batteries, while the rest of the brigade formed in front of the guns. These infantrymen formed the extreme left of the 11th Corps.[17]

Like Colonel Wainwright of the 1st Corps, twenty-six-year-old Maj. Thomas W. Osborn was a New Yorker who had begun the war with the 1st New York Artillery. Ironically, earlier in his life, Osborn had found the idea of military service repulsive, but his feelings changed after the battle of Bull Run. He quickly organized Battery D and commanded it on every eastern battlefield from the Peninsula campaign to Fredericksburg. At Chancellorsville he commanded the artillery attached to Hiram G. Berry's division. Osborn was promoted after the latter battle to major and given command of the 11th Corps artillery. In his postwar writings, he revealed a strong impulse to portray his actions in the most positive light, occasionally at the expense of his comrades and the truth.[18]

Wheeler's rifled cannon were an excellent complement to Dilger's smoothbores, and they were soon pouring an accurate fire into the two Confederate batteries near Oak Hill. Not wishing to be a mere bystander, Dilger decided to move his guns closer to the enemy. While Wheeler's guns continued belching shells, Dilger first ordered Lt. Christian Weidman's section forward about six hundred yards to a wheatfield just to the right of Carlisle Road, where it was supported by the 61st Ohio of von Amsberg's brigade. After Weidman's guns opened fire, Dilger advanced his remaining four guns. This was not easy, as the teams were brought to an abrupt halt at the edge of a five-foot-wide and four-foot-deep ditch. Spying some nearby fence rails, the artillerymen quickly filled the ditch with them while Confederate minie balls zipped and shells exploded around them. Upon reaching Weidman's position, Dilger's remaining guns resumed their firing. Wheeler's men also limbered up and followed, but they too found the going difficult as Confederate shells exploded among them. One shell killed two horses and blew off a driver's leg. A well-constructed fence also barred the battery's path, and it was not until the cannoneers knocked it down with axes that they able to continue their progress toward Dilger's battery. While waiting for the fence to go down, Wheeler saw a Confederate shell take off an infantryman's leg. It whirled in the air, finally coming to rest with a loud whack against a caisson.[19]

Realizing the danger of bringing his caissons to the front, but knowing that he would need a constant supply of ammunition, Dilger improvised. Leaving the caissons in the safety of the rear, Dilger used three limbers to shuttle ammunition to the front. The system did not work perfectly; twice Dilger's guns almost ran out of ammunition.[20]

The arrival of additional 11th Corps infantry units concerned Rodes, but he momentarily expected the arrival of Maj. Gen. Jubal Early's division. Realizing that he couldn't wait any longer, Rodes dispatched Brig. Gen. George Doles's brigade to the left to face the enemy concentrating there. Doles realized that he was badly outnumbered, so he requested artillery support, and Rodes shifted all but Charles W. Fry's battery to a position behind Doles's Georgians to provide a covering fire.[21] Carter reported that the three batteries "rendered excellent service, driving back both infantry and artillery."[22] A new target materialized when Col. Wladimir Krzyzanowski's brigade (Schimmelfennig's division, 11th Corps) moved north from Gettysburg. The batteries threw "solid shot, shells . . . and everything that could be shot out of a cannon," noted a private in the 153rd Pennsylvania. The infantrymen were ordered to double-quick toward the Almshouse, but this did not deter Carter's gunners. "The shells were coming pretty thick before we reached the barn. Some were going over us, and some did not quite reach us. A shell exploded right over the column, and every man dodged for the instant," noted the Pennsylvania private. According to Theodore Dodge, a "rebel battery, which somehow had crept up on an eminence on our right, some half-mile distance, began to pepper us with grape and canister. This was very annoying, for although the fire of a battery is much less deadly at a distance than musketry close at hand, the noises are so much more appalling that men will get uneasy under a harmless shelling quicker than under a murderous fire of small arms."[23]

According to a Confederate historian, "These batteries, by a tremendous effort, succeeded almost single-handedly in checking the Federal advance and driving back both the infantry and artillery of the enemy from the threatened point. Carter's battery, though much depleted and damaged, delivered a most effective fire with reckless daring." Unfortunately, some of the shells also hit Doles's men. He reported that his brigade "was subjected to and did receive [fire] from one of our batteries . . . several men killed and wounded." The battery was probably Page's or Carter's.[24]

This was a hot time for Dilger's and Wheeler's batteries. Carter's shells often overshot Dilger's battery and struck close to Wheeler's. Miraculously, only horses were killed or wounded.[25] Capt. Alfred Lee of the 82nd Ohio observed the duel and noted, "The return fire of the rebel guns was lively, and their shot and shell ricochetted splendidly over the open fields."[26] Artillery fire suddenly began striking the two Federal batteries from the northeast as well—Early's division had arrived from the north.

Early's Threat from the Northeast

Marching from Heidlersburg that morning, Jubal Early's division reached the battlefield between 3:00 and 3:30 p.m. As it maneuvered into position, Union Brig. Gen. Francis Barlow finalized his division's deployment on Blocher's Knoll, which was directly in the path of the newly arrived Southern division. Barlow had Lt. Bayard Wilkeson's 4th U.S. Battery of six Napoleons with him. Wilkeson, just nineteen years old, was the youngest battery commander at Gettysburg. He dropped one section near the Almshouse and positioned the other four guns on the high ground behind the infantry on Blocher's Knoll. Barlow's position, which anchored the right flank of the 11th Corps, was well advanced, and both flanks hung in the air, vulnerable to attack. Barlow advanced to this elevated piece of ground because it offered a good artillery platform.[27]

Early's division numbered about fifty-seven hundred men—far more than Barlow's three thousand, and with Doles's brigade of thirteen hundred men, it was even more of a mismatch. Early's artillery battalion was commanded by Lt. Col. Hilary P. Jones. An educator before the war, Jones's first military assignment was as a lieutenant in the Morris Artillery in August

1861. He was promoted and given command of the battery the following February and promoted again two months later. Jones commanded the artillery attached to Isaac R. Trimble's division during the Shenandoah Valley campaign and was promoted to lieutenant colonel in March 1863 and given the artillery battalion attached to Early's division. A highly critical officer paid Jones a major compliment when he described him as "a moderately good officer; no very strong points, or any objectionable ones." William Pendleton called him a "very judicious and faithful officer."[28]

Jones's artillery battalion contained sixteen guns—dramatically outnumbering Barlow's six. Quickly taking in the situation, Early ordered Jones's battalion into position on the east side of Rock Creek (to the left of Harrisburg Road), about a half mile northeast of Blocher's Knoll. From this position, Jones's guns had a clear field of fire against Barlow's infantry and Dilger's and Wheeler's guns to the southwest. The actual disposition of the batteries is not known but may have been, from right to left: Capt. Asher W. Garber's Staunton Artillery (four Napoleons), Capt. Charles A. Green's Louisiana Guard Artillery (two ordnance rifles and two 10-pounder Parrotts), and Capt. William A. Tanner's Courtney Battery (four ordnance rifles). Eight of these guns were rifled and four were Napoleons. Capt. James McD. Carrington's Charlottesville Battery (four Napoleons) remained in reserve. The three batteries immediately opened an enfilading fire against the Federal infantry's right flank and rear on the high ground. Jones later wrote that the artillery fire had a "considerable effect" on the enemy.[29]

Spying several companies of the 17th Connecticut of Brig. Gen. Adelbert Ames's brigade forming a skirmish line on the east side of Rock Creek near the York Pike to their left, Jones's guns opened fire with "shot, shell, grape, and canister."[30] It was a hot time for the Nutmeggers. The nearby Benner house caught fire and burned to the ground, driving out Federal infantry that had been using it for protection. Jubal Early now ordered Brig. Gen. John Gordon's brigade against Col. Leopold von Gilsa's brigade of Barlow's division on Blocher's Knoll, while George Doles's brigade advanced from the northwest. William Wheeler reported seeing "two great gray clouds" moving toward Barlow's infantry, so Dilger ordered the right section of the New York battery to fire into the advancing column with canister. Some of the right guns of Hilary Jones's artillery battalion opened fire on Dilger's and Wheeler's batteries, and shells from both Oak Hill and north of Blocher's Knoll added a converging fire on the ten Federal guns. Men and

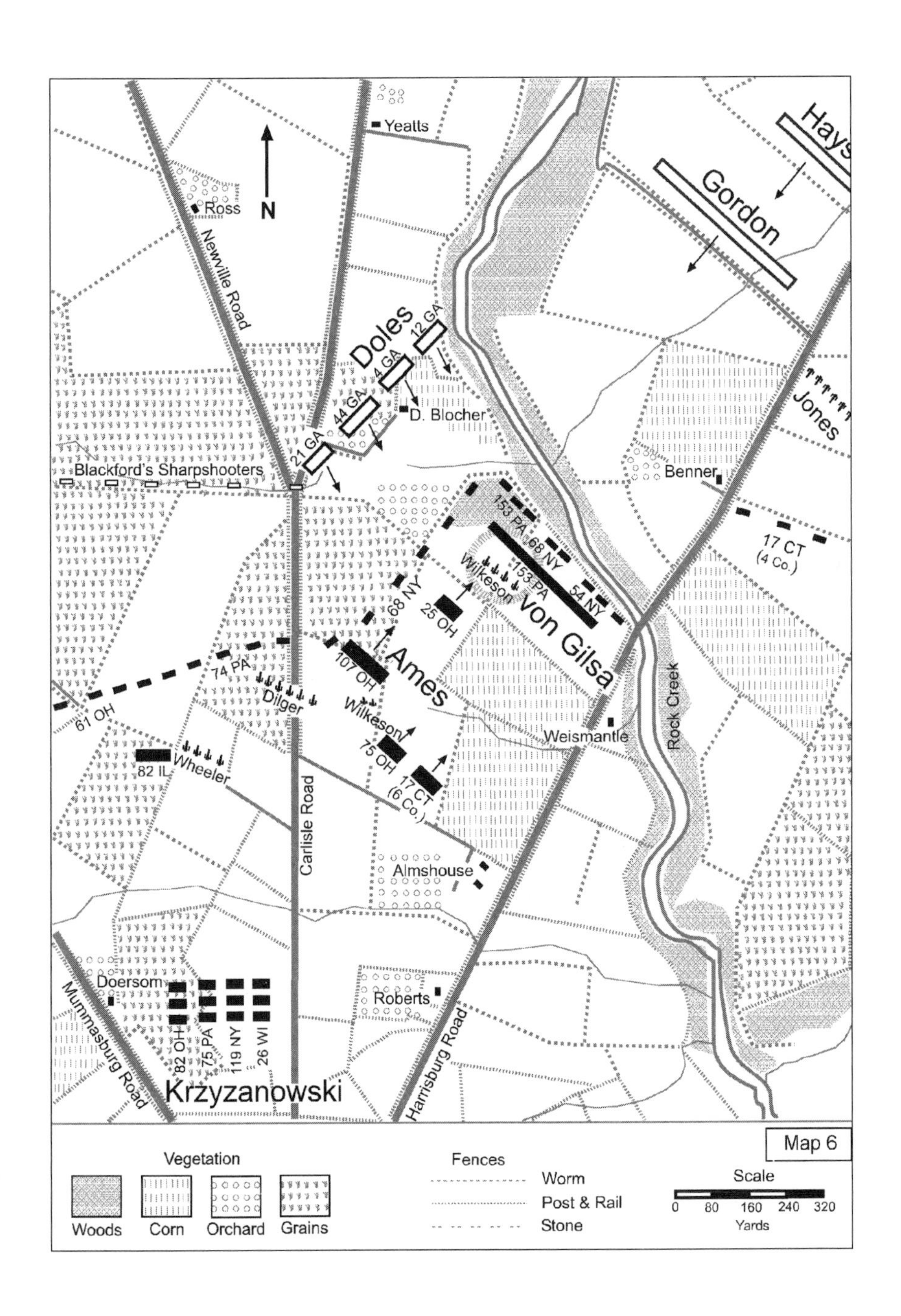

Yealts
Ross
N
Newville Road
Hays
Gordon
Doles
12 GA
4 GA
44 GA
21 GA
D. Blocher
Jones
Blackford's Sharpshooters
Benner
17 CT
(4 Co.)
153 PA
68 NY
153 PA
54 NY
Wilkeson
Von Gilsa
68 NY
25 OH
107 OH
Ames
74 PA
61 OH
Dilger
Wilkeson
75 OH
Weismantle
Rock Creek
82 IL
Wheeler
17 CT
(6 Co.)
Carlisle Road
Almshouse
Doersom
Roberts
Mummasburg Road
82 OH
75 PA
119 NY
26 WI
Krzyzanowski
Harrisburg Road
Map 6
Vegetation
Woods
Corn
Orchard
Grains
Fences
Worm
Post & Rail
Stone
Scale
0 80 160 240 320
Yards

horses went down in quick succession, and a cannon from each Federal battery was disabled.[31]

Some of Jones's guns also pounded Barlow's infantry. A private in Ames's brigade noted that as his unit left its reserve position and moved to Blocher's Knoll to reinforce von Gilsa's brigade, "We were obliged to pass through a raking crossfire from a rebel battery on our right, but most of their shells passed over us as we kept pretty close. After changing our position several times more in as many minutes, most of us . . . stretched ourselves on the ground to . . . escape the shells which were continually flying over us."[32] A soldier in the 75th Ohio of the same brigade added that it was a "deafening and furious bombardment! The ground shook."[33]

While the right guns of Jones's battalion pounded Dilger's and Wheeler's batteries, two batteries pounded Bayard Wilkeson's battery. Major Osborn, the 11th Corps artillery chief, was not impressed with Barlow's positioning of Wilkeson's four guns on the knoll, saying that they were "unfortunately near the enemy's line of infantry . . . as well as two of his batteries, the concentrated fire of which no battery could withstand."[34] Wilkeson was one of the battery's first casualties when his leg was blown off at the knee, and he died later that day. Lt. Eugene A. Bancroft assumed command of the battery and ordered his men to continue firing into the attacking Confederate infantry.[35] Pvt. Reuben Ruch of the 153rd Pennsylvania of von Gilsa's brigade noted, "A battery in the rear of us, on higher ground than we were, opened fire over us and we could feel the heat of the balls as they passed over us. About this time the Rebels made a charge, away to the left of us, and the battery in the rear turned guns on them, taking them by flank. . . . We could see balls plowing up the ground along the rear of the line, and if ever Johnnies ran for cover, those fellows did."[36]

Realizing how severely he was outgunned, Bancroft called up Lt. Christopher Merkle's section, which had been left northwest of the Almshouse. Forming on the western edge of Blocher's Knoll, Merkle opened fire on one of Carter's batteries to the west. He noted in his report that he engaged the battery "for a few moments, with solid shot."[37] To his right, the Confederate counterbattery fire was so great that Bancroft ordered his four guns, after occupying their original position for about thirty minutes, to move closer to Merkle's position.[38]

Wilkeson's battery had only a few moments for counterbattery actions before the Confederate infantry launched its attack. As Doles's brigade bore down on Blocher's Knoll from the northwest, Merkle repositioned his two Napoleons and opened fire. He reported, "I used shell and spherical case shot at first, and, as the line of the enemy came closer, and I ran out of shot, shell, and case shot, I used canister; the enemy was then within canister range."[39] This was the ideal situation for this type of cannon: hitting enemy infantry at short range. The problem was that the Federal supporting infantry was giving way after being hit on their front and flank by the fast-moving Confederate infantry attack. Merkle noted, "Our infantry fell back rapidly, and left me almost without support."[40]

Bancroft quickly realized his battery's precarious position. "Our infantry giving way in great disorder, the want of support compelled me to withdraw the guns," he wrote in his report.[41] The young commander was able to get his six Napoleons to safety despite the loss of twelve men killed or wounded and seventeen horses disabled. The retreat toward Gettysburg was not without some excitement, as some Confederate infantry rapidly followed, causing Merkle to halt and redeploy his guns. "I fired two double rounds of canister, with prolonge [rope] fixed, at their line at the end of town; then limbered up and retired," he recounted.[42]

With no further resistance, the Confederate artillery opened fire on the "disordered masses of the enemy that were rapidly retreating before our troops," Hilary P. Jones noted in his report.[43] So rapidly did Early's men chase Barlow's that they masked Jones's guns, forcing him to halt their firing lest they hit their own infantry in the back. Jones's losses were lower than Wilkeson's—one killed and one wounded. Jones, however, admitted that three of his guns were "temporarily disabled" and a fourth permanently so. During the excitement of the battle, ordnance rifle rounds were accidentally rammed into three smoothbore cannon. The rounds lodged halfway down the barrels, resisting all efforts to ram them home. This was an example of why artillery commanders preferred to form batteries with the same type of cannon. A fourth gun was permanently disabled when a Federal solid shell hit the face of its muzzle, bending it in the process.[44]

Help was on the way for Barlow's beleaguered division. Although Krzyzanowski's brigade (Schimmelfennig's division) was too late to help defend Blocher's Knoll, it distracted the victorious Confederate infantry, permitting more of Barlow's men to escape. Seeing this new body of Federal troops, the two Confederate brigades changed direction to take them on, while Carter and Jones swiveled their artillery battalions into position and opened fire. According to Alfred Lee of the 82nd Ohio, "The enemy's batteries swept the plain completely from two or three different directions, and their shells plunged through our solid squares, making terrible havoc. Yet the line swept steadily on, in almost perfect order. Gaps made in the living mass by the cannon-shot were closed again as quickly and quietly almost as though nothing in particular had happened."[45] The German American troops soon had other things to worry about as the Confederate infantry now unleashed a furious attack, sending Krzyzanowski's men in full retreat toward Gettysburg. Wheeler's battery continued firing into Doles's ranks with canister, but its commander soon realized that his guns were ineffective in halting this tidal wave. When the enemy was within a hundred yards and closing in around his rear, Dilger reluctantly gave the order to limber up and move to safety.[46]

The Retreat to Gettysburg

As Barlow's men streamed past, Capt. Lewis Heckman, commander of Battery K (four Napoleons), 1st Ohio Artillery, deployed his guns near Gettysburg College, just north of town. His mission was to stem the Confederate tide and buy time for the thoroughly beaten and demoralized 11th Corps infantry to reach the safety of Cemetery Hill. According to Maj. Thomas W. Osborn, the battery was put into position "with a view of holding the enemy in check until the corps had time to retire through the town."[47] It was almost a suicide mission as two fresh brigades of Early's division (Isaac Avery's and Harry T. Hays's) closed in on Gettysburg while Heckman quickly deployed his battery between the Carlisle and Harrisburg roads. This was not a time for precision, as the guns had to be ready to open fire immediately. To their right, the Federal gunners could see Col. Charles Coster's fresh brigade (Adolph von Steinwehr's division) in position near Kuhn's brickyard, and in front of them, they could see disorganized masses of their own infantry closely followed by the enemy. Their guns masked, the cannon remained silent. The battery opened fire as soon as its front cleared and continued to do so for about thirty minutes. The four guns expended a total of 113 rounds during this period, firing primarily canister. The cannoneers threw themselves into their work as the enemy steadily closed in on them. Brig. Gen. Harry T. Hays reported that the "fire to which my command was subjected from the enemy's batteries . . . was unusually galling."[48]

Realizing that he could wait no longer, Heckman ordered his guns limbered and sent to the rear. His order was, however, too late, and the Louisiana Tigers of Harry T. Hays's brigade and North Carolinians of Col. Isaac Avery's brigade overwhelmed two of his pieces before they had a chance to make their way to safety.[49] Major Osborn did not blame Heckman for the loss, writing in his report, "Though [Heckman] worked his battery to the best of his ability, the enemy crowded upon it, and was within his battery before he attempted to retire. . . . I think no censure can be attached to this battery for the loss of the gun[s]."[50] The two guns that survived the ordeal were no longer in fighting trim, so they were sent to the rear, and in the words of Osborn, he lost "the benefit of it during the fight of the second and third day." Heckman lost thirteen men killed and wounded, and nine horses went down.[51] One of the captured cannon was sent to Jones's battalion to replace the gun that had been permanently damaged by a Federal solid shot.[52]

Unlike some of the other battery commanders, Dilger was still full of fight. Although he was forced to pull back toward Gettysburg, he did not go easily. Instead, he halted one section of each battery and wheeled them into position just north of town while the remaining guns made their way to safety. The four cannon immediately banged away at the approaching Confederate infantry and helped hundreds of Federal troops successfully escape the clutches of the enemy. When the enemy was too close for Wheeler's rifled guns to be effective, Dilger ordered them to the rear while he continued firing canister from his two Napoleons, almost until the Confederates were at their muzzles. Dilger then ordered his two cannon withdrawn. He quickly realized that the main road through the town was awash with infantry, and he would probably lose his guns if he tried to navigate it. Instead he ordered a loop around the town, and the small, but lethal, fighting force finally rejoined the rest of the battery on Cemetery Hill.[53]

Wheeler lost one of his cannon during the retreat when a shell struck its axle, breaking it and dismounting the piece. With the enemy closing in, Wheeler quickly ordered his men to throw a prolonge around the barrel and attach it to a limber. This done, the barrel was dragged toward safety. Unfortunately, the rope broke, and Wheeler had no choice but to leave it behind. He was relieved when he rode north of town after the enemy had retreated on July 5 and found the barrel. Retrieving it, Wheeler was able to make the cannon whole again before the battery left Gettysburg.[54]

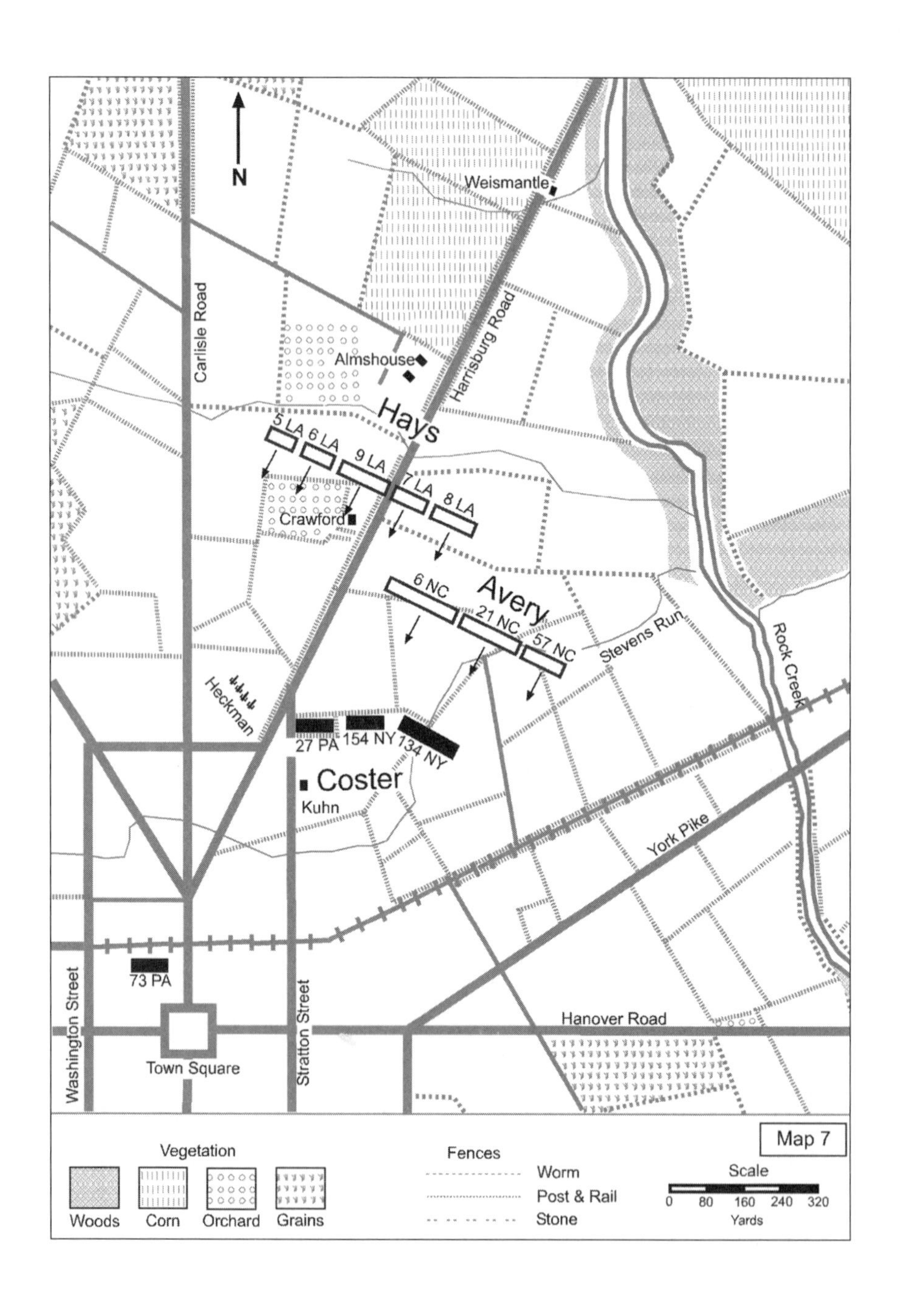
N
Weismantle
Carlisle Road
Harrisburg Road
Almshouse
Hays
5 LA
6 LA
9 LA
7 LA
8 LA
Crawford
6 NC
Avery
21 NC
57 NC
Stevens Run
Rock Creek
Heckman
27 PA
154 NY
134 NY
Coster
Kuhn
York Pike
73 PA
Washington Street
Town Square
Stratton Street
Hanover Road
Map 7
Vegetation
Woods
Corn
Orchard
Grains
Fences
Worm
Post & Rail
Stone
Scale
0 80 160 240 320
Yards

The honor of being the last organized unit to leave the town probably went to Wilkeson's battery. Relieving Dilger's section on Carlisle Road with its gun limbers replenished with ammunition, it covered the retreat by firing into the streets of Gettysburg.[55]

While Early's division was routing the 11th Corps with help from Doles's brigade, the rest of Rodes's division, south of Oak Hill, prepared to renew its attacks. The initial attacks had been disjointed, and each brigade was repulsed independently. Getting all of his brigades into line, including Alfred Iverson's and Edward O'Neal's shattered ones, Rodes's coordinated attack now smashed through the remnants of John Robinson's division on Oak Ridge. The Confederate infantry followed the Federal troops fleeing toward Gettysburg, and Charles W. Fry's battery was ordered to quickly limber up and closely follow the infantry. The battery halted periodically to fire at Federal troops attempting to make a stand. Its gunners exhibited "energy, coolness, and skill" as they discouraged further resistance.[56]

With all of the Federal troops north of town in full flight toward Cemetery Hill, Carter's battalion took up the chase with, as he put it, "a few pieces unlimbering from time to time to break up the formations of the enemy as they endeavored to rally under cover of the small crests near the town."[57] The men of Capt. James McD. Carrington's Charlottesville Battery (four Napoleons) of Jones's battalion had not been happy to be in reserve during the initial attack on Blocher's Knoll. When Early rode by, they expressed their desire to enter the fight and were told to be patient for they would see plenty of action before long. They got their wish when Carrington was ordered across Rock Creek to follow the victorious Confederate infantry down Carlisle Road toward the town. This was easier said than done, for Union artillery was shelling the bridge, and Carrington realized that if a horse was hit while crossing, it could effectively halt the remainder of the battery. There was no time to waste, however, so Carrington crossed one piece at a time and quickly followed the infantry toward Gettysburg. Carrington claimed that his was the only battery to unlimber in the streets of Gettysburg and throw shells at the retreating Federal troops. After about twenty minutes, Carrington ordered his battery back to Jones's battalion, which was still east of Rock Creek.[58]

As Major Osborn's four defeated batteries wheeled to a halt on Cemetery Hill, they reunited with Capt. Michael Wiedrich's Battery I (six ordnance rifles), 1st New York Light, which had been held in reserve atop East

Cemetery Hill. As the battery initially took position, General Howard rode up and said, "Boys, I want you to hold this position at all hazards. Can you do it?" The cannoneers yelled, "Yes sir!" Just then a shell screamed overhead, causing some of the men to instinctively duck. Closely watching them, Howard said, "Don't be alarmed, boys, that was an elevated shot, fired at random." Seeing masses of cavalry in the distance near York Road, Wiedrich's battery opened fire. The horsemen turned out to be Federal cavalry from Col. Thomas Devin's brigade (Buford's division), which had been assigned to guard Barlow's right flank. The friendly fire cost the cavalry some horses, but no casualties, and caused the troopers to quickly abandon their position of guarding the 11th Corps' right flank. The battery's six rifled cannon also exchanged fire with one of Jones's batteries on the opposite side of Rock Creek.[59]

At about 5:00 p.m., Michael Wiedrich could see the beaten Federal troops streaming into town, closely followed by Confederate infantry. Shifting his guns' positions, he opened fire, first with shells, then with canister, which dampened the Confederates' ardor. Isaac Avery's brigade of North Carolinians quickly sought the protection of the railroad cut near the town. After marching eastward about four hundred yards, Avery's men were ordered out of the cut and again exposed to Wiedrich's artillery fire, so they were ordered to lie down in a depression behind a hill. Wiedrich maintained this fire until late into the evening. His cannon fire, coupled with the fatigue of the Confederate troops who were tired from their march to the battlefield and the subsequent fighting, caused their commanders to delay the attack on Cemetery Hill until July 2. The Confederates also considered taking Culp's Hill that evening, but again they passed on the opportunity, and the battle was lost.[60]

Cemetery Hill

Who commanded the Federal artillery on Cemetery Hill has been hotly debated for many years. Col. Charles S. Wainwright, commander of the 1st Corps artillery, claimed that General Howard ordered him to assume command of the guns of both corps. Maj. Thomas W. Osborn of the 11th Corps disagreed, stating that, during the evening of July 1, Wainwright commanded the artillery on the northeastern side of Baltimore Pike, and he commanded the rest.[61] Artillery commander Brig. Gen. Henry Hunt clarified the issue in his report: "The batteries at the cemetery, under command of Colonel Wainwright, remained in position . . . and Major Osborn, chief of artillery of the Eleventh Corps, was directed to take command on the south side of the road."[62]

Osborn was angered when asked to restock Wainwright's caissons with his own ammunition because the 1st Corps ammunition train was far to the rear. "This of necessity caused considerable annoyance later in the engagement, on account of the difficulty in procuring a supply of ammunition sufficient to cover the great expenditure we were compelled to make through the engagement," Osborn noted in his report.[63]

Gen. Winfield Hancock, leader of the 2nd Corps, arrived in the late afternoon and assumed command of the hill. Wainwright insisted that he never saw him, but his men did. Hancock set about raising the men's morale and posting the guns.[64] He asked James Stewart how many serviceable guns he had, and when told four, Hancock ordered him to position three on the pike and the other at a right angle to them. "I want you to remain in this position until I relieve you in person," Hancock said before riding off. This was Hancock's first battle as a corps commander, and he did more positioning of the artillery than any of his counterparts. This would lead to confusion and confrontation later in the battle.[65]

During the retreat, Greenleaf T. Stevens's and James H. Cooper's guns became intermingled, but all arrived safely on Cemetery Hill. Hancock called for the "captain of that brass battery." When Stevens approached, he was told to "take [his] battery on to that hill," pointing to Culp's Hill to the southeast and "stop the enemy from coming up that ravine." Stevens asked, "By whose order?" and was told, "General Hancock's," and then he rode away. Separating his guns from Cooper's, Stevens led his command across Baltimore Pike toward Culp's Hill, ultimately positioning it on a knoll that commanded the valley between Culp's and Cemetery hills. It later became known as Stevens's Knoll in honor of the battery and its commander who performed so gallantly during the battle.[66]

Col. Charles S. Wainwright found Michael Wiedrich's battery deployed on the northern end of the hill when he initially arrived on Cemetery Hill. With this battery as the foundation, Wainwright began positioning the 1st Corps batteries as they arrived. Cooper's battery was placed to the right of Wiedrich's and somewhat to the rear. One section of Stewart's battery, and all of Reynolds's, now commanded by Lt. George Breck, were placed to the right of Cooper's battery. They were not actually in line with Cooper's guns but about twenty yards behind it, "owing to the nature of the ground." The remaining two sections of Stewart's battery unlimbered in front of the cemetery's gatehouse, facing northwest so they could fire down Baltimore Pike.[67] Wainwright thus had twenty-three cannon on East Cemetery Hill (thirteen rifled cannon and ten Napoleons). Breck wrote that his position was "high and commanding," but that did not stop the gunners from building lunettes for protection.[68]

Similarly, when Osborn arrived on Cemetery Hill, he found the three serviceable guns of Hall's 1st Corps battery just to the right of Taneytown Road, facing west. At about 5:00 p.m., Osborn directed Dilger's battery to take position behind a stone wall on the northwestern edge of the cemetery, just to the left of Stewart's battery. Wilkeson's battery, now under Lt. Eugene A. Bancroft, was positioned on Dilger's left. Upon taking position, Bancroft ordered a few shells thrown at the enemy, but they did not elicit a response. Bancroft's other section, under Lt. Christopher Merkle, also arrived and was positioned by Osborn to the right of Wiedrich's, about a hundred yards north of the cemetery. He was given a third gun from Bancroft's battery, which he positioned on Baltimore Pike.[69] Merkle's three guns did not remain here long, for Osborn soon ordered them to the rear of Cemetery Hill, where they remained all night. Merkle returned the guns to Bancroft the next morning.[70]

William Wheeler's battery was stationed on the left of the line, to the right of James A. Hall's. Later that evening the left section of Michael Wiedrich's battery was shifted to the west side of the cemetery to fill a gap between Hall's right and Wheeler's left, where it remained for the rest of the battle. Lewis Heckman's battery, which valiantly attempted to hold the line at Kuhn's brickyard, saw no further action, as its two remaining guns were in such bad shape that they were sent to the rear for the rest of the battle. Thomas W. Osborn had about twenty guns in his sector. Eight were rifled, and twelve were smoothbore Napoleons.[71]

With still an hour of daylight left in what had been a very long day, Wainwright worried about an attack by the Confederate troops massing in the town. His biggest fear was that his shells would put civilians and wounded troops in the town in harm's way. Seeing some of his artillery firing, he quickly told the officers to save their ammunition and "not to take orders from any man with a star on his shoulders who might choose to give them."[72]

These activities defiled a once lovely and serene cemetery. Lt. George Breck wrote, "A beautiful cemetery it was, but now is trodden down, laid a waste, desecrated. The fences are all down, the many graves have been run over, beautiful lots with iron fences and splendid monuments have been destroyed or soiled, and our infantry and artillery occupy those sacred grounds where the dead are sleeping. . . . [I]t is enough to make one mourn."[73] Hartwell Osborn of the 55th Ohio added, "The artillery of the corps was massed on the summit and created havoc in the burial grounds as the teams and heavy guns crashed over the sodded hillocks, or sent gravestones flying, regardless of everything save the necessity of placing the guns to meet the enemy."[74] Ironically, a sign posted at the cemetery's entrance read, "All persons found using firearms in these grounds will be prosecuted with the utmost rigor of the law."[75]

Perched on Cemetery Hill, the men took stock of the day's events. Lt. George Breck, now commanding Reynolds's battery, probably summed it up best when he wrote to his wife on July 2, "We were never in such a severe fight before, never, for here we were subjected to a close and heavy fire of musketry, most all the time we were engaged, which has never been the case heretofore in our encounters with the enemy, only at intervals and for a brief season."[76]

Given the mismatch, it could have been much worse. Of the fifty-four guns in the ten batteries of the 1st and 11th Corps that arrived on the battlefield that day, only a total of eighty-three men and eight officers were killed or wounded. The Confederates had captured and retained three guns (one from Gilbert H. Reynolds's battery and two from Lewis Heckman's), and seven guns were disabled (three of James A. Hall's; one from William Wheeler's; one of James H. Cooper's guns broke an axle but managed to reach Cemetery Hill; two of James Stewart's pieces had lost their pointing rings and could not be used until they were replaced). Their guns replenished with ammunition, the Federal artillerymen were ready for a renewal of the enemy's onslaught. None came, and the men rested or strengthened their defensive positions.[77]

The Confederate batteries were likewise given a well-deserved rest. Casualties were high, but not as high as the Federals'. Probably because of Early's fast-moving engagement with the Federal troops north of town, Lt. Col. Hilary P. Jones's battalion sustained only modest casualties: one killed and one wounded. Thomas H. Carter's battalion sustained the highest losses. According to Col. J. Thompson Brown, the 2nd Corps artillery chief, the battalion lost six killed and thirty-six wounded. Most of these losses—two killed, two mortally wounded, and twenty-six wounded—were sustained by Richard Page's battery during its terrible duel with Dilger's and Wheeler's batteries.[78]

None of the Confederate batteries lost guns to the enemy, but several were put out of action for varying lengths of time. For example, Maj. David G. McIntosh reported that two guns in his battalion were disabled on July 1. One of Lt. Samuel Wallace's ordnance rifles was struck on its face and sent to the rear, presumably not to return, but a broken axle on a Whitworth was repaired and quickly put back into action. William J. Pegram's losses are less clear, as the battalion's report contains its total three-day losses, which were three guns and one caisson disabled and two caissons exploded. Most of these losses probably occurred on the first day. Carter did not include an accounting of his ordnance losses during the battle, but given the human toll, it may have been fairly high. While Lt. Col. Hilary P. Jones's battalion's human losses were modest, it sustained four ordnance losses—three guns were lost when the wrong ammunition was used, and a fourth was disabled by an enemy shot. The three former guns were put back into action before the battle ended, and the fourth was replaced with one of the guns captured from Lewis Heckman's battery.[79]

While the Federal artillerists were preparing their defenses on Cemetery and Culp's hills, and most of their Confederate counterparts were resting, the men of the Fredericksburg Battery had other things in mind, such as filling their stomachs. Camped near a large "country residence," the men could not resist exploring it and found a "well-appointed household." When the food was exhausted, the men turned their attention to the wine closet. Adjourning to the "drawing room," one soldier played the piano while the others engaged in a "stag dance."[80]

Overall, July 1 was not a good day for the Federals, particularly their artillery. In the absence of their chief, Brig. Gen. Henry Hunt, the infantry commanders exercised undue influence over the batteries, often placing them in untenable positions for the sole purpose of supporting the infantry. According to two historians, there was "little coordination, no choice of terrain, no cross-fire." This would change with the arrival of Hunt later that evening.[81]

3

JULY 2

THE TWO SIDES TAKE POSITION

FEDERALS TROOPS ARRIVE AND FORM THE DEFENSIVE "FISH-HOOK"

Brig. Gen. Henry Hunt, commander of the Army of the Potomac's artillery, did not get much rest on the night of July 1–2. At forty-three years old, Hunt was a mere four years younger than Maj. Gen. George Meade, who now commanded the army. The army commander's entourage arrived at Gettysburg at about 2:00 a.m. on July 2. Meade immediately ordered the West Point graduate to examine the positions on Cemetery Hill "so far as the darkness would permit" and then accompany him on an inspection of the rest of the army's positions.[1] Probably the leading authority on the use of artillery, Hunt had distinguished himself in the war with Mexico. He learned how to effectively deploy large aggregations of cannon when he commanded the artillery around Washington, D.C. early in the Civil War. By the Peninsula campaign, Hunt was commanding the artillery reserve. His abilities were most evident at the battle of Malvern Hill during the Seven Days' campaign. Maj. Gen. George B. McClellan, then the army commander, rewarded Hunt by giving him command of all the artillery during the Antietam campaign. He later performed superbly at Fredericksburg, but then a feud with Maj. Gen. Joseph Hooker, who was the army commander at the time, relegated him to an administrative position with little authority over the artillery. The folly of this move proved evident at the battle of

Chancellorsville. When he realized his mistake, Hooker restored Hunt to command the army's artillery.[2]

Hunt spent the next few hours surveying the artillery's positions, scrutinizing their condition and assuring that they had adequate ammunition. He excelled in this activity, and the next two days would illustrate Hunt's effectiveness.[3]

The morning of July 2 was quiet as units from both armies continued arriving on the battlefield. The cannoneers of both armies spent the time resting and, in some cases, building defensive works for their guns. Some units opened fire from time to time. For example, Greenleaf T. Stevens's battery threw some shells at Confederate infantry taking position near the town. Lt. Edward N. Whittier reported, "On advice received from General Adelbert Ames [commanding Barlow's division, 11th Corps], the battery opened at 1,500 yards on columns of the enemy infantry passing within range, and inflicted some considerable damage."[4] A few soldiers believed that their artillery's sporadic fire could be a way of goading the enemy into attacking them. For example, Pvt. Stephen Wallace of the 153rd Pennsylvania recorded in his diary on July 2, "Our forces have a good position and would rather the Rebs to make an attack. We fired cannon every once in a while in order [to] get them to come out."[5]

Stevens was severely wounded during the day. Shot through both legs below the knees, probably by skirmishers from Isaac Avery's brigade, Stevens was replaced by Lt. Edward N. Whittier.[6]

At about daylight, Maj. Gen. Howard W. Slocum informed Meade that he was concerned about the gap in the infantry line between his 12th Corps positions and those of the 1st Corps on Culp's Hill. If left unattended, the enemy could drive a wedge through it and gain the army's rear. Meade told Hunt to remedy the situation by positioning some artillery to cover the vulnerable area until infantry could be found to fill it. Hunt complied by sending Lt. Sylvanus T. Rugg's Battery F, 4th U.S. (six 12-pounder Napoleons), Lt. David H. Kinzie's 5th U.S. (four 12- pounder Napoleons), Lt. Charles E. Winegar's Battery M, 1st New York (four 10-pounder Parrotts), and Lt. Charles A. Atwell's (Knap's) Battery E, Pennsylvania Independent (six 10-pounder Parrotts), all of Lt. Edward D. Muhlenberg's 12th Corps Artillery Brigade to the threatened area. Muhlenberg told a slightly different story: "The density of the growth of timber, the irregularity and extreme broken character of the ground, studded with immense bowlders, prevented the artillery from taking position in the line proper of the corps. It was, therefore, held in reserve and readiness to answer all calls which might be made upon it by the future movements of the opposing forces."[7]

Over on the east side of Baltimore Pike, Col. Charles S. Wainwright adjusted some of his batteries. James Stewart's battery shifted to the right of the road, facing northwest, with its left close to Hubert Dilger's battery. Michael Wiedrich's battery was slightly to the northeast, and R. Bruce Ricketts's, which had replaced James H. Cooper's guns, was to his right. Gilbert H. Reynolds's battery was deployed farther to the right, and even farther to the right was Greenleaf Stevens's battery, deployed on the knoll between Cemetery and Culp's hills. With the exception of Stewart's battery, which faced northwest, all of Wainwright's guns faced northeast. Most of the cannoneers had built shallow lunettes to shield them from enemy fire.

Later in the afternoon, Hunt visited Maj. Thomas W. Osborn's sector on the west side of Baltimore Pike. Osborn outspokenly expressed his concerns that the area to be defended north and west of town was just too large for the seventeen cannon he had in position. Could Hunt supplement the five 11th Corps batteries (actually four, as Lewis Heckman's was sent to the rear) on Cemetery Hill with additional ones? Hunt quickly reacted, sending thirty-two guns from his reserve. From the 1st Regular Brigade came Lt. Chandler P. Eakin's Battery H, 1st U.S. (six Napoleons), from the 2nd Volunteer Brigade came Capt. Elijah D. Taft's Independent Battery, 5th New York (six 20-pounder Parrotts), and from the 3rd Volunteer Brigade came Capt. Wallace Hill's Battery C, 1st West Virginia (four 10-pounder Parrotts), Lt. George W. Norton's Battery H, 1st Ohio (six ordnance rifles), and Capt. Frederick M. Edgell's 1st New Hampshire (four ordnance rifles). These guns swelled Osborn's total to forty-nine. This was a formidable collection of ordnance that would show its effectiveness on the afternoon of July 3 against the Pickett-Pettigrew-Trimble Charge.[8] According to Hunt, Osborn deployed his artillery in the following manner: Dilger's battery, with its right gun on Baltimore Pike, facing northwest, was deployed on the right. On its left was Eugene A. Bancroft's and then Chandler P. Eakin's, William Wheeler's, Wallace Hill's, and James A. Hall's batteries. Elijah D. Taft's large-bored guns were behind and perpendicular to Bancroft's battery, facing north-northeast. George Norton's battery was to his right, facing in the same direction.[9]

Earlier on July 2, the Federal 2nd Corps reached the battlefield, and its artillery chief, Capt. John G. Hazard, immediately positioned the corps' artillery battalion in the center of the Federal line. Hazard had begun the war as a lieutenant in a battery from his home state of Rhode Island, and a year later, in August 1862, he was promoted to captain and given command of Battery B, 1st Rhode Island. His star continued to rise because of his effective service, and he received command of the 2nd Corps artillery brigade after Chancellorsville. Gettysburg would be Hazard's first opportunity to command more than a battery with six guns.[10]

Hazard placed Lt. George A. Woodruff's Battery I, 1st U.S. Artillery (six Napoleons) on the right, just south of the town, in Ziegler's Grove. Next came Capt. William A. Arnold's Battery A, 1st Rhode Island Light Artillery (six ordnance rifles). Lt. Alonzo H. Cushing's Battery A, 4th U.S. Artillery (six ordnance rifles) was deployed to the south, just to the right (north) of the copse of trees. Lt. T. Frederick Brown deployed his Battery B, 1st Rhode Island Light Artillery (six Napoleons) to the left of the copse of trees. Further to the south (left), was Capt. James M. Rorty's Battery B, 1st New York Light Artillery (four 10-pounder Parrotts), which was attached to Gen. John Caldwell's 1st Division.[11]

The Federal 3rd Corps was also taking position farther south of the 2nd Corps. John Calef's battery, with the rest of John Buford's cavalry division, had guarded this sector since the night before, but all had left the battlefield for Taneytown, Maryland, when the 3rd Corps arrived. Neither Maj. Gen. Daniel E. Sickles, nor his artillery chief, Capt. George E. Randolph, found the area from Round Top to the south to the southern part of Cemetery Ridge particularly desirable. A nephew of former President William Henry Harrison, twenty-three-year-old Randolph had left his home state of Illinois to become a first sergeant of Battery A, 1st Rhode Island Light Artillery. He was wounded in both legs at the battle of First Bull Run and became a second lieutenant in August 1861 and a first lieutenant a month later. He hardly had a chance to break in his new bars when he was promoted in September to captain and given command of Battery E. He became a 3rd Corps division artillery chief during the winter of 1862 and was assigned to command the entire 3rd Corps artillery in April 1863. Few men saw such a meteoric rise in stature and fame.[12]

Randolph was most distressed about the lack of adequate artillery platforms. He dutifully deployed his old battery—Battery E, 1st Rhode Island

Light Artillery (six Napoleons), now under Lt. John K. Bucklyn—and Capt. A. Judson Clark's Battery B, 1st New Jersey Light Artillery (six 10-pounder Parrotts), near Gen. David Birney's division on the left (south) near Little Round Top. Randolph wrote in his report that the "positions of both were low, unprotected, and commanded by the ridge along which runs the road from Emmitsburg to Gettysburg."[13] As hard as he looked, he could not find a suitable place for Lt. Francis W. Seeley's Battery K, 4th U.S. Artillery (six Napoleons), so it remained in the rear. When the two remaining batteries—Capt. James E. Smith's Independent Battery, 4th New York (six 10-pounder Parrotts), and Capt. George B. Winslow's Battery D, 1st New York (six Napoleons)—arrived, Randolph reluctantly parked them near Seeley's battery in reserve.[14]

Sickles continually requested permission to advance his corps toward Emmitsburg Road. Meade was busy on Cemetery Hill and could not ride down to visit Sickles, so after numerous requests to move his corps forward, Sickles decided that he could wait no longer and advanced his entire corps to occupy the high ground west of his assigned position during the afternoon. Randolph approved, for "this new disposition seemed to me, notwithstanding the sharp angle in our line made necessary by the formation of the ground, to be a much more desirable one."[15] Because of the corps' long front, Randolph was forced to disperse his batteries widely. Making matters worse was the sharp bend in the line, or salient, at the Peach Orchard. James E. Smith's battery was to deploy on Houck's Ridge on the left, where it could support Gen. Hobart Ward's brigade. This was the leftmost battery on the Federal line. Because the space above Devil's Den could not accommodate all six pieces, Smith decided to place four there and the other two to his right and rear so that they could fire into the Plum Run Valley. Smith had a tough time positioning the four guns on the steep and rocky ridge, as his men had to physically haul the guns up the rocky slope. George B. Winslow's battery dropped trail about fifteen hundred yards to Smith's right in the Wheatfield. Rose's Woods separated these two batteries from each other's view. Randolph knew that Winslow's position in the Wheatfield was risky, as it was in the open and flanked by woods that could mask enemy movements. He was willing to gamble, noting, "This position was surrounded by woods, but, in my opinion, the line was materially strengthened by this battery of short-range guns." Sometimes cannon had to be sacrificed for the greater good, he reasoned, and what better guns than Napoleons, which were deadly against infantry at close range.[16]

Randolph was probably most concerned about the Peach Orchard salient, where the Federal line took a ninety-degree turn to the north. Hunt arrived at this time, and the young artillery commander asked his superior's advice on deploying his remaining batteries. As a result of this conversation, A. Judson Clark's and John K. Bucklyn's batteries moved to the Peach Orchard, and a short time later they were joined by Capt. Nelson Ames's Battery G, 1st New York Light Artillery (six Napoleons), of the 4th Volunteer Brigade, which had been assigned to the 3rd Corps that morning. Clark's battery, which would assume six different positions on July 2, first deployed between the Peach Orchard and the Trostle house, and then moved into the orchard, where each gun threw six or seven shells into John B. Hood's division, which was about fourteen hundred yards away on Seminary Ridge and in the process of marching toward its preattack position. When the enemy infantry column disappeared from view, Clark moved his battery to the left, along Wheatfield Road. Sickles approached Clark and said, "Hold this position while you have a shot in your limbers and a man to work your guns." Nelson Ames's battery deployed directly in the Peach Orchard, and to its right, along Emmitsburg Road, was Bucklyn's battery, which dropped trail just north of the Wentz house, between it and the Sherfy farm buildings. Randolph's last battery, Francis W. Seeley's, was assigned to Andrew Humphreys's division and not yet deployed.[17]

The Confederate batteries on Seminary Ridge opened fire as the 3rd Corps batteries approached their new positions. Ames estimated that the enemy batteries were firing from a distance of about 850 yards. The first shells fell while his battery was crossing an open field. "They got an excellent range of my battery," noted Ames, "nearly all of their shots striking my battery, but fortunately they did no other damage than killing 2 horses."[18] The harrowing situation was not yet over, for now two fences loomed in front of the battery. "I was obliged to halt in plain sight of the enemy, to clear away two fences which the supporting infantry had failed to throw down as they were ordered to do." This task completed, Ames's guns dropped trail in the orchard and immediately returned fire with spherical case shells.[19]

Hunt now had a massive and well-placed array of guns on Cemetery and Culp's hills and Cemetery Ridge, facing north and west. With the 1st, 2nd, 3rd, 11th, and 12th Corps batteries deployed, he looked forward to the arrival of the 5th and 6th Corps guns that were nearing the battlefield. In the meantime, Hunt had begun pulling units from the artillery reserve. This was an organization of 18 batteries, composed of 90 cannon, 1,717 men and 1,785 horses. Those guns not deployed were parked at four small farms over a one-square-mile area in the rear of the center of the line, near Taneytown Road.[20]

A new concern arose for Hunt. The 3rd Corps ammunition train had been left behind, and other corps were "deficient." Given the large ammunition expenditures by the 1st and 11th Corps on the first day of the battle, Meade was rightfully concerned that the army might not have enough ammunition to carry on a heated fight. Hunt received the news calmly.[21] Hunt knew he could trust Brig. Gen. Robert Tyler, commander of the artillery reserve, who could be expected to leave everything else behind—except for the cannon—to "bring up every round of ammunition in his trains, and I knew he would not fail me."[22] Unbeknownst to the army's commander, Hunt had formed a special ammunition train, attached to the artillery reserve, which carried 20 additional rounds above the 250-round quota for each gun. Hunt was not worried, and Meade was consoled. According to Hunt, "This was not the first nor the last time I was called upon to meet deficiencies under such circumstances."[23]

Hunt had every reason to put his confidence in Tyler. A career artillerist who had graduated from West Point in 1853, Tyler knew his craft well. In a fast series of events, he was promoted from lieutenant to colonel over three months during turbulent 1861. Much of his time in the early part of the war was devoted to building and maintaining the artillery defenses around Washington. He also commanded George B. McClellan's siege train during the disastrous Peninsula campaign, losing but one gun. Tyler received his general's star in November 1862 and by May 1863 was in command of the army's artillery reserve.[24]

Confederate Exterior Positions

The balance of heavy metal had shifted by the morning of July 2. After the mismatch in artillery on July 1, the Federal guns now greatly outnumbered the Confederates'. The batteries assigned to the Confederate 2nd and 3rd Corps spent the night of July 1–2 generally behind the infantry lines just north and west of Gettysburg. Col. R. Lindsay Walker, the 3rd Corps artillery chief, held William J. Pegram's, David G. McIntosh's, John Lane's, William T. Poague's, and part of John J. Garnett's battalions behind the crest of Seminary Ridge. These guns were ready to take position on the ridge at dawn. Another graduate of the Virginia Military Institute, Walker was thirty-six years old during the Gettysburg campaign. A wealthy landowner at the start of the war, his military career began as commander of the Purcell Battery, and he saw action at the First Manassas. He performed well on the battlefield and, combined with his circle of influence, realized repeated promotions. By the spring of 1862 he was commander of the artillery attached to A. P. Hill's division. In his recommendation for Walker to be a battalion commander, William Pendleton called him "justly distinguished for long and gallant service." He subsequently commanded the battalion at the battle of Chancellorsville. Several weeks later, Walker was promoted to colonel and given command of the 3rd Corps artillery. His inexperience as a commander of several battalions showed at Gettysburg, as he left so many battalions out of the fray on July 1.[25]

North of town, Col. J. Thompson Brown, 2nd Corps artillery chief, assigned R. Snowden Andrews's (under Maj. Joseph W. Latimer), Hilary P. Jones's, and Willis J. Dance's battalions to positions behind Gen. Edward Johnson's division. Thomas H. Carter's and William Nelson's battalions remained along the ridge north of town. The dearth of adequate artillery platforms north of town posed a serious problem, and Brown spent considerable time trying to find places for his guns.[26]

A twenty-eight-year-old Virginian at the time of the battle, Col. J. Thompson Brown had entered the war as a lieutenant in the Richmond Howitzers. A month later he had his own battery in this venerable organization. The battery had the distinction of firing the first hostile artillery round in Virginia, when one of his sections fired on a Federal gunboat on May 14, 1861. Brown was promoted again in September 1861 to command an artillery battalion. With the reorganization of the army after Chancellorsville, Brown was assigned to command the 2nd Corps artillery and its reserve.[27]

Sometime after 3:00 a.m. on July 2, Col. R. Lindsay Walker directed all of his 3rd Corps battalions to deploy on Seminary Ridge to the right of Fairfield Road. This placed them opposite the now well-fortified Cemetery Hill, which bristled with Federal artillery, and the rapidly arriving Federal troops on Cemetery Ridge to the south. David G. McIntosh's battalion formed behind a stone wall on the edge of Schultz's Woods just to the right of Fairfield Road. Twelve of his guns in three batteries were deployed and ready for action. Samuel Wallace's battery was to the left, and William B. Hurt's and Marmaduke Johnson's were to its right; R. Sidney Rice's Napoleons were in reserve. Next came John J. Garnett's battalion (now commanded by Maj. Charles Richardson) south of Schultz's Woods. Richardson was ordered to place all nine of the battalion's rifled pieces on Seminary Ridge, opposite Cemetery Hill. His remaining six nonrifled pieces stayed in reserve because they lacked the necessary range to hit the enemy on Cemetery Hill. To Garnett's right, Maj. William J. Pegram deployed his battalion near McMillan's Woods. Five batteries were in line, from left to right: Thomas Brander's, Joseph McGraw's, William E. Zimmerman's, Marmaduke Johnson's, and Edward A. Marye's. Although they had been actively engaged on July 1, these batteries were still full of fight.

Maj. John Lane, who was in temporary command of the battatlion because of the illness of its regular commander, Col. Allen Cutts, had never commanded the battalion before. He deployed it to the right (south) of Pegram's, but not before he sent Capt. George M. Patterson's Battery B, 11th Georgia (Sumter) (two Napoleons and four 12-pounder howitzers), along with one 12-pounder howitzer from Capt. Hugh M. Ross's Company A of the Sumter Artillery, farther south to support Gen. Cadmus Wilcox's brigade, which formed the right of the army near Pitzer's Woods. Lane deployed his remaining ten guns just south of McMillan's Woods. Ross's remaining five guns (one Napoleon, one 3-inch navy rifle, and three 10-pounder Parrotts) dropped trail to the right of Pegram's battalion, followed by Capt. John Winfield's Company C (three 3-inch navy rifles and two 10-pounder Parrotts). Capt. Joseph Graham's Charlotte Artillery (two 12-pounder howitzers and two Napoleons) and Capt. James Wyatt's Albemarle Artillery (one 12-pounder howitzer, two ordnance rifles, and one 10-pounder Parrott) deployed on Lane's right.[28]

Col. J. Thompson Brown, the 2nd Corps artillery chief, also shifted some of his units during the morning of July 2. Capt. Archibald Graham's battery of Willis J. Dance's battalion had been sent to the left the night before, where it joined Hilary P. Jones's battalion. Now Dance's four remaining batteries were shifted to the right, where they were placed on high ground under the direction of Thomas H. Carter. Capt. David Watson's 2nd Richmond Howitzers (four 10-pounder Parrotts) was placed on the left (north) of the unfinished railroad cut facing Cemetery Hill to the east. Capt. Benjamin H. Smith Jr.'s 3rd Richmond Howitzers (four ordnance rifles) dropped trail approximately 450 yards to Watson's right, near the seminary, and Lt. John M. Cunningham's Powhatan Artillery (four ordnance rifles) deployed farther to the right (south) and just to the left of Fairfield Road. Hupp's Salem Artillery (two Napoleons and two ordnance rifles) under Lt. Charles B. Griffin formed the reserve.[29]

Lt. Col. William Nelson, who commanded the other battalion in Brown's artillery reserve, was ordered to report with his batteries to Lt. Col. Thomas H. Carter. The latter apparently ordered him to move his battalion during the early morning hours of July 2 from Seminary Ridge to the left of Chambersburg Pike. This was, in Nelson's words, "in rear of [the] heights overlooking the town, and about one-fourth of a mile to the left of the Cashtown [Chambersburg] turnpike."[30] He was again ordered to move at 11:00 a.m. on July 2 to a point "immediately in [the] rear of the Gettysburg College [northwest of the town of Gettysburg]" where he was to "park my batteries, and await events."[31]

At fifty-four years old, Nelson was the oldest Confederate battalion commander at Gettysburg. He was a Virginia legislator at the outbreak of the war, and he quickly quit that post in April 1861 to become captain of the Hanover Light Artillery. Upon the unit's disbanding, Nelson became a major of artillery in May 1862. He was promoted again during the artillery's reorganization in March 1863 and given command of a battalion, one of the smallest in the Army of Northern Virginia. William Pendleton wrote that Nelson was as "gallant and efficient an officer as we have in his grade . . . [he] has exhibited courage of the highest order and fidelity undeviating." To his men he was a father figure, easily distinguished from other officers by his high silk hat.[32]

The batteries of both armies fired occasionally during the day. Sometimes the cannon fired at the infantry; at other times they fired to find the range of the enemy's positions. Up on Cemetery Hill, Lt. James Gardner of Cooper's battery recalled that his guns "fired occasional shots (scarcely exceeding twenty-five in all) at small bodies of the enemy's infantry and cavalry, which were manoeuvering in the skirting of some timber about one mile distant."[33] Col. R. Lindsay Walker, who commanded the Confederate 3rd Corps artillery, also noted that his batteries fired periodically through the day, "enfilading the enemy's guns when they were attempting to be concentrated."[34] Some of these shots fell among James H. Cooper's guns, killing and disabling a number of horses.[35]

Lee's last army corps, Longstreet's 1st Corps, arrived during the morning of July 2. Because it had not participated in the bloody fighting the day before, it would carry the brunt of the fighting on the Confederate right. As on July 1, a change in command occurred in the Confederate artillery. Col. E. Porter Alexander, a gifted artillerist who commanded a battalion in Col. James B. Walton's artillery reserve, was ordered to report to Longstreet on the morning of July 2. Alexander was a gifted artillerist.

Born in 1835, Alexander had been graduated from West Point, joined the Confederate army as a captain of engineers. His energy and organizational skills were quickly recognized, for he was next assigned to Gen. P. G. T. Beauregard's staff and told to organize a signal corps for the army. While serving as a signal officer at the battle of First Manassas, Alexander observed the initial Federal flanking movement, which helped secure the Southern victory on that battlefield. He was later named chief of ordnance for the army with the rank of lieutenant colonel, which primarily involved scrounging for equipment and ammunition. By November 1862, Alexander was a colonel with a field command in Stephen Lee's artillery battalion.[36]

Longstreet came to admire Alexander's talents and began relying on him more than his artillery chief, James B. Walton, who was about as opposite to Alexander as could be imagined. Born in New Jersey forty-nine years before, Walton attended Louisiana College and remained in the South, becoming a successful wholesale grocer. He saw action in the war with Mexico, and at the start of the Civil War, he commanded the venerable Washington Artillery and retained command of this unit up to the Gettysburg campaign.[37]

Like a jealous sibling, Walton responded negatively to Longstreet's growing dependence on Alexander, and tensions escalated. Now with a major task before him, Longstreet decided that he needed the best artilleryman available to command his batteries. Alexander's actions also had caught Lee's eye, and it was not surprising that they called upon him when they most needed him.[38]

On July 2, Walton met with Lee and Longstreet and was told that his subordinate would assume tactical oversight of the artillery. As a way of saving at least some face for Walton, Longstreet ordered him to remain in reserve behind the lines with his old command, the Washington Artillery, probably near the Herr Tavern. At a subsequent meeting at the Lutheran Theological Seminary, the two generals told Alexander to take charge of the 1st Corps batteries.[39] Alexander later wrote, "I could but feel sorry for Walton, who evidently felt himself overslaughted and that I was going to be practically put in charge of the artillery on the field."[40]

With Walton out of the way, Alexander was ordered to make a thorough reconnaissance prior to Longstreet's attack on the Federal left.[41] His first stop was apparently Fairfield Road, where he met Gen. William Pendleton, who wrote in his report, "[General Longstreet] desired Colonel Alexander to obtain the best view he then could of the front." Pendleton complied by escorting Alexander to a closer observation post that was well within range of Federal snipers.[42]

Lee told Longstreet that his corps would form the right of the Confederate line. The problem was getting his troops there. Longstreet decided to send John B. Hood's and Lafayette McLaws's two divisions (Pickett's division was still marching from Chambersburg) and the artillery on a circuitous route to avoid detection by Federal observation posts on Little Round Top. Everything went according to plan until the column approached Black Horse Tavern. Here McLaws, whose division led the column, realized that his men were ascending an exposed ridge and would soon be seen by the enemy. Alexander noted that the road "at one point passed over a high bare place where it was in full view of the Federal signal station."[43] Quick-thinking Alexander handled the situation expeditiously. "I avoided that part of the road [the exposed portion] by turning out to the left, & going through fields & hollows, & getting back to the road again a quarter mile or so beyond."[44] Longstreet, on the other hand, decided to turn his infantry around and retrace its steps, thus adding hours to the march and permitting more time for the Federals to establish their positions as reinforcements reached the battlefield. Alexander always regretted this delay, writing long after the war, "Our chances of success would have been immensely increased" had the infantry reached their stepping off points earlier.[45] Longstreet did not know it at the time, but the enemy had seen the column anyway.

As a result of Longstreet's decision, Alexander's batteries deployed long before the infantry arrived. Maj. Mathis W. Henry's artillery battalion, which was attached to Hood's division, deployed in the southern portion of Biesecker's Woods on the extreme right of the line. The guns had a clear shot to Devil's Den and the Round Tops from this position. A Kentuckian by birth, twenty-four-year-old Henry was a West Point graduate who had seen service in the Confederate cavalry, and by August 1862 he was commanding a horse battery under Maj. Gen. James Ewell Brown "Jeb" Stuart. Henry received his artillery battalion in the early portion of 1863, but there was some confusion over who really commanded the unit. Maj. John Cheves Haskell believed that he had received his promotion prior to Henry, so he contended that he was in command of the battalion.[46]

Capt. James Reilly's Rowan Artillery (two Napoleons, two ordnance rifles, and two 10-pounder Parrotts) deployed on the extreme right of the line, and Evander M. Law's brigade later halted near it. Capt. Alexander C. Latham's Branch Artillery (three Napoleons, one 12-pounder howitzer, and one 6-pounder smoothbore) dropped trail on the opposite side of the division, near the future position of Jerome B. Robertson's brigade. Like Reilly's battery, Latham's could fire on both Little Round Top and Devil's Den. It could also hit the Peach Orchard. The battalion's remaining batteries—Capt. Hugh R. Garden's Palmetto Light Artillery (two Napoleons and two 10-pounder Parrotts) and Capt. William K. Bachman's German Artillery (four Napoleons)—were placed in reserve despite a five-hundred-yard gap between the two deployed batteries.[47]

Col. Henry Cabell's artillery battalion, assigned to McLaws's division, dropped trail about 650 yards north of Latham's battery. At the age of forty-three, Cabell was considered to be an old man. An attorney by training, Cabell had raised an artillery company early in the war, and by April 1861 he was a captain in the Richmond Fayette Artillery. Promotion came frequently, until he reached the rank of colonel in July 1862, and with it came an artillery battalion that he commanded until the end of the war. Cabell was not well regarded by some historians. Jennings Wise referred to him as "an officer of great integrity and personal courage, but lacking in energy and ability as a field soldier." However, Alexander wrote, "The old man was not only a superb soldier, but a delightful gentleman also."[48]

Because of the concentration of Federal troops in front of the Confederate guns, Alexander "immediately put in Cabell's whole 18 guns, as one battery, from the edge of the woods about 700 yards from the Peach Orchard."[49] Capt. John C. Fraser's Pulaski Artillery (two 10-pounder Parrotts and two ordnance rifles) was on the right of the battalion; to its left was a section of two ordnance rifles of Capt. Edward S. McCarthy's 1st Richmond Howitzers, then came a section of 10-pounder Parrotts of Capt. Henry H. Carlton's Troup Artillery, followed by Capt. Basil C. Manly's Company A, 1st North Carolina Artillery (two 12-pounder howitzers and two 10-pounder Parrotts). The other section of Carlton's battery (two 12-pounder howitzers) formed the left of Cabell's line. Joseph B. Kershaw's brigade of McLaws's division ultimately formed behind these guns. The Federal artillery near the Peach Orchard immediately opened fire as the Southerners dropped trail and prepared for action.[50]

Alexander's own battalion, now under Maj. Frank Huger, was not deployed at this point, resting northwest of Cabell's battalion, near Millerstown Road. So Alexander had only twenty-five of the fifty-nine 1st Corps cannon deployed and ready for action. He would not deploy his own cannon until the Federal artillery opened fire later in the day.[51] Alexander wrote after the war, "Then, selecting 18 of my own 26, I put them in action at the nearest point, Warfield's House, where McLaws's line was within 500 yards of the Peach Orchard."[52] Capt. William W. Parker's Virginia Battery (one 10-pounder Parrott and three ordnance rifles) dropped trail on the right, adjacent to Cabell's guns. To its left, just south of the Warfield house, was Capt. Osmond B. Taylor's Virginia Battery (four Napoleons). Capt. George V. Moody's Madison Light Artillery (four 24-pounder howitzers) deployed on its left, and Lt. S. Capers Gilbert's Brooks Light Artillery (four 12-pounder howitzers) extended the line to the north. Gilbert's and Moody's batteries were in front of William Barksdale's brigade's right wing, between Cabell's artillery battalion and Wheatfield Road, while the other two were beyond Kershaw's left flank. Alexander's remaining two batteries—Capt. Tyler C. Jordan's Bedford Artillery (four ordnance rifles) and Lt. James Woolfolk's Ashland Artillery (two Napoleons and two 20-pounder Parrotts)—despite containing six guns that could easily reach the Federal line, were placed in reserve. The battalion had already sustained a casualty when a cannoneer fell off a limber and a heavy wheel crushed his thigh. He survived, but his days of being a soldier were over.[53]

With the combatants now in position, it was just a matter of time for the fight to begin.

4

JULY 2

THE BATTLE SOUTH OF WHEATFIELD ROAD

DEVIL'S DEN–HOUCK'S RIDGE, THE WHEATFIELD, AND LITTLE ROUND TOP

The Confederates Open the Battle

The plan for July 2 took several twists and turns before the actual attack was launched. After much debate, Robert E. Lee finally decided to attack both ends of the Federal line. While two of James Longstreet's newly arrived divisions, Lafayette McLaws's and John B. Hood's, attacked the Federal left, Richard S. Ewell's divisions would take on the Federal defenders on Cemetery Hill and Culp's Hill occupying the right of the Federal line.

Because Longstreet took so long to get his troops into position, his attack was not launched until late afternoon. The plan was for an *en echelon* infantry attack, where each brigade would attack a bit later than the one beside it. The goal was to find a weak spot in the Federal line and punch a hole through it. This approach was flawed because, while the line of attack was long, it was not deep, so once the enemy line was breached, there were few additional troops to exploit it. Prior to the infantry attack, the Confederate artillery would pound the defenders. Brig. Gen. Evander M. Law's brigade of Hood's division on the right would attack first, followed closely by Jerome B. Robertson's brigade. Like a row of dominos, the remainder of Hood's brigades, then McLaws's, and then Richard H. Anderson's of Hill's 3rd Corps would march off toward the Federal line.

As seen in the last chapter, the Confederate batteries did not wait long to engage the Federal 3rd Corps artillery approaching the Peach Orchard. E.

Porter Alexander explained, "I had hoped, with my 54 guns & close range, to make it short, sharp, & decisive." He noted, "At close ranges there was less inequity in our guns, & especially in our ammunition, & I thought that if ever I could overwhelm & crush them I would do it now."[1] While Henry Cabell's battalion, and later his own, aimed at the Peach Orchard, Mathis W. Henry's nineteen guns opened fire on the Devil's Den–Little Round Top sector. The battery commanders were probably surprised by what they saw as they deployed. Capt. Basil C. Manly explained that his orders led him to believe that his battalion would be on the enemy's left flank, so his battery was to "attempt to rake their line."[2] The Federal flank extended far to his right, however, which precluded such a flanking attack. This observation bedeviled Longstreet, whose whole plan of attack was predicated on the Federal line ending at the Peach Orchard. He vigorously protested his orders, asking Lee for permission to swing around the Federal left flank rather than take it on in a costly frontal attack. While Longstreet was pleading his case to Lee, his artillery opened fire on the Federal line.

Although the Confederate artillery was not on the Federal flank, the guns were arrayed in a line that could send a deadly enfilade fire into the enemy occupying the Peach Orchard salient. According to Lt. William J. Furlong of the Pulaski Artillery, "The firing at first was rapid, but soon became slow and cautious, the gunners firing slow, evidently making each shot tell with effect on the enemy's batteries."[3] By firing slowly, they could also wait for the smoke to clear and ascertain the effectiveness of each shot.

The Federal Batteries Respond

A. Judson Clark's battery was the first Federal battery to respond to the Confederate challenge. According to him, only one Confederate battery had opened fire on the Peach Orchard salient when Capt. George E. Randolph ordered him to "silence, or at least reply to, the fire, while I placed [Nelson] Ames's battery of light 12-pounders in the orchard to assist him."[4] Clark's guns had already limbered up and were in the process of leaving the orchard when he received these orders. He moved his guns to the left of the orchard, which his battery accomplished by a "right reverse trot," followed by an "action right," causing the guns to face in the correct direction, and "down went the trails." Gen. Daniel E. Sickles personally directed the battery into position and encouraged the men. Enemy shells fell among the guns as Sickles yelled to Clark, "Hold this position while you have a shot in your limbers or a man to work your guns."[5] Clark walked calmly from gun to gun, ensuring that the correct ammunition was being used and the fuses were being properly cut. He then moved to the middle of the battery and yelled, "Fire at will." According to gunner Michael Hanifen, the guns initially opened fire with shell and case shot. This ammunition—shells that released flesh-destroying shapnel when exploded—was chosen to quiet the Confederate battery by killing or maiming its gunners, not by dismounting its guns. The Union battery fired slowly, "first by gun, next by section, then by half battery, and once or twice by battery."[6]

While Clark's battery banged away at Longstreet's guns, Ames's battery wheeled into position. It had spent most of the time in the reserve artillery park beyond the Trostle barn before being ordered into the dense Peach Orchard. The men had anticipated action, so all were ready for combat.

As the guns rolled into the orchard, Nelson Ames realized that visibility was limited, so he opted to remain with the center and left sections, which were commanded by new officers. The right section was under the command of a seasoned lieutenant, so Ames assumed that he could give it less attention. After positioning the four guns, a sergeant from the right section rushed up to Ames and yelled, "For God's sake come and tell us where to place our guns; we have been running them up and down all over this field, no place is satisfactory to the lieutenant; all my men . . . are tired out."[7] Ames ran over to the lieutenant's section and found its officer clearly distraught. He told Ames that he could not open fire until a number of peach trees were cut down and pulled away. Losing his patience, Ames told him that if "another word escaped his lips during the battle, his sword would be taken from him and he [would be] sent to the rear." The officer became "dumb with silence."[8] Ames later learned that the officer had been stunned when an enemy shell whizzed past his head during the battery's movement toward the Peach Orchard. He finally regained his senses shortly after his guns were in action. Still nervous about his subordinate, Ames told the sergeant to report to him if there were any problems, and he promised to return periodically to observe the section's actions.[9]

On the opposite side, Lt. William J. Furlong of John C. Fraser's Pulaski Artillery noted that the "enemy replied with spirit, their fire being incessant, severe, and well directed."[10] One shot exploded among the guns, hitting Fraser and four of his men. Fraser's wound proved mortal.[11]

Alexander agreed with Furlong's assessment: "They really surprised me, both with the number of guns they developed, & the way they stuck to them. I don't think there was ever in our war a hotter, harder, sharper artillery afternoon than this."[12] The distances between the batteries, less than seven hundred yards, certainly contributed to the intensity.

The thick growth of trees on Seminary Ridge protected Henry Cabell's batteries. For example, Basil C. Manly's battery, which was probably the first Confederate unit to fire on the Peach Orchard, sustained only eleven casualties on July 2 and 3, but the Federal shrapnel flying through the woods unnerved even the toughest veteran, and at least one member of William W. Parker's battery bolted for the rear. Parker caught up with him and convinced the man to return to his post with a few swings of his sword at the gunner's head. He merely wrote in his report, "Several of the men [were] demoralized by the shot and shell passing through the trees."[13] The batteries comprising Alexander's battalion, being directly opposite the Peach Orchard and more exposed, sustained the heaviest losses. For example, S. Capers Gilbert's battery, on the left of the line, eventually faced the right section of Nelson Ames's battery, a section of James Thompson's, and all of John K. Bucklyn's and John G. Turnbull's: a total of sixteen guns. As a result, the battery lost heavily—forty men out of seventy-five, and two of its howitzers were dismounted in less than an hour. No battery, Northern or Southern, lost a higher percentage of men during the battle (see appendix A). Even Alexander's right knee was grazed by a bullet as he walked near Gilbert's guns.[14] He noted after the war, "Our losses both of men and horses were the severest the batteries ever suffered in so short a time during the war." The battery was at a disadvantage because it consisted of howitzers that fired projectiles with only one pound of powder, compared with the Napoleons' two pounds. The relative short distance between the two combatants helped the battery, but it also put it in harm's way, and the losses showed it. During the artillery duel, Ames ordered his right section to fire solid shot, and when its commander complained, he merely replied that he "would show them what could be done with solid shot, the purpose of course being to disable their guns."[15]

Capt. Nelson Ames wrote after the war that because of the high ground occupied by his battery in the Peach Orchard, virtually all of its shots were effective, adding, "Their shots directed against us generally struck in front and rebounding passed over our heads."[16] It was also a good position because it allowed the battery's caissons to remain close to the guns, under the protection provided by the reverse slope. Seminary Ridge provided the same situation for the Confederates. According to Alexander, "The ammunition expenditure was enormous, but fortunately the reserve supply was close at hand behind the ridge."[17]

What Alexander did not know was that Henry Hunt, who closely watched the duel, had reinforced the three hopelessly outgunned Federal batteries (Nelson Ames's, John K. Bucklyn's, and A. Judson Clark's) with a couple of batteries from Lt. Col. Freeman McGilvery's 1st Volunteer Artillery Brigade of the reserve. More batteries were also on the way.

McGilvery, who was destined to play a major role in the battle of Gettysburg, had a colorful past. Born in Maine, like so many others, he had tied his fortunes to the seas, rising to be a captain of merchant vessels. In January 1862, McGilvery entered the army as the captain of the newly formed 6th Battery of the 1st Maine Artillery. The battery performed spectacularly during the Second Manassas campaign, bringing recognition to its salty commander. Promotion followed, first to major in February 1863 and then to lieutenant colonel about four months later. He took command of the 1st Maine Artillery and then the 1st Volunteer Brigade of the new artillery reserve. McGilvery arrived at Gettysburg without ever commanding the brigade in battle.[18]

McGilvery accompanied his batteries, and upon arriving in the sector between 3:30 and 4:00 p.m., he was told by General Sickles to examine the ground and place the guns where he saw fit. Capt. John Bigelow, who accompanied McGilvery on his reconnaissance of the Peach Orchard area, was placed on the extreme left of the line. His 9th Massachusetts Battery (six Napoleons) was a green unit that was already undermanned because of illness and disease. The battery had been parked near the Trostle farm buildings, then galloped down the Trostle farm lane and through a gate that opened into a field to the south. Once in the field, Bigelow yelled, "Forward into line, left oblique, Trot!" and the guns fanned out. As they approached Wheatfield Road, Bigelow yelled, "Action front!" and the limbers halted and the gun trails dropped.[19] Enemy shells took their toll on the battery as it deployed—this was standard operating procedure when cannoneers saw enemy guns coming on the scene. "One man was killed and several wounded before we could fire a single gun," recalled Bigelow.[20] The battery was vulnerable at this time, and the Confederate gunners took full advantage of it.[21] Bigelow quickly employed an effective tool to protect his battery: "We soon covered ourselves in a cloud of powder smoke," he wrote.[22] The battery opened fire at about 4:45 p.m. Lt. Alexander Whitaker immediately realized that the field of fire of his left section was severely hampered by a rise in the ground just south of Wheatfield Road, so he moved these guns to the right side of the battery.[23]

The other battery with McGilvery was Capt. Charles A. Phillips's 5th Massachusetts Battery (six ordnance rifles), which followed Bigelow's and deployed on its right, next to A. Judson Clark's battery. Gunner Fred Waugh of Phillips's battery recalled that the unit momentarily halted upon its arrival in the sector. The reprieve did not last long, for an orderly soon arrived with orders to "advance with action." A "battery right oblique" maneuver was ordered, causing the piece closest to Bigelow's battery to deploy first, followed by the others to its right. The battery was already shy a man, as two gunners had been thrown off limber seats. A wheel ran over the arm of one of them, breaking it and sending him to the rear. The battery dropped trail and immediately opened fire on Cabell's battalion deployed on Seminary Ridge. "The air was full of bursting shell and minnie balls were falling thick and fast around us and many of our battery boys were 'mustered out' that day," noted Waugh.[24]

After positioning the batteries in and around the Peach Orchard, Hunt realized that James E. Smith's guns in the Devil's Den sector had not yet opened fire. Despite being a general officer, he had a very small staff, and all were away carrying dispatches. He bitterly wrote after the war that this was a fairly frequent occurrence and was "an awkward thing for a general who had to keep up communications with every part of a battlefield and with the general-in-chief."[25] As a result, Hunt was forced to personally ride to Smith's battery. Dismounting, he climbed the rocks around Devil's Den and found the battery in an "excellent position." Because of the rocky terrain, Smith had had to move his guns into position one by one, and this took considerable time. The men had put their shoulders to the wheels and axles while some pulled the guns up with ropes. All sweated profusely in the hot sun. Hunt was probably not happy about the fact that the ammunition had to be carried up the hill and placed near the guns, as the small space did not permit deployment of the limbers.[26] If a shell hit this ammunition, the nearby guns and gunners would be blown sky high.

Satisfied with the placement of the guns, Hunt decided to leave the sector, but not before telling Smith that he would probably lose his guns during the upcoming fight. Hunt climbed down to where he left his horse, but he experienced a plight that he called "ludicrous, painful, and dangerous." Enemy shells falling into the area spooked a herd of cattle in the valley between Devil's Den and Little Round Top. To his chagrin, cattle were between him and his horse. He watched as a shell exploded under an unfortunate animal, blowing it to pieces. Several others were also killed or wounded. "All were *stampeded*, and were bellowing and rushing in their terror, first to one side and then to the other, to escape the shells that were bursting over them and among them." Hunt was finally able to traverse this malestrom and was relieved to find his horse safe.[27]

Finally in position, Smith's four guns on Houck's Ridge opened fire on the Confederate batteries to his right, putting them under a deadly enfilade fire. Mathis W. Henry's batteries appeared in front of Smith and to his left, and a duel ensued. Although the Confederate battalion's losses were light (three men killed and six wounded), it lost heavily in metal—one of James Reilly's ordnance rifles burst and two of Alexander C. Latham's (a howitzer and a 6-pounder) were disabled by Smith's intense and accurate counterbattery fire.[28]

Around this time, Capt. Patrick Hart's 15th New York Independent Battery (four Napoleons) rolled up and took position on the right of A. Judson Clark's battery along Wheatfield Road. It appears that this was not the position indicated by McGilvery, for Hart reported that as he was about to follow his orders, Hunt countermanded them and placed Hart's battery near Clark's. Before departing, Hunt said to Hart, "It will be a gold chain or a wooden leg for you. Sacrifice everything before you give up that position."[29] The fight was becoming less of a mismatch, as thirty-five Federal guns in six batteries were now in position in or near the Peach Orchard. James E. Smith's four guns on the left of the line also opened an enfilade fire on the Confederates facing the Peach Orchard. Only George B. Winslow's guns in the Wheatfield and Francis W. Seeley's battery to the right had yet to open fire.[30]

Most of the battery commanders noted that they silenced the Confederate guns in front of them.[31] This was apparently not permanent, as evidenced by A. Judson Clark's report: "I was ordered by you [Capt. George E. Randolph] to go back and attack the battery. This I did, using shell and case shot, and after a pretty short fight, silenced the battery, but only for a short time, when they opened again, as did other batteries which they had brought into position on my right."[32]

The Federal batteries facing south along Wheatfield Road were in a hot spot because in addition to dueling Henry Cabell's and Mathis W. Henry's guns in front of them, they also received enfilade fire from Alexander's battalion to their right, which was brought into position to help take some of the heat off of the former battalion. The latter's shots were aimed at John K. Bucklyn's and Nelson Ames's batteries, but they were either fired high or hit in front of those guns and ricocheted into the ones along Wheatfield Road.[33] Capt. Osmond B. Taylor of Alexander's battalion reported that his orders were to dislodge the batteries along Emmitsburg Road. Rather than using solid shot to dismount the guns, he "opened upon the batteries with my four Napoleons, firing canister and spherical case until our infantry . . . began their charge."[34] Taylor was clearly trying to take out the enemy cannoneers and not the guns.

The Confederate counterfire was intense. As Charles A. Phillips's battery took position, its banner was positioned on its right, and the enemy batteries apparently directed their fire on it; two horses were killed within a short time. Phillips was not concerned about the enemy battery in front of him as much as the one to the right, which was enfilading his guns. These were probably batteries of Alexander's battalion whose shells were overshooting Nelson Ames's battery. Given his position, Phillips could merely bang away at the enemy batteries in front of him and hope that the Federal batteries on his right would take care of the problem.[35] According to Freeman McGilvery, the opposing batteries were at "canister range," and the enemy batteries' enfilade fire was "inflicting serious damage through the whole line of my command." S. Capers Gilbert's Confederate battery, forming the left of Alexander's battalion, probably did the most damage. Patrick Hart's battery, on the right of the Peach Orchard line, was probably hit the hardest. According to Hart, "They then brought a battery still farther to my right. They poured a tremendous cross-fire into me, killing 3 of my men and wounding 5, also killing 13 horses."[36]

Freeman McGilvery was very active during the afternoon and early evening hours in the Peach Orchard sector. After pinpointing the location of an enemy battery, McGilvery would order the four batteries to the left of the Peach Orchard (John Bigelow's, Charles A. Phillips's, A. Judson Clark's, and Patrick Hart's) to direct their fire on it. He claimed that he drove five or more batteries away by this approach. While the historian of the Federal artillery, L. VanLoan Naisawald, called this assertion "naive," there is little doubt that it had a positive impact.[37] Such a tactic helped provide an enfilade fire, increased the damage to each enemy battery, and reduced its men's morale. According to John Gibbon's *Artillerists' Manual*, "The moral effect produced by such a result being still more terrible than the physical." While this approach did not dismount many Confederate guns, it did cause many to stop firing or seek another location. Historian R. L. Murray postulated that the Confederate gunners may have "played possum." By ceasing fire, they may have fooled the Federal gunners into thinking the battery was disabled or had quit the field, therefore causing them to shift their target.[38]

Massive numbers of horses were hit on both sides during the duel. Effective battery commanders made sure that they always had adequate replacements nearby. One such battery commander was Capt. Charles A. Phillips, who wrote, "I send to the rear and have spare horses brought up, and then as soon as a horse is killed, the driver sets to work, takes off the harness and puts in a new horse. This takes some time, as an artillery harness is very heavy, and they generally have to take it all to pieces to get it off a dead horse."[39] Drivers were also pressed into action to take the place of felled gunners.

The batteries were not the only recipients of this deadly fire; the nearby infantry was also pounded. As James Reilly's Confederate battery opened fire from its position in the front of Jerome B. Robertson's brigade, the enemy's guns, probably James E. Smith's, responded, "knocking out a man here and there," according to Pvt. A. C. Sims of the 1st Texas. Pvt. John Wilkeson of the 3rd Arkansas looked down the line and saw "our men knocked out constantly. . . . I don't know how long we were held there under fire, but the time seemed endless." Pvt. John West of the 1st Texas noted that the "infernal machines came tearing and whirring through the ranks with a most demoralizing tendency." Casualties were heaviest in the 4th Texas, which lost fifteen men to a single shell. An officer in that regiment found that these cannonades were exceedingly difficult for the men because "one has time to reflect upon the danger, and the utter helplessness of his present condition. The men are all flat on the ground, keeping their places in ranks, and as a shell is heard, generally try to sink themselves into the earth." At the height of the shelling, a private stood up and quickly walked to the front of the 5th Texas, where he offered a prayer. Seeing the effect of the cannon fire, Robertson moved his men to a safer location and ordered them to lie down. Col. Van Manning of the 3rd Arkansas calmed his men by walking among them, reassuringly saying, "Steady men, steady."[40]

Brig. Gen. George Anderson's brigade in the second, or supporting line, was not immune to the Federal cannonade. A shell hit Pvt. Jackson Giles of the 9th Georgia, tearing off his leg and sending the rest of him flying ten to fifteen feet away. As he was carried to the rear, he asked the stretcher-bearers to tell his comrades "I died for my country."[41] He did just that the following day. According to Lt. J. C. Reid of the 8th Georgia, "We were lying down behind a skirt of small forest trees. The shells were striking among us, and I had made my men get before the fence to avoid splinters."[42]

An irritable Longstreet rode up to McLaws and demanded to know why a battery had not been placed in a prescribed position. McLaws calmly replied, "General, if a battery is placed there it will draw the enemy's artillery right among my lines formed for the charge and will of itself be in the way of my charge, and tend to demoralize my men."[43] Longstreet was not in a mood to be contradicted, and a battery soon arrived, probably from Alexander's battalion, and opened fire on the Peach Orchard. The Federal response was both rapid and deadly, as shells began ricocheting among the trees, dumping limbs on the men. The exploding shells killed and wounded a number of men as well, "producing a natural feeling of uneasiness among them."[44] McLaws could only ride among his men to reassure them and order them to lie down. William Hill of the 13th Mississippi of Barksdale's brigade and a veteran of many battles recorded in his diary that the cannonading was the "most terrific that I ever heard."[45] George J. Leftwich of the 17th Mississippi noted that the projectiles "tore the limbs off of the trees and plowed gaps through his men."[46] Those units occupying unprotected positions were especially hard hit. These men remained under this artillery fire for at least ninety minutes while waiting for orders to charge and the *en echelon* attack rolled toward them.[47]

The wounding of Maj. Gen. John Hood was perhaps the most important Confederate loss to a Federal shell. He had just launched his division's attack on the Federal left and was riding in an orchard just west of the Bushman barn when a shell exploded over his head. One of the fragments tore into his arm, shattering it so severely that amputation was required. Brig. Gen. Evander M. Law took over command of the division, but he was largely ineffective. As a result, the division's four brigades fought independently rather than as a cohesive unit, and this dramatically reduced their effectiveness.[48]

Later that afternoon, the Federals sustained an equally devastating blow when Maj. Gen. Daniel Sickles was wounded by a Confederate artillery solid shot. Harry Pfanz called it a "freak wounding" as the bounding shot nicked the general's right knee so gently that Sickles's horse was not spooked. Sickles was carried off on a stretcher while puffing on a cigar. His leg was later amputated.[49]

Shells from Henry Cabell's guns rained down on George Burling's brigade (Humphreys's division), which was in a supporting position between the Peach Orchard and Stony Hill. A shell hit the 2nd New Hampshire's colors, breaking the staff in three pieces and wounding several members of the color guard. According to Martin Haynes, "Never, in all its history, was the regiment exposed to such a terrific artillery fire as it received while lying upon the ground to the rear of this battery [Nelson Ames's]. . . . The air was fairly alive with bursting shell and whistling canister; the leaves fell in showers from the peach trees, and the dirt was thrown up in little jets where the missiles were continually striking."[50] John Burrill wrote home that the heavy artillery fire "made the earth tremble and the air shook and was so full of smoke you could not see."[51] After consulting his regimental commanders, Col. George Burling, who was inexperienced in commanding a brigade, decided to pull his troops back about a hundred yards to a more secure area as it did not look as though the Confederates were about to attack anytime soon.[52]

In and around the Peach Orchard to the northwest, the infantrymen of Brig. Gen. Charles K. Graham's brigade (Birney's division) were also being pounded by the Confederate artillery. Losses mounted as the men occupied exposed positions. John Bloodgood of the 141st Pennsylvania recalled receiving the order, "Cover," and "we all got flat on our faces, so as to give the rebel shells plenty of room to operate over our heads." According to Col. Edward Bowen of the 114th Pennsylvania, the hostile artillery fire was concentrated on their front: "We were in the midst of a terrific shower of shot and shell and every conceivable kind of missile, which made terrible havoc among us." A soldier from the 68th Pennsylvania recalled that the "artillery fire bearing upon it [the regiment], was terrific, carrying away men at every discharge." The soldiers could only press themselves against the ground and hope that their time was not up. Some of the units, particularly the 141st Pennsylvania, were protected by the sunken nature of Wheatfield Road.[53]

The infantry were not merely observers. Every time the large-caliber guns of George V. Moody's battery fired, they recoiled off Seminary Hill. At first the big strapping Irish gunners impatiently rolled the guns back into firing position, but they tired after a while, and infantry from McLaws's division were asked to assist.[54] The infantry were also called upon to help man the guns. "They killed so many of our artillerymen that some of the infantrymen had to go and help them handle the guns," recalled John Henley of the 17th Mississippi of Barksdale brigade.[55] Porter Alexander reported that the "loss of men was so great that I had to ask General [William] Barksdale . . . for help to handle the heavy 24-pounder howitzers of Moody's battery. He gave me permission to call for volunteers, and in a minute I had eight good fellows, of whom, alas! we buried two that night, and sent to the hospital three others mortally or severely wounded."[56]

On the Federal side, John K. Bucklyn's battery and others in the vicinity of the Peach Orchard took such heavy losses that infantry officers issued a call for volunteers, and some of the men from the 141st Pennsylvania helped carry ammunition, while others from the 68th Pennsylvania helped serve the guns.[57]

Eleven Confederate batteries sporting forty-eight guns battled ten Federal batteries with fifty-six guns. Not only did the latter have a numerical advantage, they also deployed more effective weapons—thirty-four Napoleons, perfect for this type of fighting. While both sides suffered heavily, it appears that the Confederates took the worst beating. All of George M. Patterson's battalion was driven from the field; S. Capers Gilbert's battery of Alexander's battalion, which was closest to Patterson's, had two guns dismounted and lost almost 50 percent of its men, and John C. Fraser's battery lost so many men that only two guns could be manned. Mathis W. Henry's battalion to the south was also hard hit and may have lost as many as three guns. George V. Moody's battery was also pounded, but its guns remained in action because of an infusion of infantrymen. Most important, the Confederate long arm had not achieved its mission: disabling and driving back the Federal artillery prior to the infantry attack. As a result, Confederate infantry losses mounted as they dashed across the open fields.[58]

Devil's Den–Houck's Ridge Sector

With his four Parrotts in position on Houck's Ridge and two others behind it guarding the Plum Run Valley, Capt. James E. Smith was ready to devote his full attention to the enemy. Some of the men were probably looking for redemption for what had been a checkered past. Smith had relinquished command of the battery after the battle of Fredericksburg to assume command of the 3rd Corps artillery. His replacement proved to be incompetent, and he was removed and the battery was transferred to another division. The gunners resented this action and refused to obey the order. An infantry regiment was called out with orders to "move or bury the battery." Feeling a strong attachment to the battery, Smith resigned as artillery chief and returned to command the battery.[59]

Smith's position on Houck's Ridge wreaked havoc on the Confederate batteries on Seminary Ridge because his guns were able to enfilade their positions. Col. Henry Cabell noted that when his battalion took position, "[We] exposed ourselves to a flanking fire from the enemy's mountain batteries."[60] So intense was Smith's fire that Cabell mistook it for several batteries, not a part of one. The Confederate guns opened fire and, according to Henry Hunt, "many guns were immediately turned on [Smith], relieving the rest of the line."[61] A twenty-minute duel ensued with Alexander C. Latham's battery. Smith noted, "The accuracy of the enemy's aim was astonishing."[62]

The duel between Henry's battalion and Smith's four guns lasted about half an hour. So intense was the fire that Col. Augustus Ellis ordered the right wing of his 124th New York, which was closest to Smith's guns, to abandon its exposed position and take cover in the Rose Woods. The artillery fire was actually more intense there, so Ellis quickly ordered his men back to their original positions.[63]

While his four guns on the high ground had fairly good fields of fire, Smith was concerned about the gorge to his left. He knew that, if the enemy tried to force a passage, his four guns on the ridge would not do much good as their barrels could not be depressed enough to be effective. While Smith had infantry support in the form of the 124th New York and 4th Maine on the ridge, there was none to the left of them. He did not like the prospects of these guns, without infantry support, attempting to stem a determined enemy thrust through the valley or gorge.[64] Smith noted after the war, "I felt anxious about our left flank and made an effort to get some infantry posted in the woods along the base of Round Top, but as the enemy gave little time for reflection, my attention was occupied in looking after the Battery, and replying to the concentrated fire of a number of guns [Aexander Latham's and James Reilly's batteries on Warfield Ridge]."[65]

The appearance of Hood's infantry, probably Jerome B. Robertson's brigade, marching resolutely toward Smith's position pushed the issue. Not able to find Brig. Gen. Hobart Ward, who commanded the infantry brigade in this sector, Smith strode up to Col. Elijah Walker of the 4th Maine with a request that he shift his regiment to the left of the guns where they could cover the gorge. According to Walker, Smith said that he could "take care of his front, but the enemy would come up in the woods on our left and I could better protect his guns from that place." Walker disagreed, replying that he could "take care of his guns whare we were, and I would not go into that den unless I was obliged to." Smith was not so easily placated, so he went off to find Ward, and Walker soon received orders from the brigade's adjutant to move his regiment to the left. Walker was so concerned about being in what he considered to be a vulnerable position on the left flank of the army that "I remonstrated with all the power of speech I could command." It didn't work, and he was forced to move near the gorge.[66]

Smith contested Walker's recollections. "Colonel Walker misunderstood me, if I am to judge by the tone of his letter," he wrote. "I certainly never said that I did not want his help [for my four guns on Houck's Ridge]; I was not fool enough to think a Battery could maintain a position such as was assigned to the 4th Maine without a strong force of infantry in support." Continuing his narrative, Smith added, "Not having such support in line, it was my belief that the best disposition ought to be made of the limited force at hand; therefore I suggested to Colonel Walker the advisability of moving his regiment from the rear of my Battery into the woods on our left, saying at the time, if he would protect the flank, the Battery would endeavor to take care of the front."[67]

Mathis W. Henry's battalion on the right of the Confederate line fell silent as infantry from Evander Law's and Jerome B. Robertson's brigades of Hood's division stepped off to attack the Federal left. Smith's gunners saw the infantry begin their assault and immediately opened fire with case shot, which "tore gap after gap throughout the ranks of the Confederate foe." Smith noted that he "never saw the men do better work: every shot told."[68] So intense was this fire that Law peeled the 44th and 48th Alabama from the right of the line and rushed them to the left to contend with Smith's battery. This deprived these two veteran regiments from the attack on Little Round Top and had a profound impact on the events at this vital area. Col. William Perry of the 44th Alabama recalled that Law "expected my regiment to take a battery [Smith's] which had been playing on our line from the moment the advance began."[69] Robertson's brigade, advancing on Law's

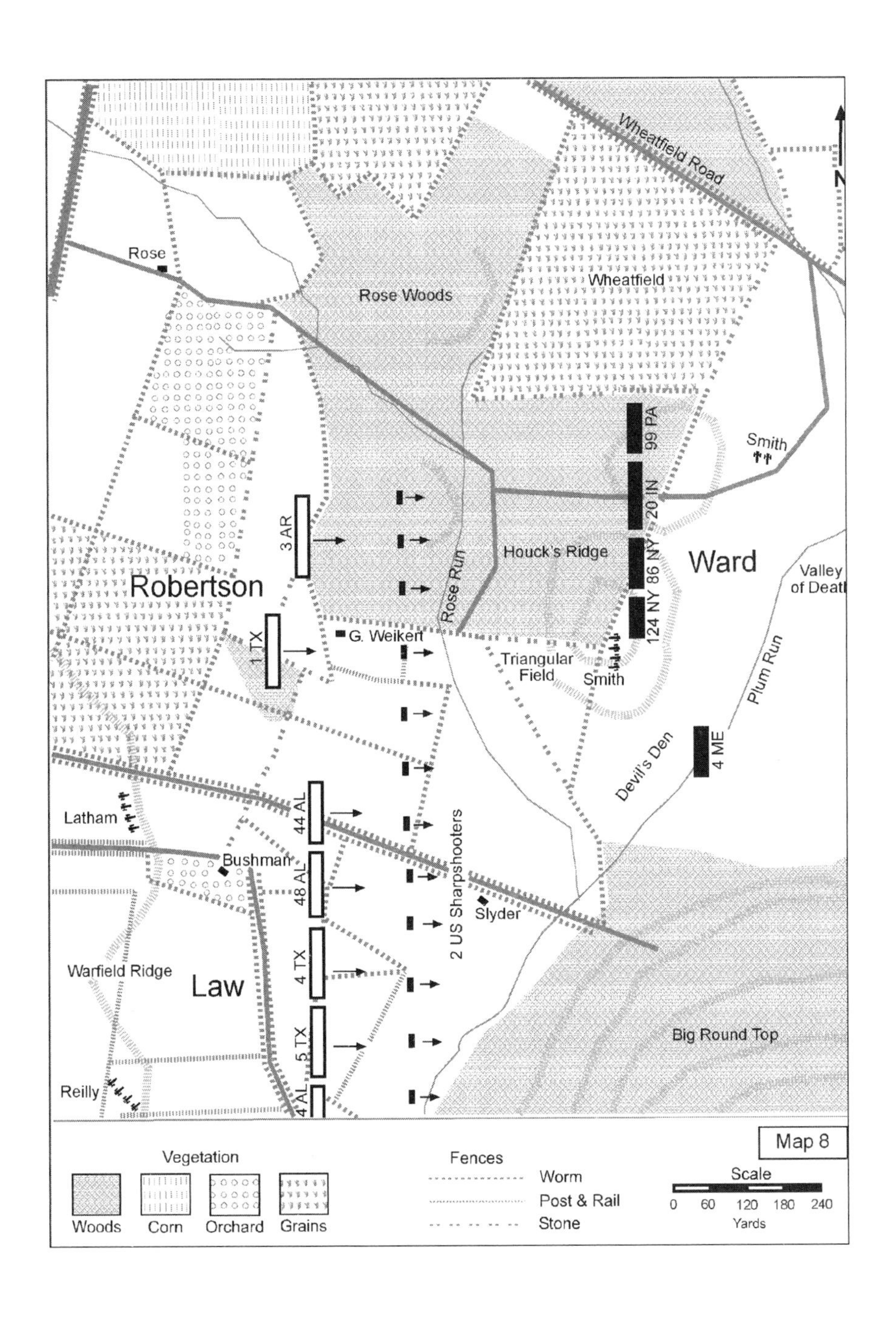
Wheatfield Road
N
Rose
Rose Woods
Wheatfield
99 PA
20 IN
86 NY
124 NY
Smith
3 AR
Houck's Ridge
Ward
Valley of Death
Robertson
Rose Run
1 TX
G. Weikert
Triangular Field
Smith
Plum Run
Devil's Den
4 ME
Latham
44 AL
Bushman
48 AL
2 US Sharpshooters
Slyder
Warfield Ridge
Law
4 TX
Big Round Top
5 TX
Reilly
4 AL
Map 8
Vegetation
Woods
Corn
Orchard
Grains
Fences
Worm
Post & Rail
Stone
Scale
0 60 120 180 240
Yards

left, was also pounded by Smith's battery. According to Robertson, "As we advanced through the field, for half a mile we were exposed to a heavy and destructive fire of canister, grape, and shell from six pieces of their artillery on the mountain."[70]

While the two regiments from Law's brigade attacked Smith's battery from the southwest, the 1st Texas of Robertson's brigade advanced on the battery from the west. As the Texans advanced, Smith's gunners progressively depressed their tubes and sent case shot screaming toward the Texans. This caused the Confederate line to quicken its pace. "The pieces were discharged as rapidly as they could with regard to effectiveness," reported Smith.[71]

The Texans reached the safety of the woods bordering the Triangular Field, about three hundred yards from Smith's position. Here they halted and opened fire on Smith's gunners before continuing their attack on Houck's Ridge. The cannoneers could see sharpshooters crawling forward to take up positions behind large rocks. According to Private Sims of the 1st Texas, "We loaded and fired, the front rank on their knees and the rear standing."[72]

One by one, Smith's men began to fall to this sniper fire. Smith ordered his guns loaded with canister, but it did little good in extracting the enemy from their hiding places. As the Texans continued their advance, Smith realized the futility of the situation, as rocks protected the enemy, and his cannon barrels could not be depressed enough to hit those not protected. Smith dashed over to Col. Augustus Ellis, commander of the 124th New York, and begged him to order a counterattack to save his guns.[73]

Ellis was happy to oblige. He had also been watching from his position behind Smith's four guns and could see his men becoming increasingly nervous. They needed to become engaged, so with the command, "Charge bayonets! Forward; double-quick—March!" the New Yorkers sprang forward. The gunners were forced to hold their fire as the infantrymen passed through Smith's cannon. The New Yorkers flew into the Texans, driving them back beyond the Triangular Field. Lt. Charles Weygant described the sound of the battle as "roaring cannon, crashing rifery, screeching shots, bursting shells, hissing bullets, cheers, shouts, shrieks and groans."[74]

Another of Hood's brigades, Brig. Gen. Henry Benning's, was now fed into the battle. It too had difficulty traversing the open terrain between Seminary and Houck's ridges because of the vicious Federal artillery fire. Although Smith's battery could not fire at the Texans, it could wreak havoc on Benning's Georgians as they advanced from Warfield Ridge. A soldier from the 17th Georgia wrote, "As soon as we came in sight a furious blast of cannon broke from the tops of the hills and mountains around and the terrific cry and scream of shells began." Although the Confederate batteries opened on their counterparts, the Federal artillery was not distracted, instead continuing to pour its destructive fire on the advancing infantry. "Down the plunging shot came, bursting before and around and everywhere tearing up the ground in a terrific rain of death," wrote the Georgian. The brigade halted after advancing about four hundred yards. The shells continued screaming overhead, but few men left the ranks, unless hit. After a few minutes in this position, the men were again ordered forward. "The ranks began to melt away, but springing forward, with a shout, the undismayed line steadily rushed on," noted a soldier in the 17th Georgia.[75]

When told that all case shot was expended, Smith screamed, "Give them shell! Give them solid shot! Damn them, give them anything!"[76] The "anything" was canister, and it was poured into the Georgians so rapidly that the gunners did not have time to sponge between blasts, thus risking premature explosion of the powder charges. Smith felt that he didn't have any choice as the enemy was rapidly approaching. Thomas Bradley of the 124th New York recalled seeing Smith's men "working as I never saw gunners work before or since, tore gap after gap through the ranks of the advancing foe."[77]

With the 15th Georgia of Benning's brigade, the 1st Texas continued its advance toward Houck's Ridge and Smith's guns. They smashed into the 124th New York and sent it reeling back to Houck's Ridge and through Smith's guns. Smith knew that the end was near. A shell hit one of his guns, knocking it out of commission. Smith immediately ordered it removed to safety. The remaining three guns opened fire again on the approaching enemy, but they were too close. The only way that Smith's guns could open an effective fire was if the Confederates were again pushed back. A desperate Smith rushed over to a portion of the 124th New York that was reforming on the ridge and yelled, "For God's sake, men, don't let them take my guns away from me!"[78] It appears that A. Judson Clark's battery in the Peach Orchard was ordered to lob some shells in the direction of Devil's Den to help support Smith's beleaguered guns.[79] This was a risky proposition, as the gunners could not ascertain the location of Smith's guns or the Federal infantry supporting them.

The New York infantry were having their own trouble. Most of their officers were killed or wounded, including Col. Augustus Ellis, so launching another counterattack against the enemy was not one of their options. As the enemy continued advancing toward the base of the ridge, Smith realized that it was too late to try to get his remaining three guns down the rocky slope to safety. All that was left for Smith was the hope that the guns would be recaptured later. He wrote in his report, "If forced to retreat, I expected my supports would have saved the guns, which however, they failed to do. I could have run my guns to the rear, but expecting to use them at any moment, and the position difficult of access, I thought best to leave them for awhile."[80] This explains why they were not spiked. Smith also believed that to have withdrawn his guns earlier would have demoralized the Federal infantry and perhaps caused them to retreat prematurely. Before abandoning the guns, Smith ordered his men to remove all equipment (friction primers, sights, sponge buckets, rammers, and sponges) so they could not be turned against the Federal troops. Through charge and countercharge, at least three regiments—the 1st Texas, 44th Alabama, and 20th Georgia—actually captured the guns at one time or another. Their efforts were significantly aided by the 2nd, 17th, and 20th Georgia and the 3rd Arkansas.[81]

The Federal infantry did not give up Smith's guns without a fight, for the 99th Pennsylvania, 4th Maine, and perhaps part of the 124th New York threw themselves at the enemy soldiers around the guns, engaging them in hand-to-hand combat. The fate of the guns hung in the balance for several minutes, but ultimately the Southerners prevailed. So intense was the fighting in this sector that Col. James Waddell of the 15th Georgia counted eighty-seven holes in his battle flag. Thirty-eight were made by minie balls; the remainder by shell fragments.[82]

As Smith was leaving the ridge with his gunners, he looked toward Plum Run Gorge and saw Confederates from the 44th and 48th Alabama of Evander Law's brigade and the 2nd and 17th Georgia of Henry L. Benning's brigade advancing toward his other two guns. Running on foot to these cannon, he ordered them to open fire. "The enemy are taken by surprise; their battle flag drops three different times from the effect of our canister. Thrice their line wavers and seeks shelter in the woods, but in a moment they return in a solid mass," noted Smith.[83]

The 6th New Jersey of George Burling's brigade and the 40th New York of Philip Regis de Trobriand's brigade had reinforced this sector, and they now threw a devastating fire into the Confederates. Watching the seesaw battle, Smith desperately hoped to garner infantry support to recapture his lost guns, so he approached Col. Thomas Egan of the 40th New York. Egan promised to help, but not right at the moment; he had his hands full trying to stem the Confederate attack. The two Federal regiments and then Smith's remaining two guns were forced back, leaving Houck's Ridge and Plum Run Gorge in the hands of the Confederates.[84]

Knowing the shame accompanied by the loss of any cannon, let alone three, Smith concluded his report, "I trust no blame will be attached to me for the loss of my guns. I did that which in my judgment I thought best."[85] The battery expended 240 rounds while defending this sector, and the gunners fought heroically, but Smith was haunted by the battle and spent the rest of his life trying to justify the loss of his guns.[86]

Smith did not receive official censor for losing the guns. His commander, Capt. George E. Randolph, called it "one of those very unpleasant, but yet unavoidable, results that sometimes attend the efforts of the most meritorious officers."[87] Despite the fact that Smith's battery was the only 3rd Corps unit to lose a cannon, it lost the fewest men—thirteen (see appendix A), all to sniper fire, as the rocks around Devil's Den shielded Smith's men during the intense duel with Mathis W. Henry's battalion.[88]

The Wheatfield

While most of John B. Hood's division took on Devil's Den, Houck's Ridge, and later Little Round Top, its fourth brigade, under Gen. George Anderson, swept to the left and headed directly for the Wheatfield, which was occupied by the 17th Maine of Philip Regis de Trobriand's brigade (David B. Birney's division, 3rd Corps), crouching behind a stone wall. A collection of other regiments from de Trobriand's and George Burling's brigades were just to the right (north) on Stony Hill. Capt. George B. Winslow's battery of six smoothbore Napoleons was ordered to deploy on an elevated piece of ground on the eastern segment of the Wheatfield, with the guns' trails facing Wheatfield Road. It was an isolated position, unsupported by infantry, although the 115th Pennsylvania would appear later. To make matters worse, strips of woods along either flank could mask enemy movements, increasing the probability of losing the guns. Capt. George E. Randolph knew the risk he was taking, but he believed it was necessary to support the undermanned infantry in the sector, and a battery of six Napoleons was just the answer. Winslow did not agree with his commander's view, noting after

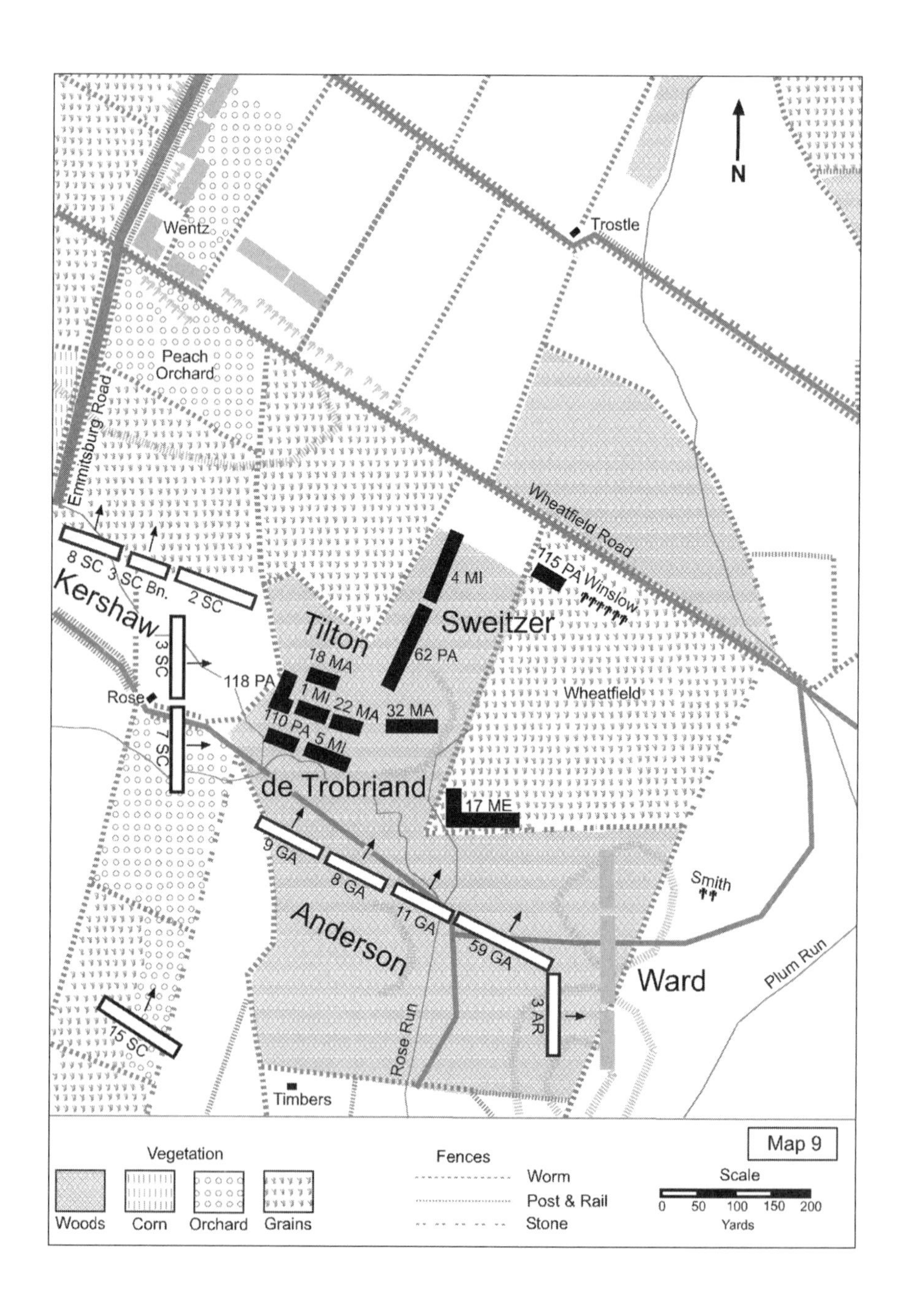

N
Wentz
Trostle
Peach Orchard
Emmitsburg Road
Wheatfield Road
8 SC
3 SC Bn.
2 SC
Kershaw
3 SC
7 SC
Rose
118 PA
Tilton
18 MA
1 MI
22 MA
110 PA
5 MI
4 MI
62 PA
32 MA
Sweitzer
115 PA
Winslow
Wheatfield
de Trobriand
17 ME
9 GA
8 GA
11 GA
59 GA
3 AR
Anderson
Ward
Smith
Plum Run
Rose Run
15 SC
Timbers
Map 9
Vegetation
Woods
Corn
Orchard
Grains
Fences
Worm
Post & Rail
Stone
Scale
0
50
100
150
200
Yards

the war that it was "not a very favorable location for offensive, much less for defensive artillery warfare." He claimed that Henry Hunt, who visited the position before it was attacked, agreed with Winslow, but the former never mentioned it in his voluminous writing.[89]

The battery had gained celebrity status because of its actions during the Peninsula campaign a year before. Under the command of Thomas W. Osborn, now the 11th Corps artillery commander, the men dug a hole and sank a gun trail into the ground, thus elevating the barrel to point skyward. Using thirty-second fuses, the gun forced the quick descent of a Confederate observation balloon. The battery had also taken on a massive Confederate cannon mounted on a railroad car, forcing it back to safety, along with its supporting battery. The unit had most recently distinguished itself at Chancellorsville, when it fired canister at the enemy at less than thirty yards until it ran out of ammunition and was forced to retreat.[90]

Although he could not see it because of the intervening woods, the smoke in the distance and the sounds of nearby explosions told Winslow that Confederate artillery had taken position on Seminary Ridge and had opened fire on the Federal line. Hunt ordered him to throw a few rounds of solid shot in the direction of the batteries. Winslow admitted that the rounds probably had "no effect, as it was evidently beyond the range of my guns."[91] The latter supposition was not true, however.

The crash of muskets a short time later told Winslow that Confederate infantry was on its way. Birney ordered Winslow to open fire. A less experienced officer might have been disconcerted, as the enemy infantry was not yet in sight and there was a distinct possibility that some of the shells could hit the 17th Maine behind the stone wall, about three hundred yards away. Winslow was nonplused. He noted, "I was unable from my obscure position to observe the movements of the troops, and was compelled to estimate distances and regulate my fire by the reports of our own and the enemy's musketry."[92]

Winslow carefully elevated his barrels so the shells arced over the Federal infantry. Because of his position on higher ground than the infantry, Winslow's guns were able to "fire over our own men more, effectively and with less danger to them, though our fire was so low as to cause them some anxiety."[93] Winslow claimed that the shells screaming toward the enemy's infantry reassured the 17th Maine. George E. Randolph observed that "the moral effect being at least equal to the physical."[94] Winslow used solid shot as he did not wish to risk using exploding shells under these conditions.

Winslow concentrated on areas "of the heaviest musketry, lessening the range as our troops fell back and the enemy advanced."[95] After repelling several charges of George T. Anderson's brigade, the 17th Maine could no longer hold its position at the stone wall and gradually fell back toward Winslow's guns. This movement masked the guns, forcing them to fall silent until the front cleared. When the guns and infantry opened fire again, Winslow's fire was so intense that Anderson's Georgians halted their advance at the stone wall and returned fire. According to Winslow, "The bullets came thick and fast, clipping the heads of wheat from the stalks as they whistled past."[96] An occasional cannoneer reeled from his gun in agony. By using shell and case shot, with one-degree elevation and one- and one-and-a-half-second fuses, Winslow was able to keep Anderson's men from advancing from the prized stone wall.[97]

Lt. Albert Ames, who commanded the left section, dashed up to Winslow with news that enemy infantry was flanking his position and had closed to within sixty yards. Not believing his subordinate, Winslow followed him back to his section and found the report to be accurate. These troops were probably from the 1st Texas (Jerome B. Robertson's brigade) and 15th Georgia (Henry L. Benning's brigade). Winslow ordered Ames to change front and open with canister. With the sound of the tin cans being shoved down the barrel, Winslow rode off in search of General Birney. Upon hearing Winslow's report, Birney affirmed his actions. Winslow then returned to his battery. One of Birney's aides soon appeared, telling Winslow to "be careful and not get cut off."[98] This caused Winslow to reconsider the situation, as he did not believe that he was in trouble yet. The guns continued barking at the enemy. Winslow's horse was shot from under him, and as he rose to his feet he noticed that the enemy, probably from Joseph B. Kershaw's brigade (Lafayette McLaws's division), overlapped the Federal line to the right. The Federal infantry was soon in full flight. Realizing that he was losing all support, and seeing a growing number of his men and horses going down each minute, he reluctantly ordered his guns to the rear. The retreat began with the left section; one piece was removed at a time as the remaining guns of the battery continued firing as best they could.[99]

The battery was not safe yet. One of the pieces had not gone more than twenty-five yards before losing several additional horses. A quick-thinking Winslow impressed two horses from James E. Smith's battery, which was passing to the rear just at this time, and he was able to save the piece.

Winslow lost ten wounded and eight missing, along with ten horses killed or disabled.[100]

George E. Randolph admitted that the battery did not occupy an optimal position, "owing to the nearness of the woods on all sides, but the result proved that the battery was able to do good service."[101] He commended Winslow, "not only for the good working of his battery, but for the handsome manner in which he withdrew under trying circumstances."[102]

Little Round Top

While a few regiments in Evander M. Law's and Jerome B. Robertson's brigades hit Houck's Ridge and the Wheatfield, other regiments headed for Little Round Top. Albeit difficult to ascend, the flat-topped hill made a perfect artillery platform that overlooked much of the battlefield. If the Confederates captured the hill, they would probably win the battle. A stroke of luck now occurred that helped seal the victory for the Federals. Watching the Confederates advancing toward Little Round Top, Brig. Gen. Gouverneur Warren, serving as the army's chief engineer during the battle, realized the tactical value of the hill and sent an aide dashing down the hill in search of reinforcements.

Marching toward the Wheatfield, Col. Strong Vincent spied an aide galloping toward him in search of the 5th Corps commander, Maj. Gen. George Sykes. Insisting that he hear the message, Vincent realized he could not wait for orders to reinforce the hill. So he turned his brigade to the left and quickly marched it up the side of the rugged hill, quickly deploying the men for action. He did not have long to wait before he was attacked by the 4th Alabama of Law's brigade and the 4th and 5th Texas of Robertson's. Three other regiments from Law's brigade—the 15th, 47th, and 48th Alabama—would later join the fight for this valuable hill. The steep and rocky west-facing slope created havoc in the Confederate ranks, and combined with the volleys from Vincent's men, each Confederate attack ended in repulse.[103]

Earlier, Capt. Augustus P. Martin, the 5th Corps artillery commander, ordered three of his batteries forward to support Gen. Joseph Barnes's division in the Wheatfield and Stony Hill. Lt. Charles E. Hazlett's Battery D, 5th U.S. Artillery (six 10-pounder Parrotts), led the column. Hazlett had initially protested being in the van as he had a premonition of this being his last battle. Martin, who greatly respected Hazlett's abilities, would have none of it, so the heavyhearted battery commander led the column toward the Wheatfield. While examining the ground, both officers spied Little Round Top and realized its value as an artillery platform. Before ascending the hill to reconnoiter, Hazlett ordered the battery to begin the climb.[104]

Martin and Hazlett encountered Warren at the crest of the hill and quickly realized that it was not as good a platform as they had anticipated. It provided but a narrow ledge that would be difficult to negotiate, and to make matters worse, the rough, rocky surface would make placing and working the guns very difficult at best. The officers also realized that the cannon muzzles could not be depressed enough to help beat off the determined Confederate attacks. Warren finally told the other officers that it was "no place for efficient artillery fire." Hazlett remained somewhat optimistic, replying, "Never mind that, the sound of my guns will be encouraging to our troops and disheartening to the others, and my battery's of no use if this hill is lost." Martin settled the issue by letting his orders stand.[105]

If climbing the steep, rocky hill was difficult for infantry, it was almost impossible for artillery. How the guns (minus the caissons) made it up the hill is not certain because of the contradictory writings of those who were there. In 1865, Capt. Augustus P. Martin wrote, "This position was gained with much difficulty, as the horses were necessarily detached from the pieces which had to be hauled up a portion of the hill by hand."[106] He may have been referring to the last, steepest portion of the hill. Lt. Benjamin F. Rittenhouse recalled after the war, "We went there at a trot, each man and horse trying to pull the whole battery by himself."[107] Thomas Scott added, "Our guns tipped over; we put them back, and somehow got them on top of the hill and trotted along on top of it to the left. I have since wondered how we ever got our guns up that hill."[108]

Some claimed that two other arms of the service aided Hazlett's gunners. According to gunner O. W. Damon, the first to help were the "sappers and engineers" who "cut a road for us through the woods on the side of the mountain, and we drew the pieces up by hand, it being too steep and too dangerous for horses on top."[109] None of the gunners mentioned the interactions with the infantry of Stephen H. Weed's brigade, particularly the 140th New York, who were ascending the hill at the same time. "Some of the guns of Hazlett's battery came rapidly up and plunged directly through our ranks, the horses being urged to frantic efforts by the whips of their drivers and the cannoneers assisting at the wheels, so great was the effort necessary to drag the guns and caissons up the ragged hillside," noted Porter Farley of the regiment.[110] It appears that several members of the 140th New York interrupted their ascent, unhitched the horses, and helped pull and push the guns up to the top of the hill. Warren also apparently assisted.[111]

Reaching the top of the hill was only the beginning of Hazlett's problems, for the large boulders there also had to be negotiated. "How to get the guns up over the boulders, where they could be brought into action was the difficult problem that confronted us. There were a few infantry soldiers, not over a dozen stragglers from Vincent's brigade, loitering behind the boulders. We decided to press them into the service and then undertake with the help of our own men to lift the guns into position," noted Martin.[112]

Hazlett positioned one cannon and then a second. It was painstaking work, particularly because the Confederate infantry were in the process of mounting their most determined attack at this time. After two repulses of the 4th Alabama and the 4th and 5th Texas, the 48th Alabama arrived and extended the line to the north. The newcomers also extended beyond the right of the 16th Michigan. Whether because of the weight of the Confederate attack or a misunderstanding of orders, some of the Michiganders began falling back, and the enemy slowly made their way toward the summit of the hill.[113] Hazlett's artillery pieces opened fire as soon as they deployed, buoying Vincent's men, who had borne the brunt of the attacks. Capt. Eugene Nash of the 44th New York confirmed Hazlett's predictions, recalling when the cannon opened fire on the surging Confederate troops, "No military music ever sounded sweeter and no aid was ever better appreciated."[114] When the din of battle subsided, the New Yorkers lustily cheered the efforts of the cannoneers. Not all the soldiers shared this enthusiasm, however. Because the cannon were fired through gaps in their line, several men in the 140th New York suffered permanent hearing loss.[115]

Warren never forgot the picture of Hazlett that early evening: "There he sat on his horse on the summit of the hill, with whole-souled animation encouraging our men, and pointing with his sword toward the enemy amidst a storm of bullets—a figure of intense admiration to me. . . . No nobler man fought or fell that day than he."[116]

The southern face of Little Round Top, which was defended by the 20th Maine, also came under attack. The 15th Alabama persistently threw itself at the defenders, but none of the trusts were successful. The artillery did not play a role in this part of the fight.

With the repulse of the last attack, Hazlett's cannoneers were exposed to deadly sniper fire from the Confederates occupying Devil's Den below. One shot hit the sponge bucket of the fourth piece, draining off the water. The gunners merely yelled, "Take that, damn you!" when they opened fire.[117]

A number of soldiers also went down, including Gen. Stephen Weed, who had been an artillery officer but left the long arm for promotion to brigadier general of infantry. As he lay mortally wounded on the rocks he beckoned his friend, Lt. Charles E. Hazlett, to his side. Those nearby could hear Weed tell Hazlett about debts he owed and about his sister. Before he could finish his statements, a sniper's bullet found Hazlett's skull. He died that night.[118] Thomas Scott called Hazlett "one of the bravest and best officers I ever served under. He always called us his boys, and, in fact, we were boys then—eight of every 10 being under 20 years old."[119] Hunt reported that Hazlett "had gained an enviable reputation for gallantry, skill, and devotion to his country and the service."[120]

Lt. Benjamin F. Rittenhouse assumed command of the battery and instantly sprang to work. He ordered the swing and wheel drivers to prepare to take the positions of disabled cannoneers; he told the wagon and forge drivers to fill kettles with water for the sponge buckets and for the men to drink; and he brought up the caissons.[121] According to Hunt, the battery "immediately opened . . . and continued it until night with marked effect, as its fire enfiladed the enemy's line." It was a close call, but the Federal troops held on to Little Round Top. The same could not be said about Houck's Ridge or the Wheatfield, but these sites were of much less tactical value.[122]

5

JULY 2

THE BATTLE FOR THE PEACH ORCHARD

PEACH ORCHARD: SOUTH SIDE

Still thinking that the Peach Orchard salient was undermanned, Henry Hunt sent a message to Gen. Robert O. Tyler of the artillery reserve for two additional batteries—one of Napoleons and one of rifled cannon. Tyler soon arrived in person with even more batteries than requested. He brought with him Capt. James Thompson's Batteries C & F, Pennsylvania Independent (six ordnance rifles) of the 1st Volunteer Brigade, which moved to the salient in the Peach Orchard and deployed to the left of Nelson Ames's battery, also facing south, where it opened fire on Henry Cabell's battalion. Capt. John W. Sterling's 2nd Connecticut Battery (four James rifles and two 12-pounder Parrotts) of the 2nd Volunteer Brigade also arrived, as did Capt. Dunbar R. Ransom's 1st Regular Artillery Brigade. The Sterling's battery had initially formed on the crest of Cemetery Hill but was sent north to Brig. Gen. Andrew Humphreys's division's sector along Emmitsburg Road.[1]

The Federal line facing south along Wheatfield Road now bristled with an impressive array of ordnance totaling thirty-two guns ready to take on the looming Confederate attack. Two batteries deployed south of Wheatfield Road. Nelson Ames's battery was on the right (west), in the Peach Orchard (adjacent to Emmitsburg Road), and to its left were four guns of James Thompson's battery (the other two were farther north, facing west along Emmitsburg Road). Ten guns thus formed this first line (four ordnance rifles and six Napoleons). Four additional batteries also faced south, but they were deployed on the north side of Wheatfield Road. The first of these was Patrick Hart's battery. Behind and to the left (east) of Hart's battery was A. Judson Clark's battery. Charles A. Phillips's battery was to Clark's left, and then came John Bigelow's. This added another twenty-two tubes (six 10-pounder Parrotts, six ordnance rifles, and ten Napoleons). This massive number of cannon was needed to compensate for the low number of Federal infantry units that Daniel E. Sickles was able to station in this sector. Charles K. Graham's brigade (Birney's division) of Pennsylvanians was in this sector, along with scattered regiments of George C. Burling's brigade (Humphreys's division). William R. Brewster's brigade (Humphreys's division) was in reserve, and farther north, along Emmitsburg Road, Joseph B. Carr's brigade (Humphreys's division) awaited the Confederate onslaught.[2]

Capt. Nelson Ames noted that he dueled with an enemy battery on Seminary Ridge in front of him for half an hour before finally silenced it. Then another enemy battery rolled up, and a second arrived on his right. The latter was probably from E. Porter Alexander's battalion, now wheeling into position. Ames responded by turning his right section toward the new threat, while ordering his other four guns to fire at the battery in front of him, probably from Henry Cabell's battalion. He called the ensuing duel "as sharp an artillery fight as I ever witnessed."[3] Seeing one of the enemy's guns from the battery on his flank dismounted and the other pieces pulling back, Ames concentrated all six guns on the batteries in front of him. The rapidly dwindling supply of spherical case and solid shot concerned him, forcing him to order the gunners to slow their rate of firing.[4]

Clark's battery, to Ames's left, was also being pounded. Michael Hanifen noted, "Our left and front was a sheet of flame. The air was dotted with little balloons of white smoke, showing where the shells had burst, and sent their deadly messengers to the fighting lines below. . . . [T]he enemy's artillery was actively engaged, as shown by the white steam-like clouds of smoke arising from their battery positions."[5]

One shot exploded below the trail of a gun, sending two cannoneers airborne. Both scrambled to their feet, dazed but still alive. A lieutenant yelled over to one of the men, "Riley, why the bloody h—1don't you roll that gun by hand to the front." Riley, turned his mangled thigh toward the lieutenant and said, "Lieutenant, if your hip was shot off like that, what the bloody h—1would you do?"[6] A shell fragment had also pierced the gun's sponge bucket, forcing the gunners to scrounge for a replacement, which they filled with the precious water from their own canteens. Now back in action, the men yelled, "Take that for Riley!" every time they pulled the lanyard.[7]

A new type of threat materialized when George T. Anderson's brigade of Hood's division broke from the cover of the ridge and made for the Wheatfield, to the left (south) of the Peach Orchard. Lt. Col. Freeman McGilvery described this movement in his report: "A heavy column of rebel infantry made its appearance in a grain-field about 850 yards in front, moving at quick time toward the woods on our left, where the infantry fighting was going on."[8] The Federal cannon along Wheatfield Road opened a massive fire, which disordered the Confederate infantry's line of battle. McGilvery watched many enemy soldiers fleeing to the rear, but he admitted, "The main portion of the column succeeded in reaching the point for which they started, and sheltered themselves from the artillery fire."[9] Lt. J. C. Reid of the 8th Georgia wrote, "Had our advance been slow they would have swept all of us away. We understood that too well to loiter, and so we dashed on through small wheat fields and over stone fences, filling up every gap made by a hit, and maintaining a line which would have delighted Ney himself."[10]

Michael Hanifen of Clark's battery described this attack after the war: "A cloud of skirmishers covered their front. We opened fire on them immediately with shell and shapnel, and every shot tore gaps in their ranks."[11] A. Judson Clark noted that while the "fire seemed very destructive, opening large gaps in their ranks, it only temporarily checked them, and they pressed steadily on."[12]

Back on Clark's right, Nelson Ames sent all of his remaining spherical case ammunition to his left section, while the remaining four guns continued dueling Alexander's guns with solid shot. The guns on the left had momentarily fallen silent when they exhausted their supply of infantry-destroying ammunition. The gunners now waited patiently until the enemy's line of battle came within canister range. It probably never did, but they would not have long to wait for new targets to appear.[13]

Henry Hunt rode over to Ames to ask how long he could hold his pivotal position in the Peach Orchard salient. Ames responded that although his ammunition was running low, he believed he could continue to hold the position as long as it lasted. No sooner had Hunt galloped away than a Confederate battery to the right opened fire with canister, "sweeping our front from right to left," noted Ames.[14] The right section's lieutenant dashed up to Ames and asked him if it wasn't time to pull back, as the battery could not contend with the crossfire of so many guns. Ames growled at him to get back to his guns and to "keep his mouth shut and make no more suggestions of that sort; if his guns were to fall back the lieutenant would hear of it at the proper time."[15] He ordered the right section to shift to meet this new threat and open with solid shot. The lieutenant was not through making suggestions, however. If the enemy was using canister, shouldn't his guns being using that as well? Ames sarcastically replied, "As long as his eyes remained good, he was able to observe what the enemy were using and hoped they would continue to use canister, but we would show them what could be done with solid shot, the purpose of course being to disable their guns."[16]

Ames's first shot passed under the axle of one of the enemy's guns. The second passed over it. The third struck under the barrel of the cannon, sending it flying high into the air—Ames thought as high as the treetops. Through his fieldglasses, Ames could see the dead and severely wounded enemy gun crew scattered around the cannon. Another Confederate piece was disabled when a solid shot hit its wheel.[17]

Charles A. Phillips's battery was also exposed to a hail of shells and small-arms fire at this time. A bullet struck the top of a cannon, just as it was being sighted, and ricocheted into the gunner aiming it. A wheel on the left piece lost seven spokes to a Confederate shell, which embedded itself into the hub. What bothered Phillips's gunners most was the enemy battery to the right that was enfilading their position. One gunner recalled that the battery was "throwing case shot at us very carelessly, and every minute a shower of bullets would come in, *whoosh*,—just like a heavy shower of hailstones."[18]

Cannoneers on both sides performed valiantly. A Number 3 gunner's foot in Ames's battery was severed just as his cannon was preparing to fire. The Number 2 gunner, whose job it was to load the gun, hesitated, but the wounded man yelled out, "D—you, what are you waiting for; put your charge in; I am going to have one more shot at them, leg or no leg."[19]

While the Federal gunners now paid most of their attention to the approaching enemy infantry, the Confederate artillery maintained their attention on their Yankee counterparts. Charles A. Phillips recalled after the war that the enemy "distributed very liberally amongst us the shells intended for [James] Thompson's battery, so that we were kept busy changing horses."[20] George Randolph noted that a Confederate battery of "six light 12-pounders" in their "front and a little to the left of its position [A. Judson

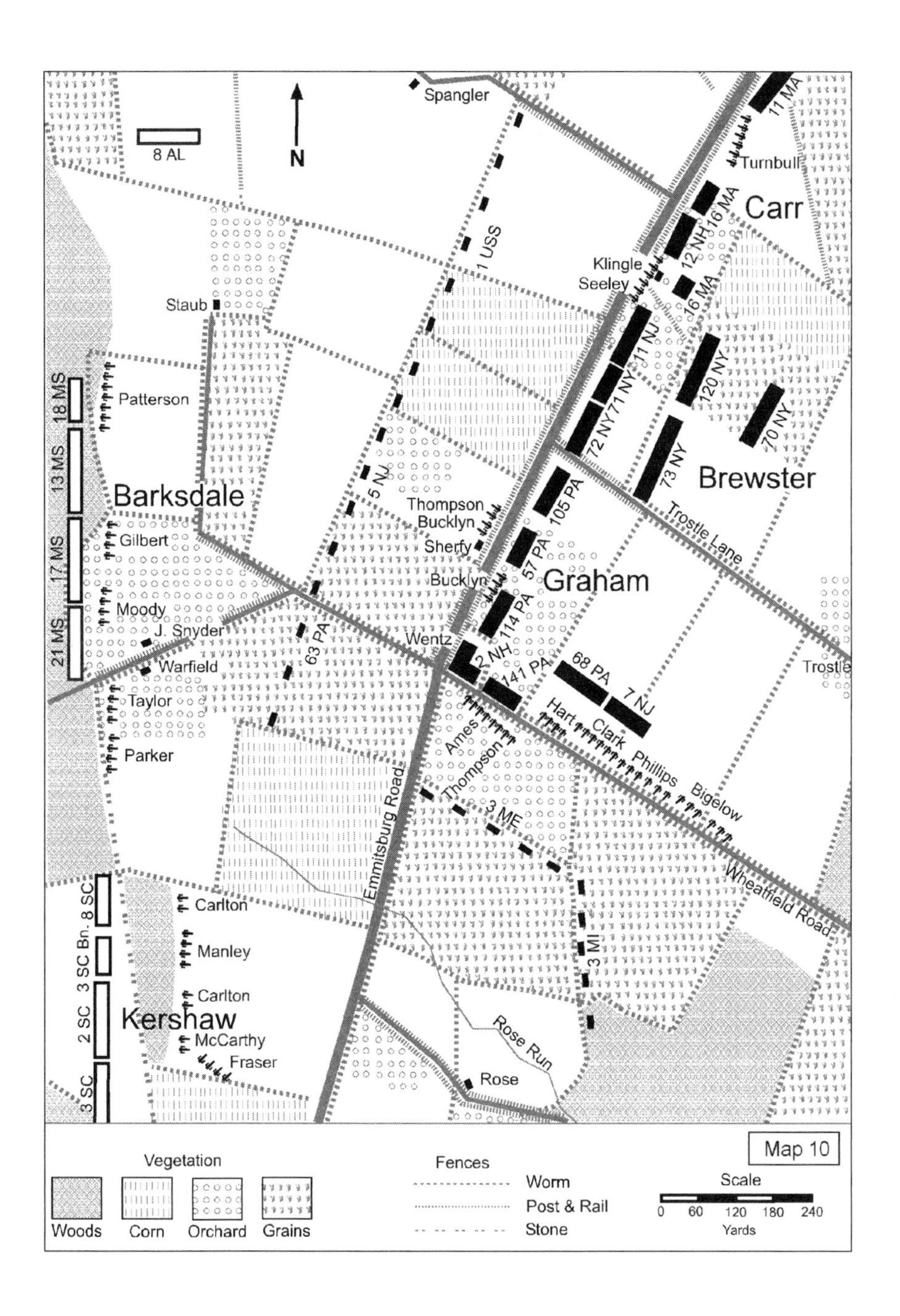

Spangler
8 AL
N
11 MA
Turnbull
Carr
Klingle
Seeley
12 NH
16 MA
16 MA
1 USS
Staub
11 NJ
120 NY
71 NY
72 NY
70 NY
73 NY
Brewster
18 MS
Patterson
13 MS
Barksdale
5 NJ
105 PA
Thompson
Bucklyn
Sherfy
57 PA
Trostle Lane
17 MS
Gilbert
Bucklyn
Graham
Moody
114 PA
21 MS
J. Snyder
63 PA
Wentz
2 NH
Warfield
141 PA
68 PA
7 NJ
Trostle
Taylor
Ames
Hart
Clark
Thompson
Phillips
Bigelow
Parker
Emmitsburg Road
3 ME
Wheatfield Road
8 SC
Carlton
3 SC Bn.
Manley
3 MI
3 SC
Carlton
2 SC
Kershaw
McCarthy
Rose Run
Fraser
3 SC
Rose
Map 10
Vegetation
Woods
Corn
Orchard
Grains
Fences
Worm
Post & Rail
Stone
Scale
0 60 120 180 240
Yards

Clark's] . . . did great damage to our lines."[21] These guns were probably from John C. Fraser's, Edward S. McCarthy's, and Henry H. Carlton's batteries of Henry Cabell's battalion.

The men could hear the explosion of small-arms fire to their left and knew that the Georgians of George T. Anderson's brigade, who had navigated the open fields pounded by the Federal artillery around the Peach Orchard, were now attacking the infantry in the Wheatfield. Without an infantry target, the Federal batteries in the Peach Orchard resumed their duels with the Confederate batteries on Seminary Ridge. Cabell's battalion fired three shots in rapid succession, and suddenly another attack column surged out from the woods on Seminary Ridge and advanced toward the Federal position.[22] Freeman McGilvery also watched this charge and described it in his report as a "larger column appeared at about 750 yards, presenting a slight left flank to our position. I immediately trained the entire line of our guns upon them, and opened with various kinds of ammunition. The column continued to move on at double-quick until its head reached a barn and farm-house [the Rose farm] immediately in front of my left battery about 450 yards distant, when it came to a halt. I gave them canister and solid shot with such good effect that I am sure that several hundred were put *hors de combat* in a short space of time."[23]

A. Judson Clark agreed with this assessment, writing in his report, "I began firing canister, doing great execution, throwing the column wholly into confusion, and causing it to seek shelter behind the slope of a hill just behind them."[24] With a stick in his hand, Clark "passed from gun to gun, animating and encouraging the men, as cool and calm as if it was a battery drill," recalled the battery's historian.[25] He noted that "the guns themselves seemed full of life; dogs of war, nearly red hot; how they roared and thundered! Shells of the enemy's guns were shrieking overhead, or throwing up clouds of dust and dirt where they exploded, bullets were zipping from front and flank."[26]

After the battle, McGilvery walked the fields and counted "120 odd dead, belonging to three South Carolina regiments."[27] These were men from Joseph B. Kershaw's brigade whose destinations were Stony Hill and the Peach Orchard. William Barksdale's brigade was to advance on its left, but it had not done so as yet. As a result, the left of Kershaw's line presented a tempting target to McGilvery's line of guns near the Peach Orchard. John Bigelow recalled, "Our case shot and shell broke beautifully."[28] He initially did not want to open fire as he believed the troops might be from the defeated Federal 3rd Corps. However, when the wind unfurled one of the battle flags, revealing a cross instead of the Stars and Stripes, Bigelow screamed for his men to open fire.[29]

The 2nd and 8th South Carolina and the 3rd South Carolina Battalion took the brunt of the metal from the thirty-two Federal cannon that opened fire at a range of only three hundred yards. The brigade's entire line was to hit Stony Hill, but when Kershaw realized that his left was so exposed, he diverted these three units toward the heavy concentration of Federal artillery deployed near the Peach Orchard, while the 3rd and 7th South Carolina continued their advance on Stony Hill.[30]

Kershaw noted after the war that his regiments advanced "majestically across the field . . . with the steadiness of troops on parade."[31] When the Federal artillery opened fire, Kershaw admitted that it "rendered it difficult to retain the line in good order."[32] Alex McNeill of the 2nd South Carolina called this "the most terrible fire to which they ever were exposed."[33] Pvt. John Coxe of the same regiment would never forget the "deathly surging sounds of those little black balls as they flew by us, through us, between our legs, and over us!"[34] Pvt. William Shumate of the 2nd South Carolina graphically described what it was like to charge across an open field in the face of massed Federal cannon fire: "Kershaw's Brigade moved . . . in perfect order and with the precision of a brigade drill, while upon my right and left comrades were stricken down by grape and canister which went crashing through our ranks. It did seem to me that none could escape. My face was fanned time and again by the deadly missiles. We had arrived within one hundred yards of the battery and had not fired a shot. The artillerists were limbering up their pieces to retire."[35]

According to several Southern soldiers, a fatal mistake now occurred. An officer from the 2nd South Carolina, seeing the 3rd South Carolina move farther to the right to attack Stony Hill, ordered his men to conform to this movement. W. Johnson of the regiment wrote, "Guess they [the Federal artillerymen in and near the Peach Orchard] thought we had had enough sight-seeing from the front, and now we were to have a side view."[36] W. T. Shumate noted that the men were stunned when they heard the order, but "true to our sense of duty we immediately obeyed the command." The Confederate veterans believed that the Federal gunners were also stunned, but it was for a different reason. Kershaw recalled that as a result of his left wing's charge, the Federal "cannoneers had left their guns and the caissons were moving off" when the fatal mistake was made.[37] Seeing the change in the infantry's movements, the Federal gunners returned to their cannon, quickly loaded them, and poured a destructive fire into the ranks of the almost helpless South Carolinians. Lt. Col. Frank Gaillard of the 2nd South Carolina recalled, "We were in ten minutes or less, terribly butchered. . . . I saw half a dozen at a time knocked up and flung to the ground like trifles. . . . [T]here were familiar forms and faces with parts of their heads shot away, legs shattered, arms torn off, etc."[38] Despite these ardent claims by Kershaw and his men, historian Eric Campbell found nothing to support their assertion that the Federal gunners had abandoned their guns, even momentarily. The destructiveness of this fire against human flesh was not disputed and it was terrible.[39]

Soon after the battle ended, infantryman Robert Carter of the 22nd Massachusetts (William S. Tilton's brigade) walked the fields that Kershaw's men had charged across. He noted, "Masses of Kershaw's and Woffords's brigades had been swept from the muzzles of the guns, which had been loaded either with double-shotted, or spherical case, with fuses cut to one second, to explode near the muzzles. They were literally blown to atoms. Corpses strewed the ground at every step. Arms, heads, legs and parts of dismembered bodies were scattered all about, and sticking among the rocks and against the trunks of trees, hair, brains, entrials, and shreds of human flesh still hung, a disgusting, sickening, heart-rending spectacle to our young minds."[40]

After regrouping, the 8th South Carolina and 3rd South Carolina Battalion continued their charge against the Peach Orchard. While these regiments approached from the south, William Barksdale's brigade finally attacked the Peach Orchard salient from the west. Its attack would prove to be almost unstoppable, and with Kershaw's renewed attack from the south, Charles K. Graham's position in and around the Peach Orchard disintegrated, jeopardizing the safety of the Federal artillery in this sector. McGilvery watched Kershaw's renewed attack on his guns from the south: "At about a quarter to 6 o'clock the enemy's infantry gained possession of the woods immediately to the left of my line of batteries, and our infantry fell back both on our right and left, when great disorder ensured on both flanks of the line of batteries."[41]

The 73rd New York of William R. Brewster's brigade was quickly dispatched toward the Peach Orchard to help stem the Confederate tide. Capt.

Frank Moran recalled how Barksdale's brigade attacked the salient and its effect on the Federal line:

> Showers of branches fell from the peach trees . . . in the leaden hurricane that swept it from two sides. . . . The rebel infantry entered the orchard and we received their fire almost in our very backs. . . . A glance to the left [along Wheatfield Road] at that moment revealed a thrilling battle picture. The shattered line was retreating in separated streams, artillerists heroically clinging to their still smoking guns, and brave little infantry squads assisting them with their endangered cannon over the soft ground. The positions of these batteries showed broken carriages, caissons and wheels, while scores of slain horses and men lay across each other in mangled and ghastly heaps.[42]

With their infantry support gone, it was now just a matter of time before the batteries would be forced to limber up and make for the rear. The Federal cannoneers learned firsthand what was contained in the basic training manual: "Artillery cannot defend itself when hard pressed, and should always be sustained by . . . infantry."[43] Few of the latter still remained in the area, and the Federal artillery would pay the price. The south-facing Federal cannon had already fired into George T. Anderson's brigade and had repulsed the initial charge of Joseph Kershaw's brigade's left regiments. Now the pressure was just too great. The artillery's movement to the rear was expedited by the small-arms fire that ripped through the batteries from the approaching enemy infantry, particularly from the strong skirmish line that picked off men and horses. McGilvery noted, "All of the batteries were exposed to a warm infantry fire from both flanks and the rear."[44] He ordered those batteries that had an adequate supply of ammunition to reassemble about 250 yards to the rear to cover the retreat of the others and attempt to stem the Confederate tide. As A. Judson Clark's battery began its movement to the rear, a Confederate soldier yelled, "Halt, you Yankee sons of—; we want those guns!" One of Clark's men yelled back, "Go to h—l! We want to use them yet awhile."[45] Remnants of the 63rd Pennsylvania were near and helped reduce the enemy's ardor, thus buying time for Clark's guns to make their way to safety. A single Confederate gun took position on the J. Wentz farm and fired into the battery, killing ten horses and forcing a caisson to be abandoned. The Federal guns in the Peach Orchard sector had fired 1,342 rounds that afternoon and evening.[46]

Charles A. Phillips was greatly relieved when Freeman McGilvery rode up and ordered him to take his battery to the rear. While the center section immediately complied, the left section fired one last round of canister before withdrawing. However, Kershaw's skirmishers set upon Phillips's right section before it could withdraw.[47] Phillips noted after the war that he erred when he ordered these last two pieces to be pulled to safety with prolonges. "Had not the order to fix *prolonge* been given I am of the opinion that the Right section would have left the field without loss," he wrote.[48] All of the horses of one piece were disabled, so the surviving men took hold of the prolonge and pulled it by hand. Seeing his men's dedication, Phillips dismounted and grabbed hold of the prolonge over his right shoulder while holding his horse's reins with his left hand. The four men dragged the gun about halfway to the Trostle house, when a limber arrived and took over the task.[49] In a letter home, Phillips noted, "I went into position on Thursday under a hot fire, and came out under a hotter one."[50]

With Barksdale's brigade rapidly approaching from the west and Kershaw's from the south, and almost out of ammunition, Nelson Ames finally ordered his men in the Peach Orchard salient to redouble their fire then begin limbering up by section, beginning with the right section. Ames claimed that the Confederate infantry were so close that he could hear their officers yelling "hoarsely and coarsely, 'surrender you yankee—.'"[51]

Ames's losses were seven men wounded (one mortally and two severely) and eleven horses killed or disabled. Phillips's were much higher—seven killed, twelve wounded, and about fifty horses disabled. Clark's losses numbered twenty men and twenty-two horses.[52]

Patrick Hart's battery had withdrawn earlier because it ran out of ammunition. It was a self-inflicted situation. When initially deploying, Hart brought up only half of his caissons, leaving the rest in the safety of the rear. When the forward caissons began running low on ammunition, he quickly sent an aide to bring up the additional caissons, but they were nowhere to be found.[53]

Tempers flared during the retreat. While Phillips was pulling one of his pieces to safety, three of his other cannon moved to the rear unsupervised. They soon were tangled up with other batteries near the Trostle house, so Patrick Hart took charge of them and brought them farther to the rear with his battery. Phillips became enraged when he learned what had occurred.[54]

On the extreme left of the Wheatfield Road line, John Bigelow's battery was ordered to hold its position to support the other batteries' departure, so it

GETTYSBURG CYCLORAMA

Above: Alonzo Cushing's guns in action

Below: James A. Hall's battery in action

BATTLES AND LEADERS OF THE CIVIL WAR

Above: Canons and limbers on the battlefield

Above: Napoleon

Above: 3-inch ordnance rifle

Below, 10-pounder Parrott

Below, howitzer

Below: The barrel in the right front is from Calef's battery and fired the opening round of the battle.

Alonzo Cushing's battery near the copse of trees

Below: Howitzer and Napoleon contrast

Below: Sterling

Below, Whitworth rifle

Below, limber and caisson

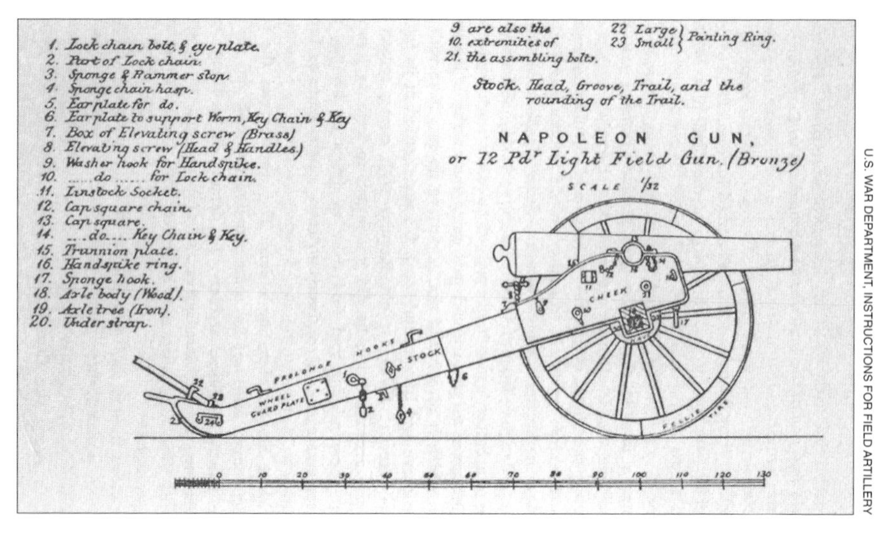

U.S. WAR DEPARTMENT, INSTRUCTIONS FOR FIELD ARTILLERY

Side view of cannon

When horses were killed during combat, the gunners had to remove the guns by manhandling them. Near the Trostle house, Capt. Charles A. Phillips and the men of the 5th Massachusetts Battery pull their gun across the farm yard.

LIBRARY OF CONGRESS

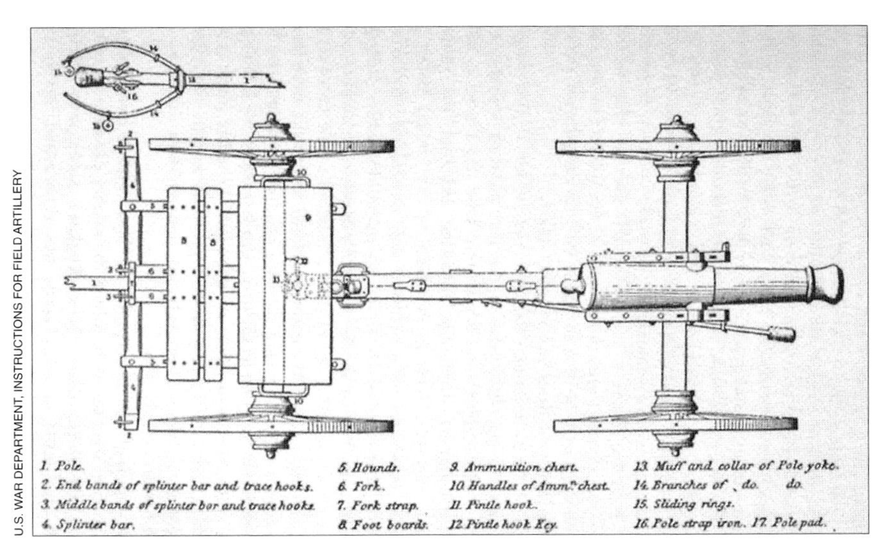

U.S. WAR DEPARTMENT, INSTRUCTIONS FOR FIELD ARTILLERY

Cannon and caisson from above

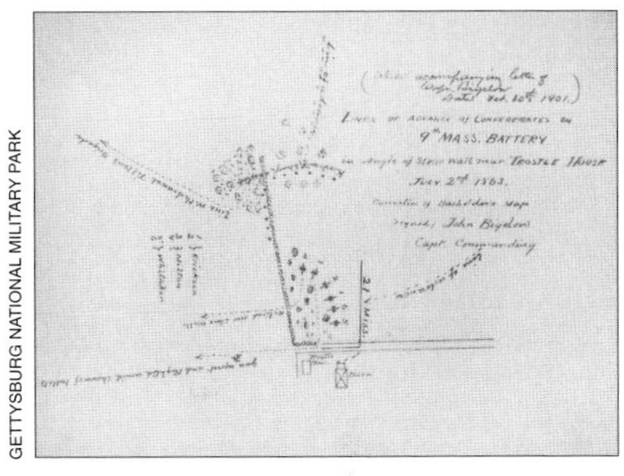

GETTYSBURG NATIONAL MILITARY PARK

Right: A sketch of John Bigelow's stand against the 21st Mississippi.

Below: John Bigelow's battery retreats over a wall.

MASSACHUSETTS COMMANDERY MOLLUS AND U.S. ARMY MILITARY HISTORY INST

WISE, THE LONG ARM OF LEE

William Pendleton

WISE, THE LONG ARM OF LEE

E. Porter Alexander

WISE, THE LONG ARM OF LEE

Frank Huger

WISE, THE LONG ARM OF LEE

James B. Walton

WISE, THE LONG ARM OF LEE

John Cheves Haskell

WISE, THE LONG ARM OF LEE

Thomas H. Carter

WISE, THE LONG ARM OF LEE

J. Thompson Brown

WISE, THE LONG ARM OF LEE

William Nelson

WISE, THE LONG ARM OF LEE

Joseph W. Latimer

BATTLES AND LEADERS OF THE CIVIL WAR

Mathis W. Henry's battery in action

WISE, THE LONG ARM OF LEE

R. Linsey Walker

David G. McIntosh

WISE, THE LONG ARM OF LEE

William J. Pegram

WISE, THE LONG ARM OF LEE

William T. Poague

WISE, THE LONG ARM OF LEE

LIBRARY OF CONGRESS

Henry Hunt

MA COMMANDERY MOLLUS AND USAMHI

James Dearing

MA COMMANDERY MOLLUS AND USAMHI

George E. Randolph

MA COMMANDERY MOLLUS AND USAMHI

Augustus P. Martin

MA COMMANDERY MOLLUS AND USAMHI

Charles H. Tompkins

MA COMMANDERY MOLLUS AND USAMHI

Freeman McGilvery

John Bigelow's battery moves into action on July 1.

MA COMMANDERY MOLLUS AND USAMHI

was the last to pull back to safety. Bigelow watched the worsening situation with growing anxiety: "Glancing toward the Peach Orchard on my right, I saw that the Confederates [Barksdale's brigade] had come through and were forming a line 200 yards distant, extending back, parallel with Emmitsburg Road as far as I could see."[55] While the left section attempted to keep Kershaw's skirmishers at bay, the center and right sections swiveled to the right and poured canister into the 21st Mississippi of Barksdale's brigade, which was rapidly approaching from the west. Col. Benjamin Humphreys, the regiment's commander, "discovered some guns at the front of the slope, to my right firing rapidly into Kershaw's line. [I] immediately wheeled the 21st Regt. away from the Brigade and to the right."[56] The regiment's line of battle was so long that Bigelow thought it was an entire brigade. When McGilvery gave the order to "limber to the rear and get out," Bigelow was concerned that the enemy's infantry was so close that he would lose his guns. "I shall lose all my men in limbering up, but would retire by prolonge instead," he recalled saying to McGilvery.[57]

One of Bigelow's men explained, "The limbers of the battery, with their teams harnessed in, were aligned in rear of the guns and attached to the trails of the pieces by long ropes. At each discharge of a gun the recoil would send it back several yards, and the limber teams would start ahead far enough to take up the slack of the rope."[58] Bigelow noted, "My battery retired by prolonge, but I should perhaps more properly say by the recoil of its guns, for the prolonges were only used to straighten the alignment and for keeping an effective front to both skirmishers as [well as] line of battle."[59] Bigelow noted that his "left section [kept] Kershaw's skirmishers back with canister, and the other two sections [bowled] solid shot towards Barksdale's men."[60] The battery retired about four hundred yards in this manner, firing all the while. Casualties mounted. In one instance, Confederate bullets cut down three men before the fourth could finally fire the cannon.[61] The situation became even worse when Alexander's battalion advanced to the Peach Orchard and opened fire. According to Bigelow, "The air was alive with . . . breaking shells . . . and the ground was being swept with canister."[62]

Pulling guns back with prolonges was a difficult task for veterans and would seem impossible for Bigelow's inexperienced troops, but he had trained his men well, and they were up to the task. Among the dangers of such a movement were uneven ground, which could cause the guns to tip over, and maintaining the alignment of adjacent guns so they did not crash into each other.[63] The officers played a major role in getting the guns to safety as they issued a steady stream of orders as well as words of encouragement. The journey was a long one, and its route could be easily seen by the line of dead and wounded men and horses.

Remarkably, all of the Federal guns along Wheatfield Road made their way to safety, but not all were in a condition to continue the fight. As Federal infantry abandoned the sector, Confederate infantry almost immediately fell on one of James Thompson's south-facing guns and captured it. Their success was short-lived, however, for some Federal infantry counterattacked, recapturing the gun in the process. It appears that during the melee around the guns, Thompson may have ordered all but two of his guns spiked.[64]

While all of the guns facing west along the Peach Orchard salient were making their way to safety, their fate was still not secure, as another Confederate threat bore down from the west.

Peach Orchard: West and North Side

Because of the strong threat to the south, most of the Federal batteries around the Peach Orchard were deployed in that direction. Fewer guns faced west along Emmitsburg Road—initially, only John K. Bucklyn's six guns and two from James Thompson's battery faced Seminary Ridge. The batteries were supported by parts of William R. Brewster's, Joseph B. Carr's, and Charles K. Graham's brigades of the 3rd Corps.

John K. Bucklyn's battery, George E. Randolph's old unit, was the first to arrive. The Confederate batteries on Seminary Ridge were already pounding the sector as the guns rolled up. Finding Nelson Ames's battery posted in the Peach Orchard, Bucklyn ordered his battery into position to the right of it, between the Sherfy and Wentz houses. The battery immediately opened fire and was periodically exposed to frontal and enfilade fire, the latter often from two sides. Losses in both men and horses mounted. No sooner had a clerk rushed over to fill a vacated spot on one of the gun crews than an enemy shell severed his head cleanly from his body. A section sergeant reeled in the saddle when struck by a shell fragment. Feeling first the back of his skull and then the front, he realized that the fragment was spent, so he continued spewing orders to his gun crew.[65]

After dueling the enemy's guns for a while, Bucklyn decided to create his own enfilade fire, detaching his right section and sending it to the right of the Sherfy house, where it deployed in a flower garden. Enemy sharpshooters had taken refuge in a house (probably Staub's) in front of them and opened fire on the battery. This threat was quickly minimized when a few of Bucklyn's shells slammed into the structure. Thompson's two guns arrived and took position to the right of the detached section.[66] It was almost immediately hit by intense Confederate artillery fire. Capt. James Thompson recalled after the war, "The first discharge swept the right section out of position like a whirlwind."[67]

The outgunned Federal artillery in this sector had the misfortune of being up against a Confederate infantry brigade that was anxious to put an end to the war. Brig. Gen. William Barksdale, a fire-eating Mississippi politician, led a crack brigade that had fought valiantly in virtually every engagement since First Manassas. Now he begged his commanders to allow him to launch his attack. Dashing over to James Longstreet, Barksdale blurted out, "I wish you would let me go in, General; I would take that battery [Bucklyn's] in five minutes." Longstreet was as patient: "Wait a little—we are all going in presently."[68]

Barksdale was overjoyed when he finally received orders to charge Bucklyn's battery. "Yelling at the top of their voices, without firing a shot, the brigade sped swiftly across the field and literally rushed the goal," noted John McNeily.[69] It was actually not as easy as that. John Henley of the 17th Mississippi recalled, "We went in perfect line. They would knock great gaps in our line. Then we would fill up the gaps and move on."[70] George Leftwich of the same regiment agreed: "When a solid shot tore a gap in your ranks it was instantly closed up, and the Brigade came on in almost perfect line."[71]

Most of Brig. Gen. Charles K. Graham's Pennsylvania brigade defended this sector, but not for long. John Lloyd of the 13th Mississippi noted within moments that the Confederates were "in among the enemy and literally running over them."[72] Henley of the 17th Mississippi wrote, "We commenced firing on them, and they ran in crowds. You could not shoot without hitting two or three of them."[73] Barksdale's brigade, "firing and shrieking like Indians, barreled into the infantry, forcing them and the artillery in the sector to flee," admitted Francis Moran of Brewster's brigade.[74]

Lt. Benjamin Freeborn of Bucklyn's battery noted that the threat to the left, probably from Kershaw's brigade and the 21st Mississippi of Barksdale's brigade, caused the supporting Union infantry to rush in that direction to deal with this threat. But this movement left the guns without support.

After about twenty minutes, Freeborn could see the troops on his left retreating: "The enemy appearing within a few yards of us and delivering a heavy musketry fire, from which we suffered severely."[75] Freeborn did not indicate whether the Confederate infantry was approaching from the left or the front, but it is likely that they were coming from both directions.

Lt. John K. Bucklyn gave a more complete version after the battle: "We shelled it [Barksdale's brigade] until it was near the road [Emmitsburg] [;] here we received its fire at short range and used some canister. General Graham, lying in the rear with the 'Pennsylvania Reserves' charged through our battery and we retired two hundred yards."[76] Bucklyn was incorrect about the unit that helped buy time so his battery could safely withdraw; it was actually the 114th Pennsylvania of Graham's brigade. Although wounded, George E. Randolph, the 3rd Corps artillery chief, had ridden up to these Pennsylvanians and yelled, "If you want to save my battery, move forward. I cannot find the general, I give the order on my own responsibility."[77] It appears that the 57th Pennsylvania and then the 105th Pennsylvania also advanced at this time to help stanch Barksdale's attack.[78]

Bucklyn pulled the detached section on the right back a handful of yards before ordering it to the safety of the rear. According to the battery's historian, this was done to "allow the infantry a better chance to manoeuvre in our front."[79] With its front masked by the Federal troops, the gunners could not fire anyway. Because of the proximity of the enemy and the short distance to be traveled, the gunners used prolonges to reposition the guns. The battery occupied this new position for a short time, as George Lewis noted, "Our rear being threatened [by Kershaw's brigade], and our front about to be attacked by infantry [Barksdale's brigade], and there being no infantry of ours as yet in front, it was considered best to withdraw the artillery and leave the field to our infantry."[80] So many horses were hit that one caisson was left behind. John K. Bucklyn had three horses shot from under him during this action, and he was wounded in the chest as the battery finally pulled back to safety. He was replaced by Benjamin Freeborn. When Bucklyn returned to action six months later, he wrote, "My battery is torn and shattered and my brave boys have gone never to return."[81]

The situation for the Federals was growing more tenuous by the minute as the Confederates had captured Houck's Ridge and the Wheatfield and had crushed the Peach Orchard salient. Complete victory was within their grasp.

6

JULY 2

STEMMING THE CONFEDERATE TIDE

Seeley's Battery Takes on Wilcox's Brigade

The situation looked bleak for the Federal troops occupying the southern portion of the line. David B. Birney's division and portions of Andrew A. Humphreys's division of Daniel E. Sickles's 3rd Corps were routed, and Houck's Ridge was now in Confederate hands, as were the Wheatfield and Peach Orchard. The Confederates were poised to attack the rest of Humphreys's division along Emmitsburg Road.

The 3rd Corps' last battery, Lt. Francis W. Seeley's Battery K, 4th U.S. Artillery (six Napoleons), had been assigned to Humphreys's division earlier in the day and was deployed about four hundred yards to the right of John K. Bucklyn's battery. Humphreys indicated that he positioned the battery at 4:00 p.m. to the right of the Klingle house and then moved it to the opposite side of the house. Regiments from William R. Brewster's and Joseph B. Carr's brigades of his division flanked it on either side. Upon arriving at the position, Seeley saw a Confederate battery hidden in Spangler's Woods on Seminary Ridge, pounding the Federal infantry in this sector. He ordered his men to open a counterbattery fire as soon as the cannon dropped trail.

Seeley's guns were up against Capt. George M. Patterson's Company B, 11th Georgia "Sumter" artillery battalion (four 12-pounder howitzers and two Napoleons) of A. P. Hill's 3rd Corps. The battalion's commander, Maj. John Lane, sent an additional howitzer to Patterson from Hugh M. Ross's battery, swelling his armament to seven guns. The battery deployed about three hundred yards north of E. Porter Alexander's battalion, where it supported Brig. Gen. Cadmus Wilcox's brigade of Richard H. Anderson's division. The battery received the full attention of Seeley's battery and then John G. Turnbull's to the north. The battalion fired about 170 rounds during this portion of the battle and sustained losses of two killed, five wounded, and seven horses disabled. It had gone into battle already shy four men, who were presumably captured during the march to Gettysburg but may have deserted.[1]

The duel between the twelve guns of Seeley's and Turnbull's batteries and the seven of Patterson's battalion lasted only about fifteen minutes. The Confederate guns were outnumbered and outclassed, as they sported howitzers as opposed to the Federals' Napoleons. The outcome was not unexpected, and the beleaguered Southern guns limped to the rear. Suddenly, four Confederate batteries opened fire on Seeley's battery and Humphreys's division at about 5:30 p.m. According to Humphreys's aide, Capt. Adolphus Cavada, who happened to be near Seeley's battery, "The air was soon full of flying shot, shell and canister—and a groan here and there attested to their effect. For more than a quarter of an hour the roar of musketry and the crashing, pounding noise of guns and bursting shells was deafening. Our own immediate line suffered severely from the enemy's guns enfilading us."[2] Andrew Humphreys calmly walked along his lines, paying special attention to Seeley's gunners. The infantrymen were content to hunker down. Thomas Marbaker of the 11th New Jersey of Carr's brigade wrote, "Tons of metal hurtled over and fell around us, and it was only by hugging the ground closely that we escaped serious loss." He admitted that lying unprotected under artillery fire truly tested the men's nerves. Henry Blake of the 11th Massachusetts of the same brigade related that the men were perhaps most unnerved by the solid shots, which "struck the earth, bounded into the air, and leaped like a rock skipping upon the surface of the ocean by the powerful arm of a giant."[3] Just to the left of Carr's brigade, Col. William Brewster's Excelsior Brigade was also being pounded. Brewster recalled that the shelling came from batteries on his left, probably from Alexander's battalion, and called it "most destructive, killing and wounding many men." Henri Brown of the 72nd New York believed that this shelling was "the heaviest artillery fire the corps had ever experienced."[4]

Although tempted to return fire, Seeley now had a much better target: Confederate infantry advancing in the open plain between Seminary and

Cemetery ridges. These were the 10th and 11th Alabama from Cadmus M. Wilcox's brigade, which were continuing the *en echelon* attack on the Federal position. Seeley's battery opened on them with shot, shell, and spherical case.[5]

Seeley's guns fell silent when the Confederate infantry reached the cover of a rise of ground between the two ridges. The gunners used the reprieve to assess the situation and load their guns with canister, for they were seasoned enough to know what to expect next. When the Alabamians appeared at the top of the rise, they would receive a warm greeting. Seeley was wounded at this time. He initially retained command of the battery, giving orders while reclining on the ground. With Wilcox's brigade approaching in front of them and Barksdale's to its left and rear after crushing the Peach Orchard salient, the battery's position was becoming untenable. To make matters worse, Federal infantry was pulling back. Seeley was now forced to relinquish command of the battery to Lt. Robert James.[6] The latter noted, "I had only time to fire a few rounds of canister, which although creating great havoc in their ranks, did not check their advance, and, in order to save my guns, I was obliged to retire."[7] Capt. George E. Randolph observed that the enemy "attacked it [Seeley's battery] almost with impunity . . . as the supports had fallen back."[8] Humphreys noted that the battery "remained to the last moment, withdrawing without difficulty, but with severe loss."[9] In his report's acknowledgments, Humphreys wrote of Seeley's "gallantry, skill, good judgment, and effective management of his battery [which] excited my admiration, as well as that of every officer who saw him."[10]

The situation just north of the Peach Orchard was nothing less than hellish. Infantry falling back in disarray, men lying all about in various states of agony, horses screaming and dying, and on top of it all, Confederate shells screeching into and over the area.[11]

After retiring about four hundred yards, Robert James ordered the battery to drop trail again. Enemy skirmishers appeared thirty yards away, almost as soon as the guns had unlimbered. Fleeing Federal infantry blocked the field of fire, so James was forced to quickly relimber and head for safety, but not before losing several men to enemy fire; others were captured.[12]

Lt. John G. Turnbull's Batteries F & K, 3rd U.S. Artillery, had also arrived earlier in the day and formed along Emmitsburg Road, facing west, approximately four hundred yards northeast of Seeley's position. According to Humphreys, he requested the battery from the artillery reserve, probably because

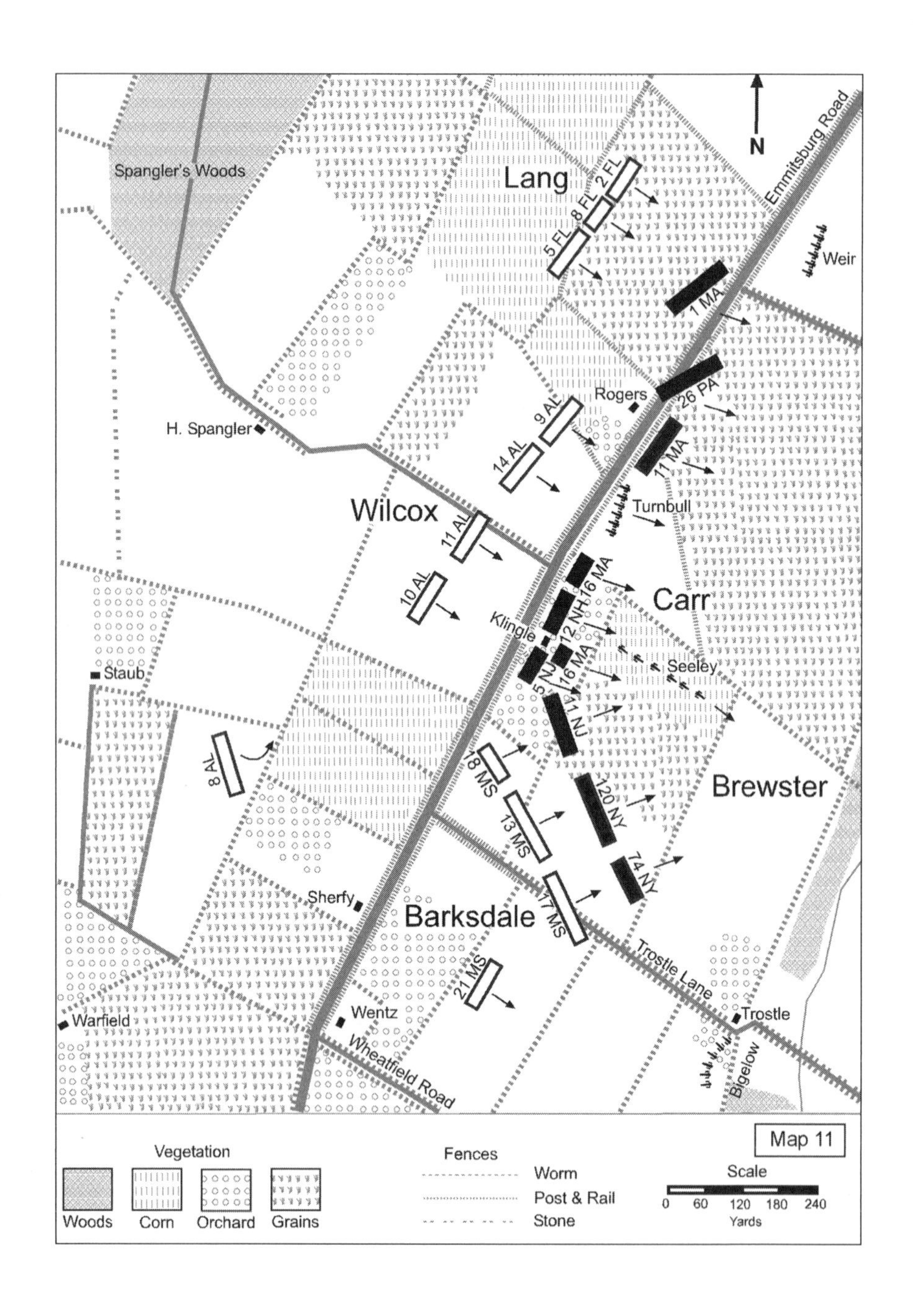
Spangler's Woods
Lang
5 FL
8 FL
2 FL
Emmitsburg Road
N
Weir
1 MA
Rogers
26 PA
H. Spangler
9 AL
14 AL
11 MA
Turnbull
Wilcox
11 AL
16 MA
Carr
10 AL
12 NH
Klingle
16 MA
Seeley
Staub
5 NJ
11 NJ
8 AL
18 MS
120 NY
Brewster
13 MS
74 NY
Sherfy
Barksdale
17 MS
Trostle Lane
21 MS
Trostle
Wentz
Warfield
Wheatfield Road
Bigelow
Vegetation
Woods
Corn
Orchard
Grains
Fences
Worm
Post & Rail
Stone
Map 11
Scale
0 60 120 180 240
Yards

he was concerned about his division's right flank. When the battery arrived, Humphreys deployed it at Seeley's first position along Emmitsburg Road, between the Klingle and Rogers houses. The battery opened fire at a "left oblique" on the Confederate guns, probably George M. Patterson's battery on Seminary Ridge. The 14th Alabama (Wilcox's brigade) advanced soon after, forcing Turnball's guns to respond with canister. With their infantry support in full retreat, the battery was forced to do the same, but it did so by prolonge so that it could continue throwing canister at the enemy.[13] According to Gen. Robert Tyler's report, the battery "was compelled to retire, with the loss of 1 officer and 8 men killed, 14 men wounded, and 45 horses killed."[14] Turnbull apparently halted four of his guns near Plum Run, near Gulian V. Weir's battery and opened fire on Col. David Lang's Florida brigade, which was approaching just north of Wilcox's Alabamians. Lang was only in temporary command of the brigade, so he was ordered to conform to Wilcox's movements. The Floridians approached so closely that Turnbull emptied his revolver into their ranks.[15] This rapid approach, coupled with the excessively high loss of horses, forced Turnbull to leave four of his guns and two of his caissons behind, while the remainder of his unit scrambled to safety.[16]

Turnbull also lost his brigade commander here. Capt. Dunbar R. Ransom, commander of the 1st Regular Artillery Brigade, accompanied the battery and took an active role in positioning it. An enemy shell exploded nearby with a tremendous clap that made the men flinch. When they looked up, they saw Ransom sliding from his horse, severely wounded.[17]

The Confederate Batteries Advance

Just prior to the infantry attack on the Peach Orchard, two of Alexander's battalion's reserve batteries—Tyler C. Jordan's and James Woolfolk's—were ordered up to reinforce the other batteries on Seminary Ridge. Watching the Peach Orchard through his field glasses, Alexander wrote after the war, "When I saw their line [Charles K. Graham's] broken & in retreat, I thought the battle was ours. Of course, I had known it was going to be all along, but now the hard part of it was over. All the rest would only be fun, pursuing the fugitives & making targets of them. I rode along my guns, urging the men to limber to the front as rapidly as possible, telling them we would 'finish the whole war this afternoon.' They were in great spirits, cheering & straining every nerve to get forward in the least possible time."[18]

Maj. James Dearing, the commander of an artillery battalion assigned to George E. Pickett's division, had ridden ahead of his unit and was now put to work, commanding Jordan's and Woolfolk's batteries. It is not surprising that Dearing was on the field ahead of his command as he was one of those leaders who thrived on battle. Only twenty-three years old, Dearing had already gained a reputation. He had dropped out of West Point as the storm clouds of war enveloped the nation and saw action as a lieutenant in the Washington Artillery at the battle of First Manassas. He commanded a battery in 1862 and assumed command of his battalion prior to the Chancellorsville campaign. According to Alexander, Dearing "was noted for his gallantry, his fine athletic figure & his attractive personality."[19]

As the two batteries advanced toward Emmitsburg Road, they encountered a rail fence that obstructed their movement. Seeing a group of Federal prisoners being taken to the rear, Dearing waved his sword and yelled, "Pull down those fences."[20] Col. Jennings Wise noted, "Never was an order executed with more alacrity. Every prisoner seemed to seize a rail, and the fence disappeared as if by magic."[21] The eight guns advanced to a position just west of Emmitsburg Road, and facing northeast, they poured an enfilade fire into Humphreys's division.

The rest of Alexander's battalion also limbered up and headed for the Peach Orchard.[22] Wise stated, "Perhaps no more superb feat of artillery drill on a battlefield was ever witnessed than this rapid change of position of Alexander's Battalion. For 500 yards the foaming horses dashed forward, under whip and spur, the guns in perfect alignment, and the carriages fairly bounding over the fields."[23] Alexander added, "Everything was in a rush. The ground was generally good and pieces and caissons went at a gallop, some cannoneers mounted, and some running by the side—not in a regular line, but a general race and scramble to get there first."[24]

Jennings Wise explained what Alexander must have been feeling that evening:

> There is no excitement on earth like that of galloping at the head of a rapidly advancing line of artillery, with the awe-inspiring rumble of the wheels, mingling with the clatter of innumerable feet close behind. The momentum of the great mass of men, animals, and carriages almost seems to forbid the thought of attempting to check the force which has been set in motion. With his mount bounding along almost as if borne on the breeze of the pursuing

> storm, the eye of the commander instinctively searches the terrain for his position, while a hundred, perhaps five hundred, human beings, and as many dumb warriors, joyfully laboring in the traces, watch his every movement. At last the leader's right arm shoots upward, then outward. No words are necessary, and if spoken would be superfluous. In that dull roar of the onrushing mass no voice but that of Jove could be heard. . . . Every man and horse knows his part and must perform it, for mistakes at such a moment are fatal. . . . Small wonder then that Alexander cherished no regret over having declined the command of a brigade of infantry.[25]

What Alexander saw next remained with him for the rest of his life: "When I got to take in all the topography I was very much disappointed. It was not the enemy's main line we had broken. That loomed up near 1,000 yards beyond us, a ridge giving good cover behind it & endless fine positions for batteries. And batteries in abundance were showing up & troops too seemed to be marching & fighting every where. There was plenty to shoot at. One could take his choice & here my guns stood & fired until it was too dark to see anything more."[26]

Upon reaching Emmitsburg Road, one of Alexander's batteries, possibly Osmond B. Taylor's, apparently turned north and trotted toward its assigned position. The cannoneers pulled up suddenly when they encountered dead and wounded from both armies on the roadway. Capt. A. W. Givin of the 114th Pennsylvania marveled at the chivalry of the battery's commander as he "halted his battery to avoid running over them and his men carefully lifted our men to one side, and carried the wounded into [the] cellar of a house, supplied them with water, and said they would return and care for them when they had caught the rest of us."[27]

Long after the war, Alexander tried to pinpoint the forward positions of his batteries. He recalled that George V. Moody's battery occupied the Peach Orchard, with its right piece near Wheatfield Road, and Osmond B. Taylor's battery was the leftmost unit; the others deployed between it and Moody's.[28] Thinking he could make a difference if he brought up additional guns, Alexander rode back to Henry Cabell's battalion, "but he had suffered so heavily in horses that I don't think any of his batteries moved forward before night."[29] This was not true, as at least Basil C. Manly's battery galloped forward. The batteries opened fire, and "a spirited duel now ensued with their new line," he wrote.[30]

Bigelow's Heroic Fight

Lt. Col. Freeman McGilvery quickly formed a new defensive line near the Trostle house, about four hundred yards in the rear of the Peach Orchard salient. John Bigelow's battery, which had been pulled back via prolonges because the enemy was following so closely, formed on the left, in front of a corner of a stone wall, and Charles A. Phillips's and James Thompson's guns formed on its right. Facing southwest, the guns immediately opened fire on the onrushing Confederate infantry from Joseph Kershaw's brigade and the 21st Mississippi (William Barksdale's brigade). The guns had fired but a few rounds when they were forced back again. Thompson's battery lost so many horses during this short, sharp encounter that the enemy recaptured a cannon and a caisson. This was not done without a fight. Thompson's gunners had tried to pull the cannon by hand with the assistance of some Federal infantry, but the latter panicked as the enemy approached. The few remaining gunners had little choice but to leave the gun and head for safety. A second gun was almost lost when the approaching Confederates shot many of the horses. Pvt. Casper Carlisle, the gun's driver, hopped down, freed the gun, and brought it to safety. For this act of heroism he received the Medal of Honor.[31]

According to McGilvery, "The crisis of the engagement had now arrived."[32] He knew that if he withdrew from this position, there would be no way of stopping the victorious Confederates from cleaving the Union line and defeating the army. It was a hopeless situation, and he had few options. Seeing Bigelow's battery preparing to withdraw, he rode to it, but not before his horse was wounded four times. While anxious about the current danger, Bigelow's men were relieved to be finally getting out of harm's way. This emotion died when McGilvery approached and yelled, "Captain Bigelow, there is not an infantryman back of you along the whole line from which Sickles moved out; you must remain where you are and hold your position at all hazards, and sacrifice your battery, if need be, until at least I can find some batteries to put in position and cover you. The enemy are coming down on you now."[33]

This was a situation that all artillerymen dread: trying to stem a determined infantry attack without support. The problems for Bigelow's battery were compounded by the fact that it was a green unit that had already sustained heavy losses. This was a suicide mission: the enemy was closing in on three sides and the battery's ammunition was almost depleted. In response to the order, Bigelow deployed his six guns in a semicircle so they could have multiple fields of fire.[34] "The ground on our front and right was much higher, and we could not see more than fifty or sixty yards in those directions; neither was there room enough to work six guns at the usual intervals; and the ground was broken by bowlders, with heavy stone walls in our rear and left," noted the battery's historian.[35] Bigelow called the task "superhuman," for the enemy was already almost on top of his battery.[36]

Although Bigelow could not see the enemy because of a knoll about fifty yards in front of his position, he knew they were approaching. He ordered the guns loaded with solid shot and aimed low so that the balls ricocheted over the hill and hit the unseen enemy. This was a desperate measure that might slow the enemy advance, but it would not halt it.[37]

The 21st Mississippi appeared on the swell of ground to the front and right of the battery in a matter of minutes. These were the same troops that had helped crush the Peach Orchard salient and forced McGilvery's guns into their initial retreat. Now the Mississippians again descended rapidly upon Bigelow's six guns. They opened fire, and cannoneers and horses fell. Some of the gunners changed to canister, while others used case shot and shell, cutting the fuses so they would explode near the guns' muzzles. The enemy approached so rapidly that Bigelow ordered his men to remove the ammunition from the limber chests and place it near the guns. This was a risky proposition, but Bigelow had no choice. During those times when the smoke cleared, the gunners could see the enemy's line was "torn and broken" but not dissuaded.[38] The Confederate artillery near the Peach Orchard opened fire, and combined with the small-arms fire from the Mississippians, they made the area around the guns a living hell.

The battery's position became more cramped with every discharge of the guns. Lt. Richard S. Milton's left section recoiled so far into the stone wall that its two guns were no longer effective, so Bigelow ordered them to the rear. This movement was not without some moments of excitement, though. One gun snagged a stone as it passed through an opening in the wall and flipped over. The exertions of the men righted it, but the exercise was harrowing in the face of the enemy, with bullets zipping about. The other cannon became stuck while going over the low wall, but with perseverance, the gunners pushed it over.[39]

This was the stuff of legends. Whitelaw Reid, a former minister to England and now a correspondent for the *Cincinnati Gazette*, happened to see the desperate fight between the battery and the 21st Mississippi and described it to his readers: "Reserving his fire a little, then with depressed guns opening with double charges of grape and canister, he smites and shatters, but cannot break the advancing line. . . . He falls back on spherical case, and pours this in at the shortest range. On, still onward, comes the artillery-defying line, and still he holds his position. They are within six paces of the guns—he fires again. Once more, and he blows devoted soldiers from his very muzzles. . . . They spring upon his carriages and shoot down his forces."[40] John Bigelow wrote that the "enemy crowded to the very muzzles . . . but were blown away by the canister . . . the air was dark with smoke . . . the enemy were yelling like demons, yet my men kept up a rapid fire, with their guns each time loaded to the muzzle."[41]

The battery, firing double-shotted loads, was doing an effective job against the enemy troops in front of it. The gunners could not, however, contend with the enemy bearing down on their flanks. Richard S. Milton wrote that the "rebel line of battle could not be checked, although its center was badly broken by our canister fire. Its flanks closing in on either side of us, obtained a cross-fire, which silenced the four pieces on my right."[42] The battery's historian merely wrote, "Our cannoneers were driven at the point of the bayonet, and were shot down from the limbers."[43] They did not go down without a fight. Accouterment-wielding cannoneers killed or disabled several Confederates in hand-to-hand combat. It was a desperate time, particularly for the battery's officers. Bigelow was shot off his horse; Lt. Christopher Erickson, already wounded, was hit at least five additional times and died instantly; Lt. Alexander Whitaker was shot through the knee. Milton took command of the battery. As the men raised Bigelow's head, he could see enemy soldiers standing on the limbers, shooting the horses to prevent the guns from being moved to the rear.[44] In the end, the 21st Mississippi captured four of Bigelow's guns. Photographs of the Trostle farm after the battle show the many dead artillery horses lying around Bigelow's position.

The Plum Run Line

While Bigelow was buying time, Freeman McGilvery was assembling as many guns as he could find in a new defensive line near Plum Run, about four hundred yards to the rear of the Trostle house. Lt. Edwin B. Dow's 6th Maine Light Artillery, Battery F (four Napoleons) of the 4th Volunteer Brigade was the first to arrive. When Charles A. Phillips's three guns appeared around 7:00 p.m., they were placed to Dow's right. Two guns from James Thompson's battery—the rest had been lost or spiked—took position on Phillips's right. Lt. Malbone F. Watson's Battery I, 5th United States (four ordnance rifles), also arrived and was placed about 150 yards to the left (south) of Dow, on the opposite side of the Trostle farm lane. Capt. James M. Rorty's Battery B, 1st New York (four 10-pounder Parrotts), of the 2nd Corps also arrived and formed to the right of the line.[45] McGilvery now had at least seventeen guns to take on William Barksdale's victorious Mississippians. He reported that another volunteer battery manned the line, swelling the number of guns to twenty-three, but "I have never been able to learn the name of [it]."[46] Equally important was the arrival of Col. George Willard's fresh infantry brigade (Alexander Hays's division, 2nd Corps). This brigade had been posted near the Bryan farm on Cemetery Ridge but was marched rapidly south to help stem the Confederate tide.

Confederate sharpshooters quickly occupied thicket-surrounded Plum Run, which flowed between the two forces. While Edwin B. Dow's guns sprayed the thickets with canister, Charles A. Phillips's opened fire on the Confederate batteries near the Peach Orchard that were pounding the Federal line.[47]

McGilvery could see Bigelow's battery being overwhelmed by the 21st Mississippi, and he could see the rest of Barksdale's victorious Confederate infantry reforming in an open field about seven hundred or eight hundred yards in front of him. Phillips's battery opened fire on Bigelow's guns after they were captured, creating a no-man's-land around them. Bigelow purportedly told Phillips that his battery "covered them [his captured guns] so completely that the enemy could not take them off."[48]

The 21st Mississippi, operating well to the south of the rest of Barksdale's brigade, surged ahead. Malbone F. Watson's isolated battery was directly in its path. A 3rd Corps staff officer had apparently placed the battery at this highly vulnerable position in order to plug a hole in the line.[49] Watson was hit by a Confederate bullet and fell heavily to the ground, forcing Lt. Charles C. MacConnell to take over. Fifth Corps artillery chief Capt. Augustus P. Martin noted in his report that the "battery was without support of any kind. The enemy appeared shortly . . . nearly in front, at a distance of about 350 yards . . . as they approached nearer, the battery poured in canister, some twenty rounds, until men and horses were shot down or disabled to such an extent that the battery was abandoned."[50]

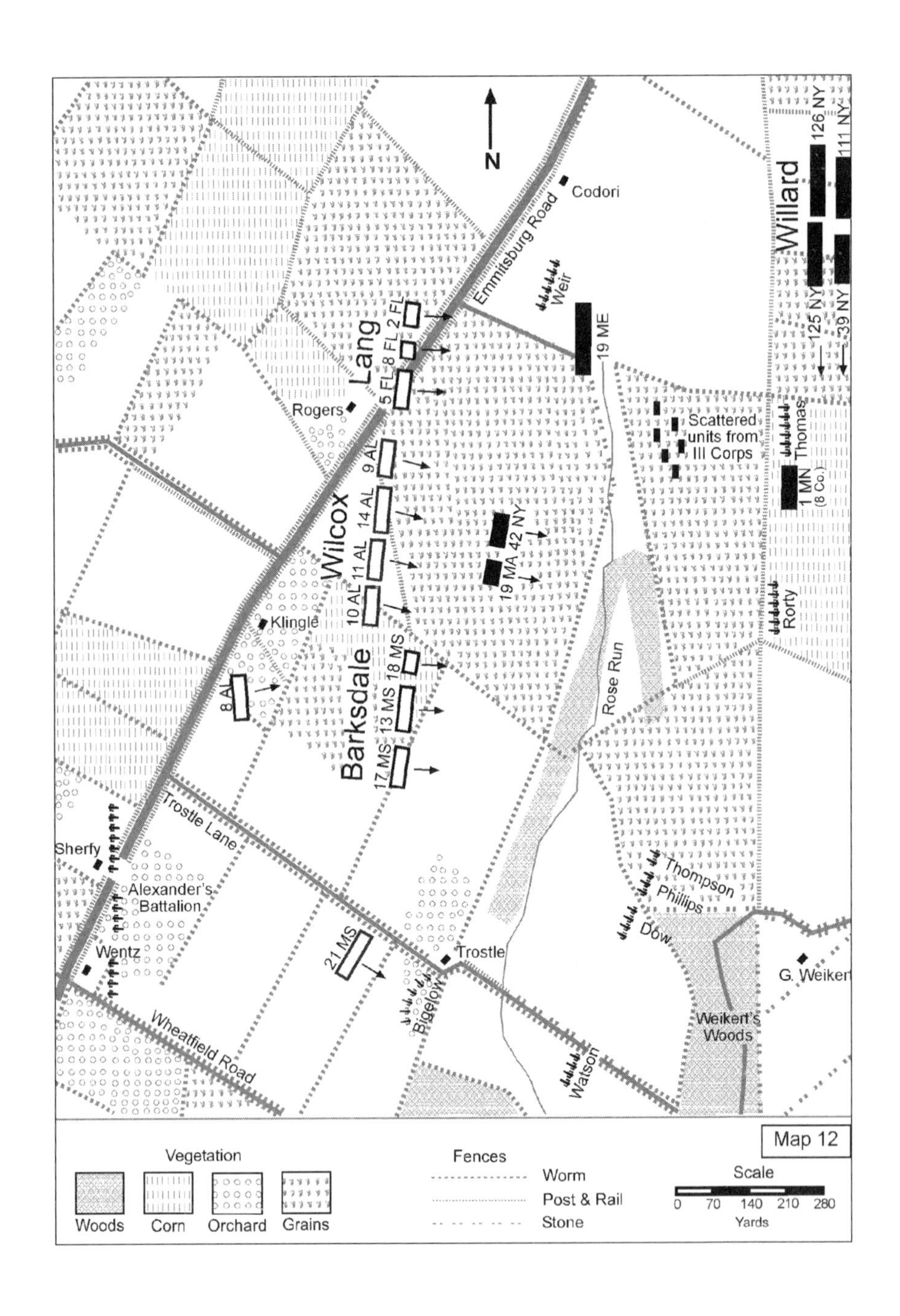

THE PLUM RUN LINE

In actuality, the Mississippians probably attacked so rapidly that they overwhelmed the battery before it could fire a shot. The 21st Mississippi had now captured eight Federal cannon within minutes. After the war, Andrew Humphreys recalled this moment: "From the position I occupied then, no enemy could be seen or heard in my front. Nor a gun was being fired at me. The federal army was cut in twain. Eight hundred yards, to my right a confused mass was retreating, driven by McLaws, and Hood. I attempted to turn the guns just captured on them but no rammers or friction wires could be found. Eight hundred yards to my left, the enemy's line was kept busy by Barksdale."[51]

Seeing the Mississippians capture Watson's battery and attempt to turn the guns on the Federal forces still in the area, a quick-thinking aide to Birney, Capt. John Fassett, rushed over to the 39th New York of George L. Willard's brigade, which was resting nearby. Fassett ordered the regiment's commander, Maj. Hugo Hildebrandt, to recapture the guns. "By whose orders?" demanded the major. "By order of General Birney," replied Fassett. "I am in General Hancock's Corps," replied Hildebrant. Fassett immediately replied, "Then I order you to take those guns, by order of General Hancock." Probably noting what was going on around him, Hildebrant nodded and ordered his regiment forward. It subsequently crashed into the 21st Mississippi, forcing it to abandon Watson's guns and then Bigelow's as it retreated back toward Seminary Ridge. Fassett was awarded the Medal of Honor for his heroics.[52]

To Watson's right, the rest of McGilvery's line fired into Barksdale's three regiments as they continued their surge. McGilvery watched as his patchwork defensive line began unravelling. James Thompson's guns ran out of ammunition and withdrew, and to his dismay, James M. Rorty's did as well. He remarked in his report that the latter "remained only a few minutes."[53] According to a member of Rorty's battery, "We opened with terrible effect, for the enemy was close up in solid ranks. Being without infantry support, it was impossible to hold our position . . . there was no time to file out of our position, so we went back in a solid battery front."[54] So anxious were the men to reach safety that they drove their teams over the stone fences that stood between them and sanctuary.

The Plum Run line was now down to seven guns—Edwin B. Dow's battery and three guns of Charles A. Phillips's. They continued blasting away at Barksdale's men "until they [the enemy] had ceased firing."[55] Dow remarked in his report that while his men were mowing down Barksdale's men, the Confederate artillery, "to which we paid no attention, had gotten our exact range, and gave us a warm greeting."[56] Dow later wrote that McGilvery told him to "hold my position at all hazards until he could re-enforce the position and relieve me."[57] The orders were reminiscent of those he had issued to Bigelow a short time before.

These heroics would not be needed because the Confederate attack finally ground to a halt. The infantry reinforcements of Willard's brigade and then Brig. Gen. Henry Lockwood's (Thomas H. Ruger's division, 12th Corps) barreled into the enemy. Barksdale was mortally wounded just prior to this attack. Leaderless, disorganized from the long charge, and with night falling, the Mississippians decided to break off the attack. They always regretted not being able to bring off the two captured batteries, but they did spike one of Bigelow's Napoleons. The Plum Run line had been very active that evening. For example, Dow's battery expended 244 rounds in a short period of time.[58]

Not convinced that the threat was over, McGilvery ordered Charles A. Phillips to ride back to Taneytown Road and bring up Francis W. Seeley's battery, which subsequently deployed near Edwin B. Dow's with orders to open fire on any enemy column that appeared. So undermanned was the battery that two guns were sent to the rear and only four deployed near Dow's cannon. Their ammunition chests depleted, Phillips's three guns finally withdrew.[59] When McGilvery realized that the threat was over, he turned his attention to Bigelow's and Watson's eight recaptured cannon. Details from Dow's battery, augmented by infantrymen from Henry H. Lockwood's brigade, ventured forward and pulled the guns to safety.[60]

The sector had held because of the bravery of the Federal gunners who had performed spectacularly. Most notable were the actions of Freeman McGilvery, who understood the Union army's predicament and, although unsupported, almost single-handedly held the Confederate infantry at bay until Federal infantry reinforcements arrived. According to a contemporary, "McGilvery fearlessly exposed himself all over the field . . . escaping without injury. His horse was hit four times by musket-fire, once by a shell and once by a spent solid shell, of which wounds he soon after died."[61]

As might be expected, Bigelow's battery sustained the heaviest losses: eight men killed, seventeen wounded, and two captured. Sixty horses of its eighty that went into combat were killed or disabled. Seeley lost two killed, nineteen wounded, and two missing. He also lost twenty-eight horses. Dow's battery, which was not engaged near the Peach Orchard or along Emmitsburg Road, lost only eight men wounded. It did shoot off 244 rounds.[62]

Overview of the Actions on the Southern Part of the Federal Line

It is clear that the Confederate guns received the worst of it during the afternoon and early evening hours in the southern part of the field. According to Southern historian Jennings Wise, "The fire of the Confederate Artillery was most effective, but it was hopelessly outmatched in numbers."[63] An examination of the situation does not support Wise's contention, however. The Confederates put forty-eight guns in eleven batteries into action along Warfield Ridge and the southern portion of Seminary Ridge. The Federals deployed ten batteries with fifty-six guns. That is a mere eight additional pieces. The Federals did have the advantage of having more Napoleons (thirty-four vs. eleven) that are particularly effective at these ranges. According to historian R. L. Murray, the Confederates lost at least five guns and possibly more. So great were the losses of cannoneers that at least one battery, John C. Fraser's, had to consolidate gun crews, so only two guns could continue firing. George V. Moody's battery lost some of its effectiveness when infantrymen were pressed into service as gunners. The Federals, on the other hand, did not lose any guns during the counterbattery action, but they did lose seventeen to enemy infantry, and all but four were recaptured. Lost were three guns from James E. Smith's and one from James Thompson's batteries.[64]

E. Porter Alexander has been criticized for not deploying all of his available artillery. He did not have the services of the Washington Artillery, which was left on Chambersburg Pike, but of the guns available to him, only 25 percent were initially deployed. At the peak of the cannonade, only sixty-nine fired at the enemy; eighteen sat idly behind the tree line. Most of these guns were howitzers that did not have the range to duel with the Federal artillery. Alexander also wanted to use them to follow the infantry later on.[65] Alexander subsequently explained, "I hoped they would crush that part of the enemy's line in a very short time, but the fight was longer and hotter than I expected."[66] Historian R. L. Murray speculated that if Alexander had deployed all of his available guns from the start, he could have inflicted severe losses on the Federal batteries, particularly those taking position in the Peach Orchard, as they were most vulnerable when deploying for action.

Left undeployed were some of Alexander's most effective guns, which could have been positioned to enfilade the Federal line.[67]

Anderson's Division Strikes the Federal Center

As each brigade of Sickles's 3rd Corps met with defeat, 2nd Corps commander Winfield Hancock became increasingly concerned about his sector of the battlefield, which stretched from the 3rd Corps on the left to the southern edge of the town of Gettysburg on the right. A yawning gap already existed because his left division, Brig. Gen. John Caldwell's, had been sent south to the Wheatfield, where it was ultimately overwhelmed.

Few units occupied the gap. Seeing the six Napoleons of Lt. Evan Thomas's Battery C, 4th U.S. Artillery, of the artillery reserve's 1st Regular Brigade approach on its way to reinforce the 3rd Corps, Hancock diverted it to a position on Cemetery Ridge approximately midway between Emmitsburg Road and the Hummelbaugh house near Taneytown Road. The 1st Minnesota of William Harrow's brigade was sent to support it.[68]

With Hood's and McLaws's divisions engaged, it was time for Richard H. Anderson's division to enter the fray. Cadmus M. Wilcox's brigade and David Lang's now rushed toward Cemetery Ridge. Seeing Wilcox's brigade of Alabamians breaking through the 3rd Corps' line, Hancock galloped up to the Minnesotans and sent them into immortality. Thomas's battery was not idle for long, for its cannoneers opened a deadly fire when they saw Wilcox's brigade and David Lang's small Florida brigade approach Cemetery Ridge. Seeing the danger that Thomas's battery posed to his brigade, Wilcox sent the 9th Alabama to silence it. The unit did not go far before it encountered the 1st Minnesota, which was essentially sent on a suicide mission to seal the breach in the Federal line. After engaging the plucky Minnesotans and two other Federal regiments, and lacking reinforcements, Wilcox ultimately decided to break off the attack and pull back toward Emmitsburg Road.[69]

Hancock found another battery from the 1st Regular Brigade that was also heading south to reinforce Sickles. As with Thomas's battery, Hancock countermanded the orders given to Lt. Gulian V. Weir's Battery C, 5th U.S. Artillery (six Napoleons), and deployed it about two hundred yards northwest of the Hummelbaugh farm lane and about three hundred yards northeast of Thomas's battery. The gunners quickly dropped trail and opened fire on Lang's brigade, which was charging toward their left. Weir reported, "I was ordered by General Gibbon [2nd Division commander] to open fire to the left with solid shot at 4 degrees elevation."[70] An old artilleryman, John Gibbon remained with the guns, helping to adjust their range.

Riding around his sector, Hancock looked for James M. Rorty's battery of his corps' artillery brigade. Rorty's battery was to support Lt. T. Frederick Brown's battery, which occupied an exposed area along Emmitsburg Road near the Codori farm, but Rorty was nowhere to be found. Galloping back to Weir's battery, Hancock ordered the young officer to limber his guns and accompany him to a new position, which turned out to be about three hundred yards to the southeast of the Codori farm—well in advance of Thomas's battery.[71] Weir was dismayed to see the vulnerability of his new position and was not afraid to tell Hancock his feelings about it, and how it lacked vital infantry support. Hancock agreed with the latter, saying, "Go in here, I will bring you support,"[72] as he galloped to the rear.

Looking around him, Weir angrily noted that the enemy "met with no opposition whatever from our infantry, who were posted on my right and front."[73] About to be overwhelmed, he immediately ordered his guns to open with solid shot and spherical case, and then with canister. Col. David Lang, commander of the Florida brigade, wrote that his men were met at the first hill "with a murderous fire of grape, canister, and musketry."[74] L. B. Johnson of the 5th Florida observed that the "enemy's guns are making great gaps in our lines, and the air seems filled with musket balls, our men are falling on all sides."[75]

Continuing their charge, Lang's men drove into the right flank of Joseph B. Carr's brigade (Humphreys's division, 3rd Corps) aligned along Emmitsburg Road and sent its men fleeing to the rear. The Floridians now advanced on Weir's beleaguered gunners. Weir did not know it, but Brig. Gen. Ambrose R. Wright's brigade was also approaching on his right. Without orders to withdraw, Weir's men could only watch the 3rd Corps infantry flee to the rear, along with Francis W. Seeley's and John G. Turnbull's batteries. Weir realized that to remain would mean the certain loss of his battery, so he decided to disobey Hancock's orders and get his guns to the rear.

Help was on the way, however. Hancock found the 19th Maine of Norman J. Hall's brigade and brought it double-quicking toward Weir's position. Seeley's battery suddenly approached as the regiment crossed Plum Run. The cannoneers were in such a rush to reach the safety of the rear that they plowed into the center of the regiment, sending men and mud flying. They were followed by the six caissons of Turnbull's battery. According to Francis Heath, commander of the 19th Maine, Hancock was clearly stressed at this point, for he "gave way to a curious outbreak of temper."[76] As Heath ordered his regiment to break ranks and allow the batteries to pass, Hancock snapped, "If he commanded the regt. he'd be God Damned if he would not charge bayonets on him [Seeley's battery]."[77]

Although his battery was now making its way to the rear, Weir decided to drop trail when he saw the oncoming 19th Maine. The battery was now apparently farther south than its former position, facing Emmitsburg Road at a right angle, with the Codori barn about five hundred yards to its right.[78] Weir hoped that the infantry would "drive the enemy back, as their force seemed to be small and much scattered."[79] While Weir was correct in his assessment of the Floridians' condition, it did not stop them from continuing their surge toward Cemetery Ridge. Two sections from Turnbull's battery were in the vicinity, and the Confederates gobbled them up.[80] Now the 2nd Florida turned its attention on Weir's guns.

Looking to his right, Weir could see some troops advancing from the Codori barn. Thick smoke, however, obscured their identity. When these unknown troops were no more than thirty yards away, they let go a volley that staggered the battery. Weir immediately gave orders to limber the guns and get them to the rear. Seeing his gunners in the process of obeying his orders, he began riding to the rear. Looking back, he was stunned to see that three of his guns were about to be captured by Wright's Georgia brigade. Weir wrote, "[The] enemy were too close. I endeavored to get my guns off the field; succeeded in getting off but three."[81]

By this time, the enemy infantrymen were so close they were disabling the gunners and horses. Weir's horse was shot, throwing him to the ground. Slowly getting to his feet, Weir was struck in the forehead by a spent bullet, spinning him around and disorienting him. He noted, "Everything seemed to be much confused."[82] Although he knew that three of his guns had escaped, he could find only one of them in the rear. Wilcox's and Lang's attack did not continue much longer, because they were not reinforced and the units were disorganized.[83]

Approximately three hundred yards to the northeast, Lt. T. Frederick Brown's Battery B, 1st Rhode Island (six Napoleons) of Capt. John G. Hazard's 2nd Corps Artillery Brigade was deployed just to the northwest of the

Codori farm buildings. Seeing Sickles's movement off of Cemetery Ridge, Gen. John Gibbon felt compelled to advance a small force that would "give support to its right flank."[84] In actuality, the 3rd Corps right flank was several hundred yards to the left (south). Lt. Col. Charles Morgan, Hancock's chief of staff, wrote that Gibbon's actions were a grave mistake, "particularly the placing of the battery in front of the main line." He went on to state that "Captain Hazard was justly apprehensive concerning this battery, and General Hancock himself disapproved of its being sent out, but said nothing to General Gibbon concerning it."[85] This was Hazard's old battery, known for its gallantry at the battle of Fredericksburg when it was sent to an exposed position to support the infantry attack on Marye's Heights. Although the battery did little good in this vulnerable position and took heavy casualties, it showed the mettle of the men and their commander.[86]

Brown's battery was supported by the 15th Massachusetts and 82nd New York, both from William Harrow's brigade, which deployed along Emmitsburg Road. Both were veteran units, but their ranks were depleted.

The battery deployed at a forty-five-degree angle to Seminary Ridge, facing northwest. Its left gun was nearest Emmitsburg Road, with its right extending back toward the main Federal line on Cemetery Ridge, about a hundred yards away. Soon after the battery deployed, it opened fire on a Confederate battery that was pounding George G. Meade's headquarters behind the Federal line.[87] The identity of the battery is not known, but it was probably from John Lane's battalion. Some of the Confederate artillery also played on the Federal position along Emmitsburg Road. The men of the 15th Massachusetts quickly erected flimsy breastworks to protect themselves, but Confederate gunners soon found them and blew them apart.[88]

At about 6:30 p.m., a dense mass of men emerged from Seminary Ridge. At first the cannoneers thought that they might be their own infantry from the 3rd Corps. A Rebel Yell from fourteen hundred throats told the artillerists of their error. T. Frederick Brown's gunners sprang to their guns as Brig. Gen. Ambrose R. Wright's brigade of Georgians headed for the copse of trees on Cemetery Ridge. The Georgians presented a "solid front of two lines of battle," noted the historian of Brown's battery. Watching this advance, a Federal soldier in the Philadelphia Brigade near the copse of trees on Cemetery Ridge was awed by the Georgians' new uniforms. Unlike other Confederate units, these soldiers seemed to have uniforms that matched. John G. Hazard was on the scene, watching the actions of his former battery.[89]

A gunner from Brown's battery recalled this initial phase of the encounter: "As our artillery fire cut down their men they would waver for a second, only to soon close up and continue their advance, with their battle flags flying in the breeze, and the barrels of their muskets reflecting the sun's dazzling rays."[90] In actuality, only two of Brown's sections opened fire on Wright's men; the right section continued to shell the Confederate battery on Seminary Ridge. The four Napoleons facing Wright's rapidly moving infantry were perfect for this type of fighting, and they made the enemy pay. They opened fire with spherical case shells filled with seventy leaden or iron balls, timed to explode in four seconds. As the distance between the Georgians and the battery decreased, "Our fuses were cut at three, two, and one second, and then canister at point blank range, and finally, double charges were used," noted the battery's historian. A Federal cannoneer observed after the battle, "Our fire was very destructive to the enemy. I could see, at every discharge of our guns, a vacant space appear in the enemy's ranks, but they would immediately close up."[91]

Confederate infantryman Frank Foote recalled, "Shells around us tore our bleeding ranks with ghastly gaps. . . . We pressed on, knowing that the front was safer now than to turn our backs, and with a mighty yell, we threw ourselves upon the batteries and passed them, still reeking hot."[92]

The gunners yelled out, "Don't give up the guns!" There was only one thing left to do—Brown gave the order, "Limber to the rear!" A Confederate foot soldier noticed an officer riding along the battery, a pistol in his hand, giving a stream of orders. Halting, the infantryman took aim, and Brown slid from his horse with a serious wound to his neck. Lt. Walter S. Perrin assumed command of the battery. All of the gun commanders instantly obeyed Brown's order, except one. Sgt. Albert Straight, who commanded the forth gun, ordered it fired before being limbered. In the short time it took to obey his order, two of the gun's horses were hit. Straight noted, "The boys had to look out for themselves, as the Johnnies were all around us, and the bullets flew very lively." The gun was ultimately captured.[93]

The remaining five guns made for the safety of the Federal line behind them. They first had to cross a stone wall that marked the infantry's main line on Cemetery Ridge. The gunners aimed for a small opening in the wall that could accommodate only one cannon at a time. The first two guns made it through without a problem, but the next two tried to navigate the opening at the same time, resulting in a giant collision. Gunners, horses, and guns were

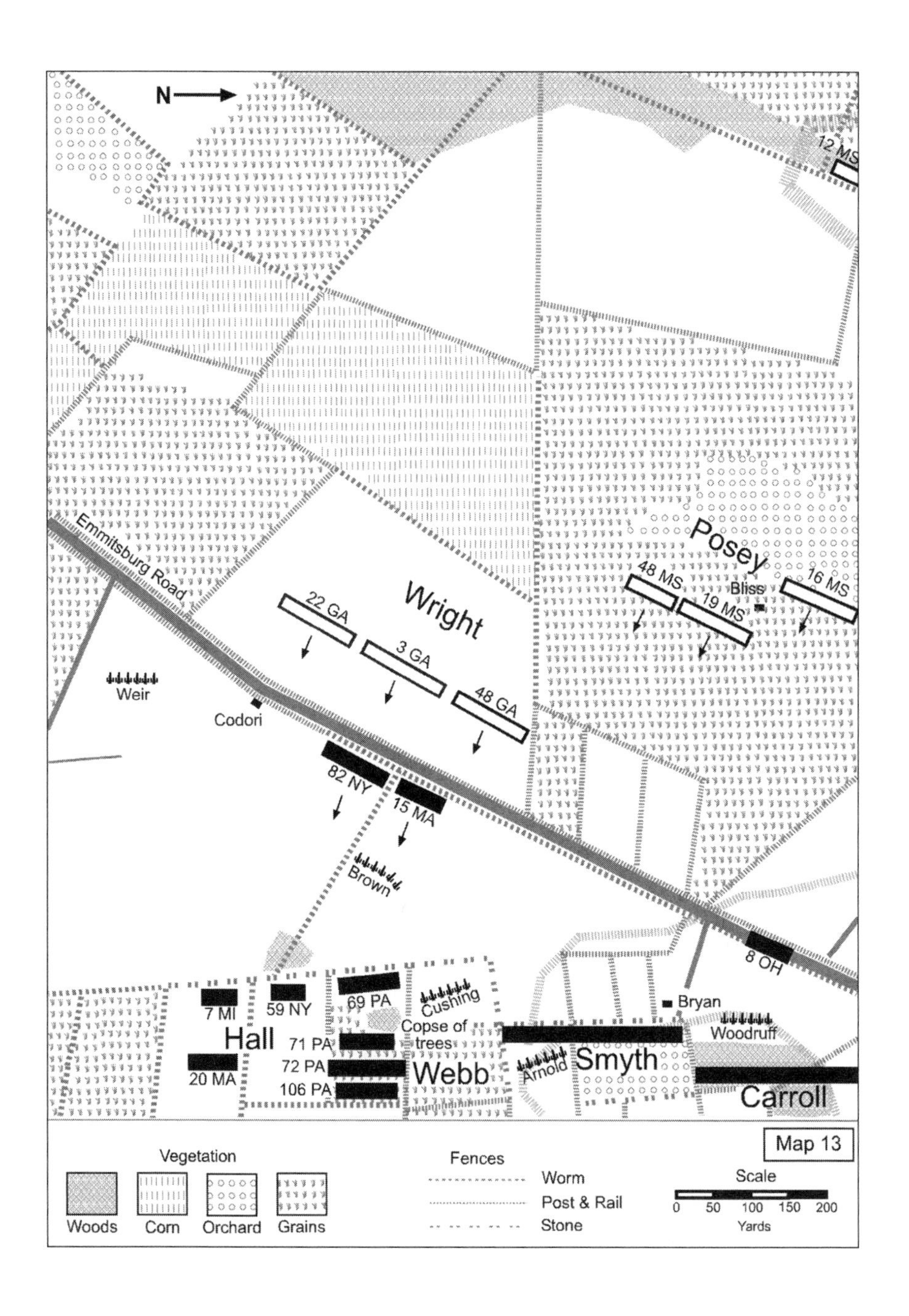

WRIGHT'S CHARGE

knocked about in disarray in front of the wall. The enemy continued firing, and that added to the confusion and suffering. The two groups of gunners crawled to safety behind the wall, leaving their guns a tangled mess. This meant that the remaining gun could only wait outside the wall for the passageway to clear.

The Irishmen of the 69th Pennsylvania of the Philadelphia Brigade impatiently waited behind the wall for the guns to clear their front and finally screamed for the cannoneers to hit the ground as Wright's Georgians approached. They did, and looking up, the cannoneers could see "a vivid flame sending forth messengers of death to the foe."[94]

Confederates surged over both of Brown's abandoned cannon, but they didn't remain there long as two other 2nd Corps batteries—Capt. William A. Arnold's Battery A, 1st Rhode Island Light Artillery, and Lt. Alonzo H. Cushing's Battery A, 4th U.S. Artillery—opened fire from their positions to the right (north) of the copse of trees, creating a no-man's-land around the guns. Not much is recorded about these actions. A member of William A. Arnold's battery merely recorded in his diary that "this battery [Cushing's] and ours fired canister in the evening."[95] Arnold's battery was apparently hard hit by the Confederate guns on Seminary Ridge earlier in the day, for one of its guns was out of action by this time. The rear wheels of a caisson had also been blown away. The veterans of Cushing's battery were more descriptive. According to Sgt. Frederick Fuger, Cushing ordered the three left guns to oblique to the left and fire single canister charges into Wright's men near the captured cannon. "This movement was so rapidly executed that it staggered the Confederates (who, by the way, were in the act of training those pieces on us), and they fled to the rear in short order, totally broken up."[96]

Most of the Georgians still advanced, at least those still standing, despite the fire being poured into them by the infantry behind the wall and the two fresh Federal batteries. Brown's two cannon that had successfully navigated the wall deployed south of the copse of trees and opened fire. Then a vicious counterattack by the 106th Pennsylvania of the Philadelphia Brigade on the Georgians' left flank, and another by the 13th Vermont of George J. Stannard's brigade on the right, staggered the Georgians and finally convinced them to abort the attack and return to Seminary Ridge. Several regiments followed, forcing the Confederates to abandon both of Brown's guns.[97]

With the battle for the copse of trees over, Brown's cannoneers reclaimed the two pieces briefly captured by the enemy and cleaned up the logjam around the stone wall. Intact again, the battery established a position to the left (south) of the copse of trees. The battery sustained about twenty casualties, including Lt. T. Frederick Brown.[98]

Counterattacks by other Federal infantry reclaimed other lost cannon. It appears that the 13th Vermont recaptured John G. Turnbull's four guns, and the 105th Pennsylvania (Charles K. Graham's brigade) retook Gulian V. Weir's pieces.[99] Col. Calvin Craig, the latter regiment's commander, reported, "Noticing at this time three pieces of artillery that had been abandoned by our artillerists and turned upon us by the advancing Rebels (and who were in turn compelled to abandon them), I sent forward my few remaining men to bring them off the field, but being unable to bring them all off, I got assistance from some men of the Excelsior Brigade with two of the pieces, and brought the third off the field with my men."[100] Craig noted that the pieces belonged to "Battery C, 5th U.S. Artillery [Weir's Battery]."

The Final Repulse Near Wheatfield Road

The Confederates had gained partial success against the Federal left flank as dusk began settling on the battlefield. They occupied Devil's Den–Houck's Ridge, the Wheatfield, and the Peach Orchard. The most important prize, Little Round Top, remained in Federal hands, however. Now, the remnants of at least five Confederate brigades (George T. Anderson's, Joseph B. Kershaw's, Jerome B. Robertson's, Paul J. Semmes's, and William T. Wofford's) approached Plum Run Valley, later called the Valley of Death, on their way toward the prize.

Capt. Frank C. Gibbs's Battery L, 1st Ohio Light Artillery (six Napoleons), had been ordered into action to the right of Little Round Top. The ground was so rocky that the gunners were forced to unhitch the horses and drag their pieces into position by hand. Two sections were placed on the right of Wheatfield Road; the other was on the left. Both overlooked Plum Run Valley, and the gunners could see the Confederates approaching just beyond it. The left section supported William McCandless's brigade (Pennsylvania Reserve Division), while the two sections on the right supported Frank Wheaton's (David J. Nevin's) brigade's (John Newton's–Frank Wheaton's division). Almost as soon as the guns had deployed, the infantry units from the 5th Corps—probably from Sidney Burbank's, Hannibal Day's, and Jacob B.

Sweitzer's brigades—"came rushing through us, but began rallying on us as soon as they understood matters."[101]

Capt. Almont Barnes's Battery C, 1st New York (four ordnance rifles), took position to the right of Gibbs's battery. Capt. Augustus P. Martin, commander of the 5th Corps artillery, apparently allowed Barnes to decide where to place his battery, telling him to "follow the Regulars and don't let [Daniel E.] Sickles get you."[102] According to historian Harry Pfanz, the battery occupied a poor position with limited fields of fire. As a result, it played but a minor role and sustained no casualties.[103]

A third 5th Corps battery, Lt. Aaron F. Walcott's Battery C, 3rd Massachusetts Light Artillery (six Napoleons), was deployed by a 3rd Corps officer. The latter apparently rode up to the inexperienced battery commander and said that he had the authority to commandeer any battery in the vicinity. Walcott believed him, so he followed and deployed his battery on a slight rise on the north side of Wheatfield Road with his left gun touching it. The position was about 175 yards in front of Frank C. Gibbs's battery.[104]

Without warning, Georgians from Brig. Gen. William T. Wofford's brigade burst out of the Rose Woods and headed directly for Walcott's battery. The six cannon responded by throwing canister into the approaching enemy, but their position was too unsupported by infantry and too advanced. Walcott realized his battery's untenable position, but the enemy was on him before he could withdraw, so he ordered the guns spiked before abandoning them. Only one could be spiked before they were overwhelmed by the surging Confederates.[105]

As soon as the last Federal infantryman cleared Gibbs's front, an "irregular, yelling line of the enemy put in his appearance, and we received him with double charges of canister, which were used so effectively as to compel him to retire," wrote the battery commander.[106] Gibbs neglected to mention that Frank Wheaton's (David J. Nevin's) infantry also attacked Wofford's brigade around this time, clearing the front of Confederates. "So rapidly were the guns worked that they became too hot to lay the hand on," noted Gibbs.[107] The infantry continued their charge, recapturing Aaron F. Walcott's guns in the process.

To the south of Wheatfield Road, the Confederates from Anderson's, Kershaw's, Semmes's, and Wofford's brigades continued their surge. As they approached Little Round Top, an officer from Gibbs's left section panicked and ran back toward McCandless's brigade of the Pennsylvania Reserves, which had taken position behind the battery. He yelled, "Dunder and blixen, don't let dem repels took my batteries." Col. S. Jackson of the 40th Pennsylvania told him to "double-shot his guns, hold his position, and we would see to their safety." Hearing this exchange, Jackson's men yelled, "Stand by your guns, Dutchy, and we will stand by you."[108]

McCandless's brigade soon launched an attack against the remnants of Anderson's, Kershaw's, and Semmes's brigades, while Wheaton's (Nevin's) brigade attacked Wofford's brigade.

Charles E. Hazlett's battery atop Little Round Top also opened fire on the disordered Confederates. The two regular army brigades of Brig. Gen. Romeyn Ayres's division (5th Corps) that had entered the Wheatfield (and been quickly defeated) initially blocked the battery's field of fire as they retreated through Plum Run Valley. John Page of the 3rd U.S. looked up and saw artillery officers on the summit of Little Round Top waving their hats "for us to hurry up. We realized that they wished to use canister, so took up the double-quick."[109] Some of the men dove for cover when the artillerymen felt they could wait no longer and opened fire on the approaching enemy. "Here we did the most rapid and continued firing we ever did during the war," recalled artilleryman Thomas Scott.[110]

Realizing the futility of their mission, the Confederates finally pulled back and the battlefield fell quiet. It was quite a day for the artillery of both armies, but the Federal's was handled much more skillfully and fully participated in the Confederate repulse. Gen. Henry J. Hunt claimed a major role. He noted in his report that the "batteries were exposed to heavy front and enfilading fires, and suffered terribly, but as rapidly as they were disabled they were retired and replaced by others." Because they were not being attacked, the Confederate artillery had merely played a supporting role. While they held Houck's Ridge, Devil's Den, and the Peach Orchard, the most important heights to the south—Cemetery Ridge and Little Round Top—were still in Federal hands.[111]

The Federals could boast of another accomplishment. With the arrival of the 6th Corps, beginning in the late afternoon of July 2, the Army of the Potomac was at full strength. The entire 6th Corps artillery was now on the field, boasting eight batteries with forty-eight guns.[112]

7

JULY 2 AND 3

CEMETERY HILL AND CULP'S HILL

Preliminary Artillery Clashes

Maj. Gen. Edward Johnson's division was slated to storm Culp's Hill during the evening of July 2, and R. Snowden Andrews's artillery battalion—now under Maj. Joseph W. Latimer because its permanent commander was wounded at Winchester—was ordered to support the effort. Beginning at 4:00 a.m., the nineteen-year-old Latimer spent the early morning hours scouting the area, ascertaining the best positions for his sixteen guns. Sometimes called the "Boy Major" and "Young Napoleon," Latimer had won the respect of the entire army for his skill and bravery. Infantry commonly cheered him as he rode by, and according to historian Harry Pfanz, this was "a distinction usually reserved for General Lee, Stonewall Jackson, and scampering rabbits."[1]

After considerable examination, Latimer realized that Benner's Hill, situated northeast of Culp's Hill, was the only suitable place for the battalion. The site was less than ideal, however, because it was lower in elevation than Culp's Hill, it was highly exposed, and it was too small to accommodate all of his sixteen guns. Latimer was ordered to deploy his artillery at about 4:00 p.m., and he packed fourteen guns atop Benner's Hill. Capt. William D. Brown's Chesapeake Artillery (four 10-pounder Parrots) occupied the right, Capt. John Carpenter's Alleghany Artillery (two Napoleons and two ordnance rifles) occupied the center, and Capt. William F. Dement's 1st Maryland Battery (four Napoleons) and one section of Capt. Charles Raine's Lee Battery (one ordnance rifle and one 10-pounder Parrott) dropped trail on the left. Raine's other two guns—20-pounder Parrotts—were placed on an eminence behind Benner's Hill, where they joined Capt. Archibald Graham's Rockbridge Artillery (four 20-pounder Parrotts) of William Nelson's battalion, which had taken position there earlier in the day.[2]

Latimer's sixteen guns soon opened on Cemetery and Culp's hills, about fourteen hundred yards away. Col. Charles S. Wainwright, whose 1st Corps artillery brigade occupied the eastern sector of Cemetery Hill, called the firing "the most accurate fire I have ever yet seen from their artillery."[3] Wainwright responded with thirteen ordnance rifles, "with good effect." John Hatton, a gunner in Dement's 1st Maryland Battery, also admired the Federal shooting. "A storm of shell greeted us the moment our first gun fired. It seemed that the enemy had gotten range of the hill even before we fired, and were expecting us to occupy the position and were watching for us." Greenleaf T. Stevens's battery, on the knoll later named for its commander, also joined the fight.[4]

Edward Moore of Graham's battery vividly remembered the initial stages of the duel between Latimer's guns and Wainwright's on Cemetery Hill: "In less than five minutes one of Latimer's caissons was exploded, which called forth a lusty cheer from the enemy. In five minutes more a Federal caisson was blown up, which brought forth a louder cheer from us."[5] Interestingly, Federal battery commander James Stewart described the same situation from his perspective: "Cooper at once replied, and about the first shot he fired blew up a caisson. I ordered my men to give three cheers for Cooper's battery. The echo had scarcely died away when one of my caissons met the same fate. Then the hurrah was on the side of the Johnnies."[6] The caisson horses bolted for the rear, and when they were finally returned, Stewart observed, "Every hair was burnt off the tails and manes of the wheel horses."[7]

Charles Coffin, a Northern newspaper correspondent, recorded what it was like on Cemetery Hill that afternoon: "Then came a storm of shot and shell; marble slabs were broken, iron fences shattered, horses disemboweled. The air was full of wild, hideous noises—the low buzz of round shot, the whizzing of elongated balls and the stunning explosion of shells overhead and all around."[8] Gen. Oliver O. Howard watched some of Latimer's shells flying high overhead before landing in the rear, which sent a "host of army followers into rapid motion farther to the rear." Others fell short and exploded, "Throwing out their fragments to trouble the artillery men and horses, or to rattle among the tombstones," he wrote.[9]

James H. Cooper's battery, in the center of the Federal line, was also pounded by Joseph W. Latimer's guns. "The shots of the enemy came thick and fast, bursting, crushing, and ploughing, a mighty storm of iron hail, a most determined and terrible effort of the enemy to cripple and destroy the guns on the hill," wrote Lt. James Gardner of the battery.[10] One of Latimer's early shots exploded directly under gun number three, killing and wounding all of the gun crew. One man was killed outright, a second was mortally wounded and died within the hour, and three more were wounded. However, even before these unfortunates could be tended to, others sprang to the gun, and it was soon back in action. A short time later, another shell struck Cooper's gun number two, breaking an axle. The cannoneers continued firing the gun until the gun carriage broke down, forcing a makeshift repair, and the gun was soon back in action.[11]

Wainwright had always been critical of the battery's comportment in camp, but now the artillery chief wrote, "Here I had a specimen of the stuff this battery is composed of, and forgave Cooper and his men their utter unmilitariness and loose ideas of discipline in camp."[12] Wainwright was less complimentary to Michael Wiedrich's battery, which was aligned near Cooper's: "Wiedrich . . . made wretched work of it: his Germans were all excitement, and stood well but were utterly ignorant as to ranges, and the old man [Wiedrich] knew little more himself." Wainwright marched over to the offending batteries and at once set to work helping the confused cannoneers.[13]

Capt. Frederick M. Edgell's Battery A, 1st New Hampshire Light Artillery (four ordnance rifles), was on the extreme left of Maj. Thomas W. Osborn's artillery line, having relieved James A. Hall's battered battery. Because they were ordered to fire slowly, the men had opportunities to take in the surrounding scenes, which included the wreckage of Hall's battery and its dead and wounded all around them, as well as the unearthed bodies from the Evergreen Cemetery that had been blown apart by Confederate shells.[14]

The Federal guns were not the only ones being pounded; apparently some Confederate shells rained down on the 11th and 12th Corps infantry as well. Wainwright described one 20-pounder Parrott shot striking the center of a regiment lying behind a stone wall on Cemetery Hill: "Taking the line lengthways, it literally ploughed up two or three yards of men, killing

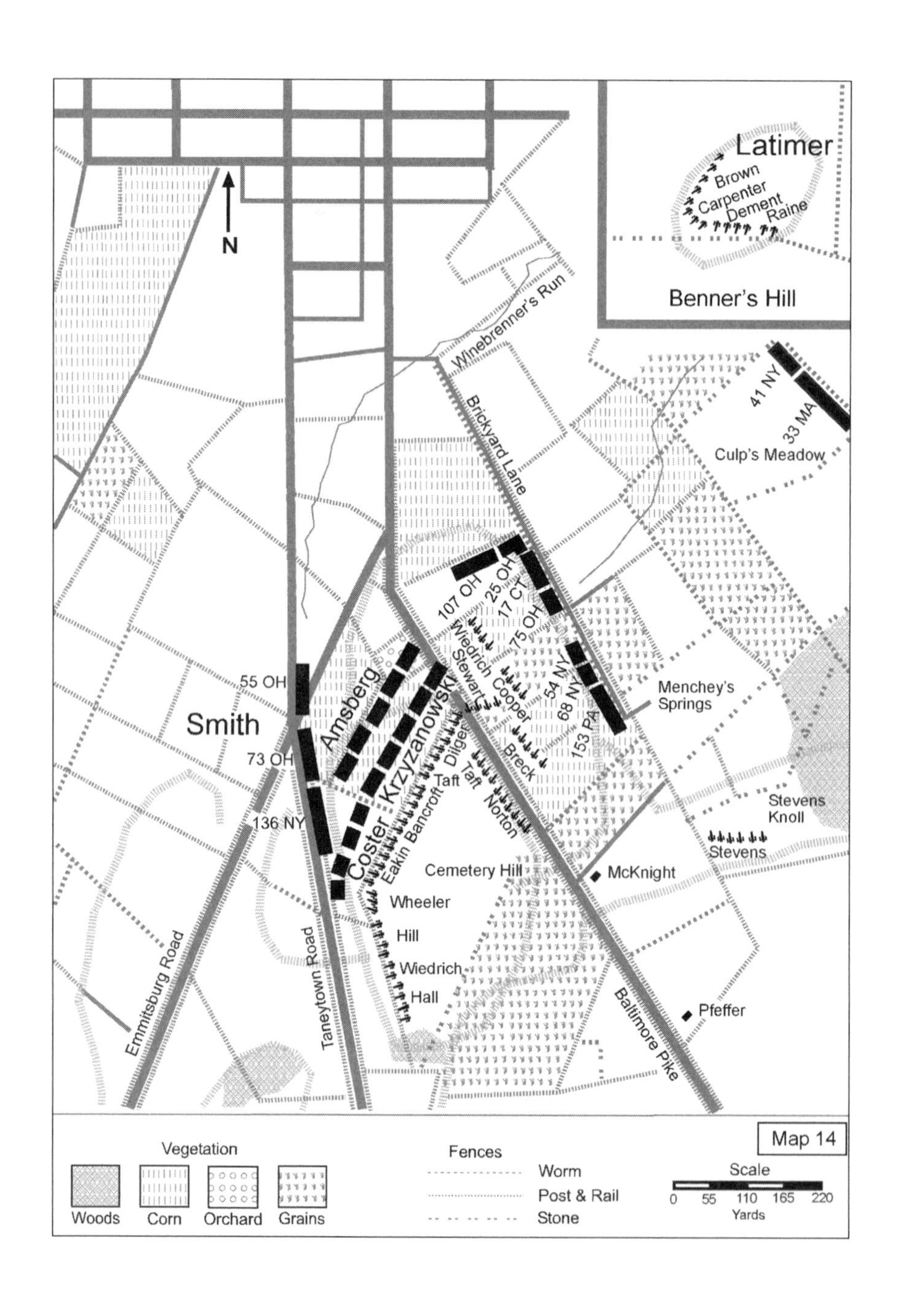

INITIAL POSITIONS ON CEMETERY HILL

and wounding a dozen or more."[15] This was probably the 153rd Pennsylvania (Leopold von Gilsa's brigade). Its historians wrote after the war that the "enemy's shot and shells which, hitherto had injured us but little, were now doing terrible execution in our ranks. Everywhere men were seen writhing in the agony of death, while the wounded were shrieking for help which no one could render them."[16] A soldier from the 55th Ohio of Orland Smith's brigade recalled that one Confederate shell killed and wounded twenty-seven men of his regiment.[17]

The 33rd Massachusetts of Smith's brigade was also exposed to this hostile fire. According to the regiment's commander, Col. Adin Underwood, "Whizz came a shot over . . . and plunged into the earth with a dull sound. A shell came shrieking and hissing in its track and exploded itself into destructive atoms; in almost a moment of time a hundred shot and shell were tearing about, bursting into fragments that hurried away many a brave man. Splinters of gun carriages, pieces of tombstones, even human legs and arms and palpitating flesh were flying about in every direction."[18] The regiment moved several times to avoid this hostile fire.

On Culp's Hill, the foot soldiers from the 12th Corps were also dodging Latimer's shells. Steuben Coon of the 60th New York of George S. Greene's brigade wrote that the "shells burst over and all around us, but only a few were hurt—for it is a fact that unless troops are massed together, shells or shot do very little damage except to scare raw recruits. They do make an unearthly noise."[19] When a number of artillerymen fell victim to sniper's bullets, men from the 60th and 78th New York sprang forward to help carry ammunition and work the guns.

Confederate infantry were not immune to this duel. William Goldsborough of the 1st Maryland Battalion of George H. Steuart's brigade (Edward Johnson's division) wrote, "[The] air is filled with exploding, crashing, screaming shells. 'Lay down!' is the command, and every man was flat on his face."[20] Even more problems were experienced by Francis Nicholls's brigade, which was directly in line with Joseph W. Latimer's artillery battalion and was pounded accordingly. The Louisianians found whatever cover they could and waited. According to one soldier, "Perhaps nothing in battle is so trying to an infantryman's nerves and patience as the preliminary artillery fire."[21]

Capt. William Seymour of Harry T. Hays's brigade, also of Johnson's division, watched the duel and left a vivid description of it: "The roar of the guns was continuous and deafening; the shot and shell could be seen tearing through the hostile batteries, dismounting guns, killing and wounding men and horses. . . . Ammunition chest[s] would explode, sending a bright column of smoke far up towards the heavens."[22]

The duel continued for close to two and a half hours, ending after 6:00 p.m. Some of Joseph W. Latimer's left guns rained shells on the left side of the 12th Corps on Culp's Hill, causing Lt. Edward D. Muhlenberg to send first a 10-pounder Parrott and, later, two more from Lt. Charles A. Atwell's battery (six 10-pounder Parrotts). They were joined by a section of Napoleons from Lt. David H. Kinzie's battery. According to Muhlenberg, these guns were placed in a "vacant space eligible for a battery . . . about two-hundred yards on the right of the First Corps."[23] Muhlenberg noted that from the "moment their [his five guns'] presence was observed, the enemy opened with eight guns."[24] Lt. Col. R. Snowden Andrews described this action in his report: "The enemy in the meantime planted some guns on the left, which partially enfiladed our batteries, which caused Capt. John Carpenter to suffer very severely."[25] Carpenter and his men weathered the storm, and Carpenter noted, "We stood our post like men, until we had exhausted all our 12 pd. Ammunition."[26] As soon as the Federal guns on Culp's Hill opened fire, eight of Latimer's guns directed their attention to them, and a duel ensued for about thirty minutes.[27]

Two batteries of Federal artillery assigned to Maj. Thomas W. Osborn on Cemetery Hill also opened fire on Benner's Hill. This included Capt. Elijah D. Taft's 5th New York Battery of six 20-pounder Parrotts, the largest guns in the army. To his right, Lt. George W. Norton's Battery H, 1st Ohio Light, with six ordnance rifles, also opened fire.

A lull in the fighting permitted both sides to pull aside the bodies of dead cannoneers and horses and tend to the wounded. After removing some of the debris, the action was renewed with a vengeance.

Robert Stiles of the Richmond Howitzers had occasion to ride by Benner's Hill toward the end of the encounter; he recorded what he saw:

> Never, before or after, did I see fifteen or twenty guns in such a condition of wreck and destruction as this battalion was. It had been hurled backward, as it were, by the very weight and impact of the metal from the position it had occupied on the crest of a little ridge into a saucer-shaped depression behind it; and such a scene as it presented—guns dismounted and disabled, carriages splintered and crushed, ammunition chests exploded, limbers upset,

> wounded horses plunging and kicking, dashing out the brains of men tangled in the harness; while cannoneers with pistols were crawling through the wreck shooting the struggling horses to save the lives of the wounded men.[28]

Joseph W. Latimer's batteries on Benner's Hill were now being pounded from their left, center, and right by as many as thirty-six Federal guns. Casualties mounted, including Capt. William D. Brown, and several guns and caissons fell victim to the hail of steel. Others were out of ammunition or so undermanned that they were unserviceable. All caused a marked reduction in the intensity of the firing. To a soldier in Steuart's brigade, Latimer was "working his guns savagely, but is being terribly handled, for three times his number of guns are concentrated . . . rending and tearing him to pieces. Caisson after caisson shoot high up in the air as they are exploded by the enemy's shells."[29] When a shell from a 12th Corps battery exploded yet another caisson, Latimer knew he was not accomplishing anything and was in danger of losing his entire battalion. He sent his sergeant major to Edward Johnson with a message: "Owing to the exhausted state of his men and ammunition and the severe fire of the enemy, he was unable to hold his position any longer."[30] Because Latimer's reputation as one of the most effective artillery fighters in the army was well known, Johnson simply replied that he could pull back "if he thought [it] proper."[31]

Latimer withdrew all but four of his guns, which remained silent until Johnson launched his division's charge at about 8:00 p.m. Being the leader that he was, Latimer stayed with these guns. As might be expected, when the attack began, all of the Federal pieces turned their attention to these four Confederate guns, and the area became perfectly hellish. A shell exploded over Latimer, shattering his right arm and throwing him to the ground, where he was pinned by his horse. He would not permit the cannoneers to pull the horse off of him but instead continued issuing a stream of orders to the men. Capt. Charles I. Raine took command of the battalion, and on Latimer's suggestion, he pulled the guns back to safety. Joseph W. Latimer died almost a month later, and all who knew him mourned his loss.[32] Col. J. Thompson Brown, 2nd Corps chief of artillery, wrote, "No heavier loss could have befallen the artillery of this corps."[33]

After the battle, Charles S. Wainwright received word that twenty-eight dead horses were counted on Benner's Hill. He lost only about a dozen. "How it was they did not kill more horses I cannot understand," he wrote, "huddled together as we were, for their fire was the most accurate I have ever seen on the part of their artillery."[34] According to Lt. Col. R. Snowden Andrews's report, the Confederate battalion lost 10 killed, 35 wounded, and 30 horses killed. One cannon and one caisson were disabled, and another exploded. Historians recently have questioned the loss of human life, ascertaining that, of the 356 men in the battalion, 22 were killed and 29 were wounded.

On the Federal side, in addition to James H. Cooper's two disabled cannon, Gilbert H. Reynolds's battery lost a cannon due to a broken axle. In addition, according to Lt. George Breck, the battery sustained another loss when "an ammunition chest of one of my caisson limbers was struck by a shell, exploding a few rounds of ammunition, which it contained, and completely destroyed it."[35] Most of the Federal battery commanders did not break down their losses by day. Cooper, however, reported that two of his men were killed, four were wounded, and three horses were killed or disabled that evening. He also indicated that he fired more than five hundred rounds of ammunition. Sgt. David Nichol reported that three men from Charles A. Atwell's battery were wounded. Gen. Alpheus Williams noted that the two 12th Corps batteries lost a total of eight men.[36]

Relative quiet returned to the battlefield, but Confederate snipers in houses and other buildings in and around the town continued firing at the cannoneers and infantry on Cemetery Hill. This did not elicit much of a response until a cannoneer from James Stewart's battery was nicked. This demanded a response. Selecting one of their newer Napoleons, which was more accurate than the others, the men loaded it with case shot and cut the fuse for six hundred yards. They targeted the house from which they believed the sharpshooters had been firing and threw the shell at it. When the smoke cleared about two seconds later, the men could see bricks flying from between the two windows of the second story as the shell exploded within the house. Michael Wiedrich's battery was also harried by enemy snipers, who were firing atop a steeple near the edge of town. The gunners responded by sending a shell into the structure. These actions helped put a halt to the sharpshooters' activities that day.[37]

Second Corps commander Lt. Gen. Richard S. Ewell's handling of his artillery during the hours immediately preceding his attack on Cemetery and Culp's hills has been severely criticized, as four artillery battalions did not fire a shot while Joseph W. Latimer's was being devastated. A major reason was the lack of adequate artillery platforms in the area north of Gettysburg. On the Federal side, the artillery operated flawlessly. The batteries were deployed in a manner to maximize fields of fire in all directions, which made Confederate troop movements difficult.[38]

The general plan of attack that afternoon was for Ewell's 2nd Corps to take on the Federal right flank on Cemetery and Culp's hills, while James Longstreet's 1st Corps assailed the enemy's left flank. Longstreet launched his attack at about the appointed time, but Ewell was content to watch the artillery duel between Latimer's battalion and the Federal guns on the two hills. This changed at about sundown, when Ewell sent Edward Johnson's division against Culp's Hill and Jubal Early's division, supported by Robert Rodes's division, against Cemetery Hill.

The Fight for Cemetery Hill

Gen. Oliver O. Howard's 11th Corps infantrymen on Cemetery Hill probably felt more secure as the sun began to set on July 2. While the fighting had been tremendous to the south, where Longstreet's divisions tangled with elements of the 2nd, 3rd, and 5th Corps, the 11th Corps sector was fairly quiet. Howard's men had been soundly thrashed the day before and were in no condition to face the enemy again. The enemy was not known for attacking at night, so the men became less concerned as the hours passed.

This quickly changed at about 7:30 p.m. when the troops on Cemetery Hill spied a long line of Confederates emerging from the town and marching resolutely toward them. These were Brig. Gen. Harry T. Hays's and Col. Isaac Avery's brigades of Early's division—the same two that had fallen on Lewis Heckman's battery at Kuhn's brickyard the day before, capturing two cannon in the process. Now Hays's brigade bore down on Cemetery Hill from the north, and Avery's brigade approached on their left (from the northeast).[39]

According to Capt. R. J. Hancock of the 9th Louisiana, "Gen'l Harry Hays, who was no man to deceive his men nor any one, rode along the line about dusk and told his men that Gen'l Early had said we must go to Cemetery Hill and silence those guns and [John B.] Gordon would reinforce us and hold them."[40] Lt. R. Stark Jackson of the 8th Louisiana received the orders to advance with dread: "I felt as if my doom was sealed, and it was with great reluctance that I started my skirmishers forward."[41] Staff officer Capt.

William Seymour recalled it differently: "The quiet, solimn mien of our men showed plainly that they fully appreciated the desperate character of the undertaking; but on every face was most legibly written the firm determination to do or die."[42]

Lying wounded in the upper floor of a church in Gettysburg, Pvt. Reuben Ruch of the 153rd Pennsylvania had a perfect view of the attack. He noted, "Between the Rebel and Union positions was a ridge about six or eight feet high. The Johnnies started stooped over, scattered like a drove of sheep, till they got to this ridge. Then every man took his place, and giving the Rebel yell . . . they closed up like water, and advanced on a double-quick."[43]

The two Confederate brigades, probably numbering about twenty-two hundred men, faced not only Francis Barlow's division (now under Brig. Gen. Adelbert Ames) but also Gilbert H. Reynolds's, Michael Wiedrich's, and Capt. R. Bruce Ricketts's Batteries F & G, 1st Pennsylvania Light Artillery (six ordnance rifles). The latter had just replaced James H. Cooper's battery, which had been soundly thrashed during the duel with Latimer's battalion. Over to the right, Greenleaf T. Stevens's battery was also firing into the Confederate ranks. Thus a total of fifteen guns in front and six on the flank took on the two Confederate brigades as they advanced across the open terrain. From the orientation of the Federal guns, it appears that Wiedrich's four guns fired into Hays's brigade and Reynolds's, Ricketts's, and Stevens's batteries took on Avery's.[44]

Driving up the hill they had been resting behind, the Confederate infantrymen reached the top and were met by cannon and small-arms fire. "But we are too quick for them, and are down in the valley in a trice, while the Yankee missiles are hissing, screaming & hurtling over our heads, doing but little damage," recalled Seymour.[45] Directly in front of them was Adelbert Ames's brigade, now under Col. Andrew Harris, protected by a stone wall. These troops opened fire on the Louisianians, while Federal artillery fired canister into their ranks. No Confederate artillery fired into the Federal infantry positions at this time; the Southern attack column was on its own.

Ruch watched the massed Federal artillery open fire on the attacking columns almost as soon as they left the town. He "could see heads, arms, and legs flying amid the dust and smoke. . . . It reminded me much of a wagon load of pumpkins drawn up a hill and the end gate coming out, and the pumpkins rolling and bounding down the hill."[46] Lt. Charles Brockway of

Ricketts's battery wrote, "We threw in their midst shrapnel and solid shot; but when they charged, we used single and finally double rounds of canister."[47]

Andrew Harris marveled at the Confederate attack: "When they came into full view in Culp's meadow our artillery . . . opened on them with all the guns that could be brought to bear. But on, still on, they came, moving steadily to the assault, soon the infantry opened fire, but they never faltered. They moved forward as steadily, amid this hail of shot shell and minnie ball, as though they were on parade far removed from danger."[48]

The order to the men serving Stevens's battery was, "Case, 2 1/2 degrees, 3 seconds time." According to one of the cannoneers, this was immediately followed by the sounds of the opening of limber chests and the cutting of fuses: "Slap went the heads of the rammers against the faces of the pieces . . . at the same moment came the order 'Fire by battery,' and at once there was the crash and roar of our six guns, the rush of the projectiles."[49] The initial shots were exceedingly accurate because, earlier in the day, one of the officers had used a "French Ordnance Glass" to ascertain the distance of conspicuous landmarks. The men checked the accuracy of these calculations by firing a round or two.[50]

While Ricketts's, Stevens's, and Wiedrich's batteries apparently switched to canister as the enemy approached, Reynolds's battery did not. Lt. George Breck noted, "This charge was mostly repelled by the infantry in support, whose presence in front prevented the use of canister."[51]

Harry T. Hays believed that the growing darkness, coupled with the dense powder smoke, reduced his losses, as "our exact locality could not be discovered by the enemy's gunners, and we thus escaped what in the full light of day could have been nothing else than horrible slaughter."[52] Capt. R. J. Hancock of the 9th Louisiana partially agreed: "The enemys cannon lighted up the heavens but most of the charges they shot over us but even at that we suffered terribly."[53]

Looking to his left, Hays could barely make out Avery's brigade that was also advancing against the enemy's line. The North Carolinians continued their advance despite the hail of bullets and shells being thrown against them. As the Confederates approached the wall, "Our colonel [Adin B. Underwood] just gave the 33d [Massachusetts], 'Fix bayonets and remember Massachusetts!' when Stevens' 5th Maine Battery to our right let go all six guns in one volley and swept our front clear of Rebels with canister,"[54] recalled John Ryder. Underwood indicated that Stevens's battery "opened on them at point-blank range,"[55] and he was stunned by this turn of events because he did not know the battery was in position. "Right over my head, it seemed to me, there was a flash of light, a roar and a crash as if a volcano had been let loose," Underwood wrote.[56] When the smoke cleared, he could not see any Tar Heels still on their feet.

In actuality, Greenleaf T. Stevens's guns had been firing at Avery's brigade from the time it left the vicinity of Gettysburg, about twelve hundred yards away. As the Confederate infantry advanced, Lt. Edward N. Whittier, now in command of the battery, changed front and fired to the left, which enfiladed the line at a distance of about eight hundred yards. Whittier noted that the guns initially fired "spherical case and shell, and later with solid shot and canister."[57] According to the cannoneers, Stevens's battery poured a "fearful blast of canister almost at right angles to the direction of their advance . . . as a matter of course, no troops that ever lived could stand that, particularly when isolated and unsupported on either flank as these were, and so they broke when so near us that if it had not been almost dark we could actually have seen the whites of their eyes."[58] Forty-six rounds of canister were fired into the enemy, which broke and retreated at about the same time that the limbers were depleted. The cannoneers pulled the battery back after the charge was repelled in order to replenish its ammunition chests. The guns returned to occupy their same positions on Stevens Knoll at about 10:30 p.m. that night.[59]

Capt. R. Bruce Ricketts's Batteries F & G, 1st Pennsylvania, apparently opened fire when the enemy was three hundred yards away. According to Joseph Todd, "We waited the advance of the charge [probably Avery's brigade] who were yelling their most demonic yells as though the infernal regions had let loose as many devils on us." He recalled that his battery's guns "mowed them down like ripe grain before the cradle."[60]

Colonel Wainwright also watched the charge: "As the enemy advanced we commenced firing canister, depressing the guns more and more, until it was one continual shower straight down the hill. The night was heavy, and the smoke lay so thick that you could not see ten yards ahead; seventeen guns vomiting it as fast as they can will make a good deal of smoke."[61] He walked away, thinking that the massed guns would halt any advance. He was wrong.

When the Confederate charge hit the stone wall, several Federal regiments abandoned their positions and scurried up the hill, followed by about seventy-five men from the 6th North Carolina and a handful from

Hays's 9th Louisiana. The other Confederate units were either repulsed or remained at the base of the hill. Ricketts watched the flight of the 11th Corps infantry with disgust, for not only were these troops breaking to the rear, they were masking the fire of his cannon. He later wrote, "When the charge was made on my position, their conduct on that occasion was cowardly and disgraceful in the extreme. As soon as the charge commenced, they, although they had a stone-wall in their front, commenced running in the greatest confusion to the rear, hardly a shot was fired, certainly not a volley, and so panic stricken were they that several ran into the canister fire of my guns and were knocked over."[62] According to Wainwright, "Not a single regiment of the Eleventh Corps exposed to the attack stood fire, but ran away almost to a man. [Capt. James] Stewart stretched his men along the road, with fence rails to try to stop the runaways but could do nothing."[63]

It was dark now, and the combatants could be seen only by the flashes from their rifles as they fired into the artillerymen. Hand-to-hand combat also erupted around Michael Wiedrich's guns. One of Hays's men threw himself at the muzzle of a gun, exclaiming, "I take command of this gun." In reply, a German artilleryman yelled, "No you don't," and pulled the lanyard, blowing away the enemy soldier.[64] The Federal artillerymen fought the Confederate infantry with whatever was handy—bayonets, clubbed rifles, pistols, rammers, and rocks. Wainwright grudgingly wrote in his diary that the primarily German battery fought "splendidly, sticking to their guns, and finally driving the rebs out with their hand-spikes and fence-rails." Earlier in the fighting, Wainwright had criticized the battery's fighting abilities, stating that the cannoneers were cutting the fuses incorrectly.[65] Despite the cannoneers' heroics, the guns were ultimately captured.

With a final rush, the Confederates threw themselves on the left section of Ricketts's battery. The fighting was hand-to-hand, as the gunners refused to abandon their guns. At least one gun was captured and another spiked during this bloody confrontation. Lt. Charles Brockway noted that the latter occurred only after the entire gun crew had been killed, wounded, or captured. The cannoneers at the other guns were still firing, probably at Avery's brigade, during this time. Being out of canister, they used "rotten shot," which was, according to Brockway, "shrapnel without fuse, the shell bursting at the muzzle of the gun."[66] The result was a canisterlike blast.

Confederate Gen. Harry T. Hays triumphantly wrote in his official account of the battle, "Arriving at the summit, by a simultaneous rush from my whole line, I captured several pieces of artillery, four stands of colors, and a number of prisoners. At that time every piece of artillery which had been firing upon us was silenced."[67] Hays did not describe the intense hand-to-hand fighting that raged around the guns. Ricketts tersely recorded the interactions in his diary: "My battery was charged by Gen. Early's division just at dusk—punished them terribly with our canister—They took my left gun, spiked it, killed six men, wounded 11 and took 3 prisoners. The boys fought them hand to hand with pistols, handspikes and rammers."[68] L. E. C. Moore recalled that the orders were to "die on the soil of your State, but don't give up your guns."[69]

A Confederate soldier demanded a Federal sergeant's surrender. Seeing this, Lt. Charles Brockway, who never carried a sword, rushed over, picked up a large stone, and knocked the Confederate down. The sergeant caught the enemy's gun and wounded him as he fell, then gripped the gun with the intention of "braining him." Ricketts passed by at this time and ordered the sergeant to desist, as the enemy soldier was obviously badly wounded. Nearby, James Riggin, the battery's guildon bearer, was accosted by a Confederate officer who demanded the flag. Riggin pulled out his pistol and shot the officer, but not before the latter fired, cutting the staff in two and mortally wounding Riggin.[70]

Help was on the way. Gen. Oliver O. Howard mobilized several 11th Corps regiments, and Gen. Winfield Hancock sent Samuel S. Carroll's brigade (Alexander Hays's division, 2nd Corps) to the rescue. Soon the remnants of the Confederates who had briefly taken the heights were either captured or making their way back to their lines near Gettysburg. While a close call, the two batteries only temporarily lost one gun, which was spiked by the enemy. Losses among the cannoneers were high, however. For example, Ricketts's losses during the short, sharp encounter were heavy: six killed, fourteen wounded, and three taken prisoner. Twenty horses also perished.[71]

In his report, Wainwright, who commanded the guns in this sector, demanded all the credit for his artillery. "I believe it may be claimed that this attack was almost entirely repelled by the artillery," he reported.[72]

The Fight for Culp's Hill

Cemetery Hill was not the only prize that Lee coveted. He also wanted possession of Culp's Hill to the southeast. The latter hill was considerably higher

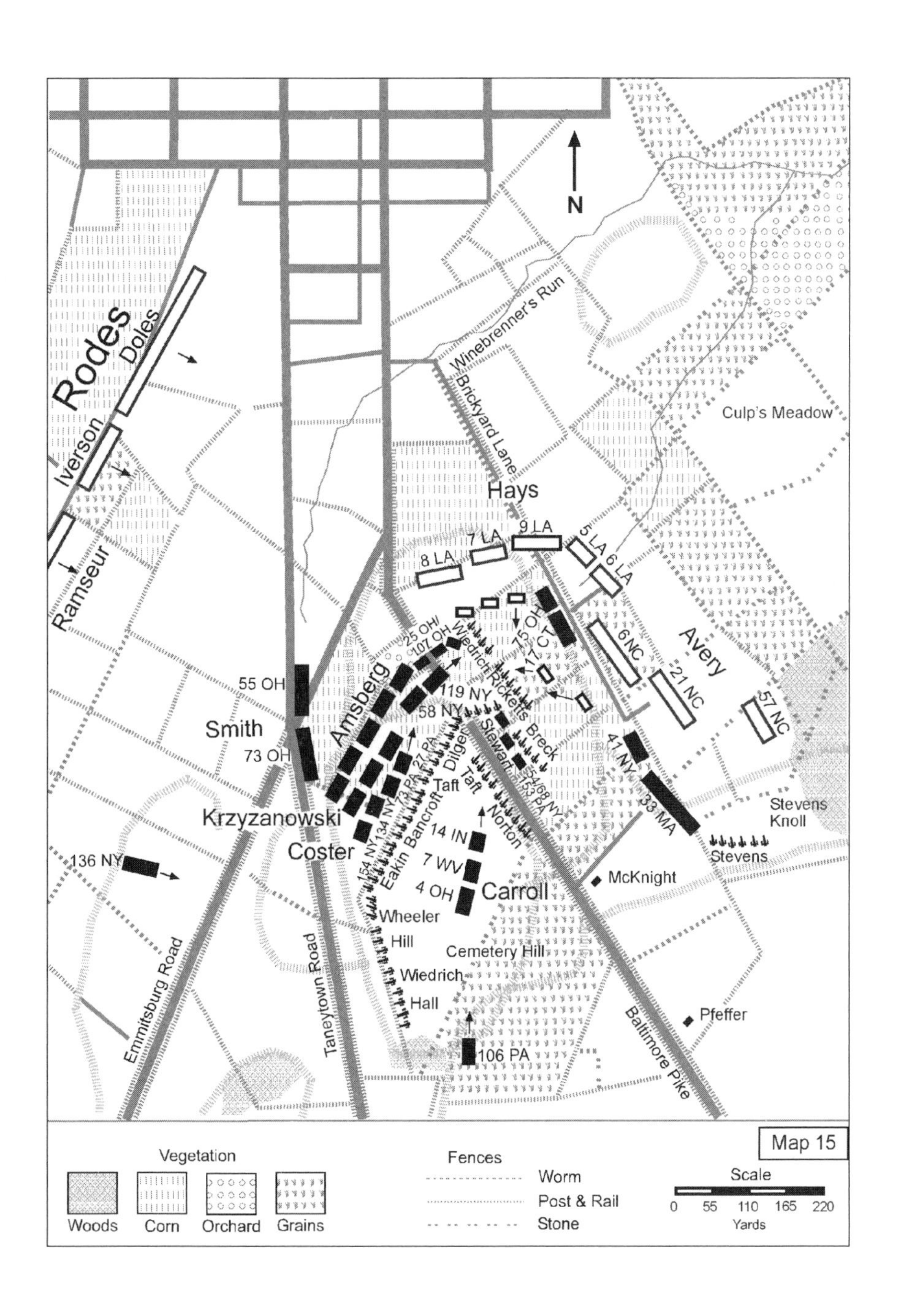

THE FIGHT FOR CEMETERY HILL

and commanded much of the battlefield, including the invaluable Baltimore Pike. While Early sent his two brigades against Cemetery Hill, Gen. Edward Johnson was to throw his entire division against Culp's Hill.

Waiting for the order to advance, William Goldsborough and his comrades in George H. Steuart's brigade lay prone to avoid the Federal shots being thrown at Joseph W. Latimer's guns on Benner's Hill. He admitted, "Perhaps nothing in battle is so trying to an infantryman's nerves and patience as the preliminary artillery fire that precedes it; and the same effect is produced upon the artilleryman by the whistle of the minnie ball."[73]

The orders to take the hill finally echoed through the fields at about 7:00 p.m., and Johnson's men were quickly on their feet and moving toward Rock Creek. Some parts of the line were almost immediately hit by Federal artillery fire, probably from the guns on Culp's Hill. This accelerated the attack as the men were anxious to reach the cover of the trees lining the creek. Lt. Col. Logan H. N. Salyer of the 50th Virginia (John M. Jones's brigade) wrote, "Our whole line moved forward in handsome order." Few casualties were sustained during this movement. There was no return fire from the Confederate batteries as none were now in position. As the four Confederate brigades crossed Rock Creek and began scaling Culp's Hill, they were finally safe from the Federal artillery, which could not depress their barrels enough to fire on them.[74]

The infantrymen in Johnson's division did not know that only one Federal brigade, Brig. Gen. George S. Greene's, manned the heights above them. The others had been called south to help relieve the 3rd and 5th Corps, which were under savage attack by Longstreet's 1st Corps. Maj. Gen. Alpheus Williams's division came under sporadic artillery fire during its two-mile trek to the southern end of Cemetery Ridge. Pvt. Abner Smith of the 20th Connecticut (Archibald L. McDougall's brigade) wrote home that the "rebs had a battery playing onto us all the way from where we started till we got where it was thought we should be needed. . . . One in our regiment was knocked down by a cannonball . . . but did not hurt him very much."[75] Sgt. L. R. Coy of the 123rd New York of the same brigade wrote home that the "shot and shell fell thick and fast around us."[76] These shots were probably fired by R. Lindsay Walker's 3rd Corps battalions, which were aligned along Seminary Ridge. Although Greene's brigade repelled most of Johnson's troops during the evening of July 2, portions of Steuart's brigade successfully gained control of some entrenchments on the right of the Federal line. Shortly thereafter, the remainder of the 12th Corps brigades began the trek back to their original positions on Culp's Hill.

The Federals planned to drive the enemy from their toehold on Culp's Hill on July 3; the Confederates wanted to expand it. Both sides decided to resume the fighting just prior to daylight, and both intended to begin with an artillery barrage. Because of the height of the hill, the Federals would be potentially more effective. The Confederates planned on sweeping up the hill again with Johnson's division, reinforced by units from Robert Rodes's and Jubal Early's divisions, while the Federals defended the heights with Geary's and Williams's-Ruger's divisions of the 12th Corps.[77]

Although Lt. Edward D. Muhlenberg was nominally in charge of the 12th's artillery brigade, Maj. Gen. Howard W. Slocum wrote that the four batteries were positioned by Lt. Col. Clermont Best, who had commanded the corps' artillery brigade but was now on the 12th Corps commander's staff.[78] Harry Pfanz believed that Best was "likely still its supervisor."[79] Lt. Sylvanus T. Rugg's Battery F, 4th U.S. Artillery (six Napoleons), and Lt. David H. Kinzie's Battery K, 5th U.S. Artillery (four Napoleons), were wheeled into position at 1:00 a.m. "parallel to and on the southwest side of Baltimore Pike, almost opposite the center of the line formed by the Twelfth, and controlling the approach of the enemy along the ravine formed by the stream known as Rock Creek," Muhlenberg reported.[80] The corps' other two batteries were already in position. Lt. Charles E. Winegar's Battery M, 1st New York Light Artillery (four 10-pounder Parrotts), was split, with one section on McAllister's Hill, about eight hundred yards south of Spangler's Spring, and the other section on Powers Hill, about seven hundred yards southwest of it. Lt. Charles A. Atwell's Pennsylvania Independent Battery E (six 10-pounder Parrots) was also on Powers Hill. The eight guns on Powers Hill were joined by Capt. James H. Rigby's Battery A, 1st Maryland Light Artillery (six ordnance rifles), of the 4th Volunteer Brigade of the artillery reserve. Thus the ten smoothbore Napoleons were closest to the action, with the rifled pieces farther to the rear.[81]

Brigade commander Col. Charles Candy recalled that the Confederate artillery opened on the Federal lines at 3:45 a.m. on July 3. Muhlenberg reported that his own twenty guns opened fire on the enemy at 4:30 a.m. and continued firing for fifteen minutes without interruption. The firing began again at 5:30 a.m. and continued at various intervals until 10:00 a.m.[82] The Confederates mounted charge after charge, but all failed. John Storrs of the

20th Connecticut (Archibald L. McDougall's brigade) vividly recalled the battle: "The sharp and almost continuous reports of the twelve pounders, the screaming, shrieking shell that went crashing through the tree tops; the deadened thud of the exploding shell; the whizzing sound of the pieces as they flew in different directions; the yells of the rebels when they gained a momentary advantage; the cheers of the men when the surging tide of battle turned in our favor."[83] Charles A. Atwell's battery was directly behind the 29th Ohio (Candy's brigade), and when it opened fire, "the blaze of the guns would almost reach us, and the concussion would lift us up bodily," noted J. R. Lynn.[84]

The Federal artillery was also asked to help discourage the pesky Confederate sharpshooters who were taking a toll from the opposite side of Rock Creek. The 2nd Virginia (James A. Walker's brigade) and 1st North Carolina (George H. Steuart's brigade) became so dangerous that an officer from Silas Colgrove's brigade approached Lt. Charles E. Winegar with a request that he fire at a stone building that was harboring many enemy sharpshooters. Winegar complied by moving one of his Parrotts from Powers Hill to a better position, where it opened fire on the enemy. Its fire was accurate, and the enemy was seen scurrying from the building. They soon returned, however.[85]

There are few accounts of the effects of the Federal artillery on the Confederate soldiers. Pvt. Abisha Gum of the 25th Virginia (John M. Jones's brigade) recalled the initial Federal bombardment at about 4:30 a.m.: "The bums [bombs] fell that thick that I did not know which way to go. They throwed the dirt all over me a time or too but did not hurt me, but [I] thought of the other world rite smart."[86] Benjamin Jones of the 44th Virginia (Jones's brigade) recalled years after the battle that "the roar of artillery and rattle of musketry was awfully severe. The mountain trembled under our feet like an aspen leaf, as the great number of artillery belched forth death in our ranks."[87]

The Confederates were not the only recipients of the Federal shells; some rained down on the 12th Corps infantry. The 20th Connecticut (Archibald M. McDougall's brigade) engaged in a seesaw battle with the enemy over possession of a stone wall. According to McDougall, the regiment "encountered great difficulty, while resisting the enemy, in protecting himself against the fire of our own artillery . . . his [Col. William Wooster's, the regiment's commander] greatest embarrassment was, the farther he pushed the enemy the more directly he was placed under the fire of our own guns." Wooster lost several men to this friendly fire.[88]

Several men from McDougall's brigade were also hit by David H. Kinzie's and Sylvanus T. Rugg's batteries, which led to hostility between McDougall and the commander of the 145th New York, E. Livingston Price. The latter described the incident in his official report: "Deeming it advisable and proper to report the facts [about the friendly fire] to my commanding officer [McDougall], I dispatched . . . to inform the colonel commanding the brigade that several of my men had been wounded by the fire of our own artillery." Price wrote that the aide returned with the message, "Tell Colonel Price 'not to fret.'"[89] Col. James Selfridge of the 46th Pennsylvania apparently went over to see McDougall and, drawing his revolver, said that he would shoot the battery commander if additional shells fell on his troops. This did not stop the carnage, and when three additional men were hit, an enraged Price and an equally angry Selfridge stormed over to the offending battery, where they encountered General Slocum. The regiments were subsequently pulled back out of cannon range. McDougall questioned the validity of this version, writing that, upon questioning Price's aide, he was told that none of the troops had been injured.[90]

The lopsided battle for Culp's Hill finally ended by 11:00 a.m. The Confederate losses were horrendous. While most of the losses were incurred by the Federal infantry firing from the crest of the hill, Edward D. Muhlenberg took his share of the credit. "The artillery was of essential service, and did excellent execution at this part of the field, no doubt contributed greatly in preventing the enemy from establishing himself in so desirable a position [Culp's Hill]. . . . The marks on the trees and immense boulders contiguous to the line of intrenchments prove conclusively that the practice of the artillery was excellent and splendidly accurate."[91]

8

JULY 3

UP TO 1:00 P.M.

The Confederates Prepare for the Renewed Battle

The artillery officers of both armies worked through the night of July 2–3, adjusting their guns' positions, looking after the men and horses, and ensuring that the batteries were restocked with ammunition, men, and horses. A full moon aided their efforts. The matter of ammunition was of greatest concern to Confederate artillery chief William Pendleton as he had what he considered an unacceptably small quantity remaining. He warned the battalion commanders that there was not much more to draw upon, so the precious commodity had to be conserved.[1]

Although the Confederates captured many artillery pieces on July 2, they were able to get only a few to safety. Capt. Sam Wilson of the 1st Texas was given the task of getting James E. Smith's guns to safety from their perch on Houck's Ridge. This was both a difficult and dangerous task as Federal troops occupied the Wheatfield nearby. In the darkness, Wilson's men crawled up "to the cannon, picked out of the path over which the guns were to be drawn, all stones and large pebbles, [and] not speaking above a whisper, wrapped the wheels with blankets and brought the guns off so carefully that the noise was not heard 100 yards away." Ironically, the three cannon were given to Mathis W. Henry's battalion—two pieces to Alexander

C. Latham's and the third to James Reilly's batteries to replace the guns knocked out by Smith's battery.[2]

After ensuring that his guns were ready for the morrow, Col. E. Porter Alexander retreated to a fence and slept between 1:00 and 3:00 a.m. While the 2nd and 3rd Corps batteries remained in their original positions, Alexander deployed the recently arriving batteries of James Dearing's battalion, assigned to George E. Pickett's division, and the Washington Artillery.[3]

Lee's plans for July 3 took many twists and turns before he finally settled on a massive infantry attack on the center of the Union line. Gen. Ambrose Wright's Georgia brigade had pierced the Union center with one brigade on July 2, so Lee decided to try it again, this time with about thirteen thousand men. After two days of bloody but indecisive fighting, it would come down to this desperate charge. James Longstreet was placed in charge of the thrust, but a more pessimistic soldier could not be found on the battlefield. He was especially concerned about Charles E. Hazlett's guns on the "high rocky hills" that could enfilade the attacking column. Lee and his staff discounted this concern by suggesting that Confederate artillery could neutralize this threat. It was a rationalization based on desperation, and all were to see the folly of it later that afternoon. "Our artillery equipment was usually admitted to be inferior to the enemy's in numbers, calibres, and quality of ammunition," Alexander noted after the war.[4]

In the words of one historian, the Confederates' planning was "thorough, detailed, exhaustive and precise."[5] Lee began preparing for the assault during the evening of July 2 and spent as many as nine hours in planning on July 3. He carefully examined the battlefield and was very visible to his men during this period. Much of the planning was for the preliminary artillery barrage that was designed to take out the dangerous Federal guns and neutralize the infantry in the center of the Federal line.[6] Then, in Lee's words, the artillery was to be "pushed forward as the infantry progressed, protect their flanks, and support their attacks closely.[7] This was a desperate measure, but Lee intended to give his infantry every chance of success. A major shortfall of the plan was that no one was placed in charge of coordinating the widely dispersed Confederate batteries, so each corps operated independently.[8]

With his ammunition trains far in the rear, Alexander apparently expected to neutralize the Federal artillery with the ammunition he had in his limbers and caissons. This meant that the cannonade would be intense but short in duration. In principle, the cannonade could accomplish its goals, as more than eleven thousand projectiles were available to be thrown at the Federal position. The number is deceptive, however, as much of this ammunition was defective.[9]

Lee assembled an impressive array of artillery to soften the Federal line prior to the charge. On the right of the line, Longstreet posted eighty-three guns on a small ridge that paralleled Cemetery Ridge. Eight from Mathis W. Henry's battalion were positioned on the extreme right, to protect that flank, while the remaining seventy-five were posted along a thirteen-hundred-yard line, which began at the Peach Orchard and extended north to the northeast corner of Spangler's Woods on Seminary Ridge.[10] Alexander's battalion (now under Maj. Frank Huger) was to the left of Henry's battalion, in the following arrangement, from right (south) to left (north): Tyler C. Jordan's, George V. Moody's, William W. Parker's, S. Capers Gilbert's. The former battery deployed just east of the Peach Orchard; the latter, just north of Trostle Lane. Farther north, near the Klingle house, Osmond B. Taylor's battery of Alexander's battalion dropped trail. Merritt B. Miller's 3rd Company of the Washington Artillery was on its left, followed northward by Capt. Charles W. Squires's 1st Company and Joe Norcom's (Lt. H. A. Battle's) 4th Company. Next came ten pieces from Henry C. Cabell's battalion. Basil C. Manly's battery formed on the right, followed by two guns from John C. Fraser's battery, which had been badly beaten up the prior day and was low in manpower. The two guns were placed under the command of Lt. J. H. Payne and attached to Manly's unit. Next came a section of ordnance rifles of Edward S. McCarthy's battery under the command of Lt. R. M. Anderson. The final guns of Cabell's battalion on this line were two more from Fraser's battery under the command of Lt. William J. Furlong. John B. Richardson's 2nd Company of the Washington Artillery completed this line. James Dearing's battalion formed just north, from right to left: Robert M. Stribling's, William H. Caskie's, Miles C. Macon's, and Joseph G. Blount's batteries. This was a prime position—suitable for a fresh, crack artillery battalion. As the gunners deployed, Lt. W. Nathaniel Wood of the 19th Virginia heard Dearing scream, "That hill must fall."[11] Finally, the remainder of Cabell's battalion formed in the left and rear, on Seminary Ridge.[12] Farther north was deployed Pichegru Woolfolk's battery of Alexander's (Huger's) battalion, Henry H. Carlton's battery, and a section of Edward S. McCarthy's battery from Cabell's battalion.[13] Cadmus M. Wilcox's and David Lang's brigades advanced to support James Dearing's batteries just before daylight.[14]

The ground around the Peach Orchard was particularly distasteful to Alexander. He found it "generally sloping toward the enemy. This exposed all our movements to his view, & our horses, limbers, & caissons to his fire. . . . I studied the ground carefully for every gun, to get the best cover that the gentle slopes, here and there, would permit, but it was generally poor at the best & what there was often gotten only by scattering commands to some extent. And from the enemy's position we could absolutely hide nothing."[15]

Not being able to ascertain the exact position of the Federal lines in the darkness, Alexander had quite a scare when daylight revealed that twenty of his guns were in danger of being enfiladed by the Federal cannon on Cemetery Hill. "I had a panic, almost, for fear the enemy would discover my blunder and open before I could rectify it. They never could have resisted the temptation to such pot-shooting," he wrote after the war.[16] "For 9 hours—from 4 A.M. to 1 P.M. we lay exposed to their guns, & getting ready at our leisure, & they let us do it. Evidently they had felt the strain of the last two days, but for all that they ought to have forced our hand."[17] Alexander hypothesized, "I can only account for their allowing our visible preparations to be completed by supposing that they appreciated in what a trap we would find ourselves."[18] Actually, the answer was much simpler—Hunt expected another Confederate infantry attack and wanted to conserve his ammunition.

Henry J. Hunt never forgot the sight of the massed Confederate artillery: "Our whole front for two miles was covered by batteries already in line or going into position. They stretched—apparently in one unbroken mass—from opposite the town to the Peach Orchard, which bounded the view to the left, the ridges of which were planted thick with cannon. Never before had such a sight been witnessed on this continent, and rarely, if ever, abroad."[19]

Alexander noted that Longstreet's guns were deployed as "virtually one battery" and were in position by 10:00 a.m.[20] After the war, he continued being surprised that the Federal batteries were "remarkably amiable all that morning, in allowing our batteries to move about in easy range, & often in columns & masses which presented the prettiest possible targets. . . . We did occasionally receive a few shots & suffered some casualties but I would not allow our guns to reply reserving everything I had for the assault."[21] The battery commanders in Dearing's battalion were told that they would fire for fifteen minutes, roll forward, then fire again in stages as they kept up with the infantry. Most of the officers expressed concern about the stout fences lining Emmitsburg Road. They would not be the only ones, for the infantry would also find them to be effective barriers during the charge.[22]

Gen. William Pendleton gave Alexander nine 3rd Corps howitzers under the command of Maj. Charles Richardson of John Garnett's battalion. These guns were useless at long distances, so Alexander rationalized that he would use them to follow the infantry charge. He accordingly placed them in a protected "hollow" behind Spangler's Woods, on the western slope of Seminary Ridge, near the Pitzer house.[23]

Col. Lindsay Walker's sixty guns of the 3rd Corps artillery were deployed a few hundred yards to the left (north) of Longstreet's guns. Their positions along Seminary Ridge on the south side of Hagerstown (Fairfield) Road had not materially changed since the day before.

Maj. William Poague's battalion, normally attached to William D. Pender's division of the 3rd Corps, had been kept in reserve for much of July 2. Twenty-seven years old, Poague had been a Missouri attorney before the war, but he returned to Virginia in 1861 to become a lieutenant in the Rockbridge Artillery. He became a captain a year later and a major in another year, and with it came command of the artillery battalion. A historian who compiled an encyclopedia of Civil War leaders called Poague "one of the most effective of the Army of Northern Virginai's battalion level artillery commanders." Contemporaries called him a "man of the most dauntless courage and of the highest Christian character" and a "superior officer, whose services have been scarcely surpassed."[24]

Toward nightfall of July 2, Poague had been ordered to gallop toward Richard H. Anderson's division on Seminary Ridge just as the final attacks were being repulsed. Capt. Joseph Graham called this the "thickest fire I ever experienced."[25] The next morning, Poague deployed his guns on William J. Pegram's right. Ten of his guns were placed in position, while the remaining nonrifled howitzers were kept in reserve, as, according to Poague, "no place could be found from which they could be used with advantage."[26] Poague positioned three ordnance rifles from Capt. James Wyatt's Albemarle Artillery and two Napoleons from Capt. Joseph Graham's Charlotte Artillery under the command of Wyatt on the left of the line, next to Pegram's guns. Farther south, in front of Anderson's division's left, Poague placed five Napoleons from Capt. James V. Brooke's Warrenton Artillery and Capt. George Ward's Madison Light Artillery to be commanded by Ward. The two groupings could not see each other because a strip of Spangler's Woods extended between them. Poague's orders were different from those of the 1st Corps batteries; he was to advance his batteries only after the Confederate infantry had taken the heights.[27]

Only thirty-three of the 2nd Corps' guns were in a position to participate in the grand cannonade because the ground did not permit more to be deployed. Lt. Col. William Nelson was ordered to reconnoiter the area to find suitable platforms for his battalion. "Finding none suitable for the purpose . . . I kept my batteries concealed during the day behind the hills, immediately in the rear of the battlefield," he reported.[28] One battery was engaged for a short time, however. Jones's battalion was also kept in reserve. Thomas H. Carter had ten of his own guns in position astride the unfinished railroad cut west of town, and he was also given Capt. William H. Griffin's 2nd Baltimore Maryland Artillery (four 10-pounder Parrotts) that was normally assigned to Jeb Stuart's cavalry division. East of town, on Benner's Hill, six large 20-pounder Parrotts from Charles I. Raine's (Andrews's-Latimer's battalion) and Archibald Graham's batteries (Willis J. Dance's battalion) were in position. Three pieces from Capt. John Milledge Jr.'s Georgia battery of William Nelson's battalion were with them.[29] Thomas H. Carter of the 2nd Corps had twenty guns north and northeast of town. According to Alexander, eighty guns from the 2nd and 3rd Corps were deployed; the remaining fifty-six stood idle. They could have been sent to other sectors of the field with better ground, but that didn't happen.[30]

The Confederate batteries wrapped around the Union line so that shells could be thrown at the Federal center from three directions.[31] According to Pendleton, Lee's artillery chief, "His [Meade's] troops were massed, ours diffused." Thus, "his fire was unavoidably more or less divergent, while ours were convergent."[32] Because most of the Confederate 2nd and 3rd Corps batteries were more than two thousand yards from the Federal line on Cemetery Ridge, it was beyond the range of their smoothbore Napoleons. Alexander's 1st Corps guns in the Peach Orchard, on the other hand, were only six hundred to seven hundred yards away.

There were other problems. The Federal artillery occupied higher ground, particularly on the opposite ends of the line at Little Round Top and Cemetery Hill. Then there was the Confederate's defective ammunition. As many as 80 percent of the shells equipped with timed fuses exploded too early, too late, or not at all. Perhaps most important was the critical shortage of ammunition. As a result, the gunners were repeatedly told to fire slowly. To further enhance their effectiveness, each battery was given a specific target.[33] One of the biggest problems was the fact that the massive numbers of Confederate guns were lined up parallel to the Federal line on Cemetery Ridge, rather than concentrated on the flanks to provide a deadly enfilade fire.

The sheer length of their lines was another problem for the Confederates. The six-mile line wrapped around the Federal line, which was only half as long. As a result, Meade's army could concentrate its artillery at a density of 120 guns per mile, compared with Lee's 45 guns per mile. Not only did this disperse the Confederate firepower, it made reinforcement from one part of the line to another very difficult. Meade's interior lines made battery reinforcement and replacement much easier. As we will see, however, there was not much of this as there was little coordination among the artillery of each Confederate corps. Each corps' artillery operated fairly autonomously, when in fact a strong coordinating hand was needed. Pendleton did not have the abilities needed for this vital task.[34]

The Federal Positions

Union Gen. Henry J. Hunt set off on an extensive examination of his artillery positions at 10:00 a.m. on July 3. His goal was to ensure that his guns were "in good condition and well supplied with ammunition."[35] July 2 had been a difficult day for his guns. About forty batteries had been engaged, including fifteen of the nineteen in his artillery reserve; only the 6th Corps batteries had not seen action. Losses were high, particularly among the horses and officers. John K. Bucklyn's battery was so scattered that it took all night and much of July 3 for it to reassemble.[36]

Charles E. Hazlett's battery, now under Lt. Benjamin F. Rittenhouse, anchored the left of the line atop Little Round Top. Its fields of fire were an artillerist's dream. A number of batteries were deployed at the base of the hill, but they were not in effective firing positions. Looking north, Hunt could see the compactly arranged guns commanded by Lt. Col. Freeman McGilvery on the southern portion of Cemetery Ridge, just north of Weikert's Woods. The 1st Volunteer Artillery Brigade commander had spent the night ascertaining the whereabouts of nearby batteries and in the morning had begun assembling them on the ridge. As the sun rose, McGilvery had only twelve cannon in line, distributed over a wide front. None were from his own brigade: two from Francis W. Seeley's battery, four from Edwin B. Dow's, and six from Evan Thomas's. With its ammunition depleted, Seeley's guns pulled out. Around the same time, Nelson Ames's battery arrived and took position to the left of the line. Dow's battery was on his right. James Thompson's battery pulled into place about seventy yards south of Thomas's battery. When Patrick Hart's battery arrived, Thompson suggested that he drop trail on his left, which he did, leaving a sixty-yard gap. Arriving next were four Napoleons, which were placed to Ames's right. Two were from John Bigelow's battery, and the other two were from John G. Turnbull's. The two sections of the different batteries were consolidated because of heavy losses and placed under the command of Lt. Richard S. Milton of Bigelow's battery. They did not remain here long, for one of Hunt's aides escorted the four guns north, where they deployed near the Abraham Bryan house. Charles A. Phillips's battery arrived and was crammed into the small space between Thompson and Hart.

Capt. William Rank's 3rd Pennsylvania Heavy Artillery (two ordnance rifles), attached to Capt. James M. Robertson's 1st Brigade of Horse Artillery, arrived next and was placed on the right of Milton's guns. The final battery to come into line was Capt. John W. Sterling's 2nd Connecticut Light Artillery (four James rifles and two 12-pounder howitzers), which was part of Elijah D. Taft's 2nd Volunteer Brigade. This strong thirty-nine-gun line was in place by 6:30 a.m. McGilvery must have been very satisfied with his position because a low ridge to the west shielded it from Confederate artillery, yet there was a clear field of fire to the northwest, from which the Confederate infantry attack was expected.[37]

The 2nd Corps batteries were farther north. James M. Rorty's four 10-pounder Parrotts were deployed on Cemetery Ridge directly opposite the Nicholas Codori house, and T. Frederick Brown's, now under Lt. Walter S. Perrin, were about a hundred yards north, on the left of the copse of trees and about sixty yards from the stone wall that concealed the Federal infantry. Alonzo Cushing's guns were to the right of the copse, and William A. Arnold's battery was deployed to the right (north) of it. What was to be called the "Bloody Angle" was between them. Brown's and Cushing's batteries had lost so many men and horses on July 2 that each was composed of only four guns. George A. Woodruff's battery was farther north, in front of Ziegler's Grove, to the right of the Bryan house. Woodruff was having his share of problems as division commander Brig. Gen. Alexander Hays insisted on posting his infantry support—the 108th New York—between the deployed cannon and their limbers. The infantry's position would interfere with the transport of ammunition to the guns and would put them in a dangerous position if the Confederate guns concentrated their efforts on this battery. Despite all of Woodruff's pleas, Hays refused to move the troops.[38]

Another concentration of forty-three guns was on Cemetery Hill under the command of Maj. Thomas W. Osborn of the 11th Corps. The batteries were deployed from left to right: George W. Norton's, Wallace Hill's, Chandler P. Eakin's (Philip D. Mason's), Eugene A. Bancroft's, two guns of Elijah D. Taft's, and Hubert Dilger's. Behind this line were Frederick M. Edgell's on the left and Taft's remaining four guns, which were farther north. Cemetery Hill was a very vulnerable spot as Confederate artillery could theoretically fire on it from three directions. The fact that so many cannon were squeezed into such a small space also increased the vulnerability of the Union gunners.[39] Commenting on this situation, Osborn wrote:

> Nearly all the guns and all the caissons were among the graves. Each battery in position as in park—fourteen feet between the guns—the limbers and caissons at proper distance in rear of the guns . . . yet they were close together. No earthworks were thrown up to protect the men, nor could there have been without digging up the dead in the Cemetery. . . . We made the best target for artillery practice the enemy had during the war. But there was another side to it. We commanded their guns as well as they did ours, with the advantage on the enemy's part of being more scattered. In addition to this we commanded the plain perfectly, with no timber intervening, over which the enemy's infantry must advance to the charge.[40]

Finally, Hunt had a number of fresh batteries in the reserve, including Robert H. Fitzhugh's, Augustin N. Parsons's, and three batteries in James M. Robertson's 1st Brigade of Horse Artillery. The eight 6th Corps batteries, sporting forty-eight guns, were also fresh and ready to move at a moment's notice. In response to a call for a rifled battery, Robertson dispatched Capt. Jabez J. Daniel's 9th Michigan Battery. This was a new unit, itching for a fight, and it galloped into position just to the right (north) of Thomas's battery, whose men had constructed lunettes around each gun to protect its gunners.[41]

While inspecting his guns, Hunt periodically stopped to gaze at the Confederate artillery positions, which were clearly being strengthened. He speculated that Lee could be withdrawing his infantry and replacing them with artillery. This meant that Lee expected a Federal attack, or was it was a prelude to a grand assault? He believed it would be the latter.[42] Hunt also spent time fretting about his ammunition supply. He told each of his batteries and their artillery chiefs "not to fire at small bodies, nor to allow their fire to be drawn without promise of adequate results." He further instructed them to "watch the enemy closely, and when he opened to concentrate the fire of their guns on one battery at a time until it was silenced." Above all, the batteries were to "fire deliberately, and to husband their ammunition as much as possible."[43]

Daytime Activities

Despite the cautions of their commanders to conserve ammunition, some batteries on both sides opened fire almost as soon as the sun appeared. Lt. Robert James, now in command of Francis W. Seeley's battery, deployed near the Jacob Weikert house, noted that Confederate gunners opened fire on his battery at daybreak. So rapid was James's return fire that he quickly expended all the ammunition in his limbers. Confederate Col. E. Porter Alexander recalled that the Federal guns opened fire first on the Washington Artillery, wounding some of the gunners. Capt. Joe Norcom was slightly wounded, and Lt. Henry A. Battles took over the 4th Company.[44] "I would never allow more than one or two shots in reply," Alexander reported.[45] His guns worthless without ammunition, James requested orders from General Birney, who told him to pull back and occupy a reserve position. The battle was finally over for Seeley's battery.[46]

Sgt. Frederick Fuger of Alonzo H. Cushing's battery estimated that his guns engaged the enemy "three or four times, lasting a few minutes each," up to 11:00 a.m., but their fire did not sustain any casualties other than the loss of some valuable ordnance.[47] During the morning, Cushing was chastised by his commanding officer, Capt. John G. Hazard, for firing too many shells at the enemy.

"Young man," Hazard said, "are you aware that every round you fire costs $2.67?"

"Yes sir!" was Cushing's response, and soon the guns fell silent—at least temporarily.[48]

At one point, the right section was ordered to swing toward the right and throw shells at the William Bliss house.[49]

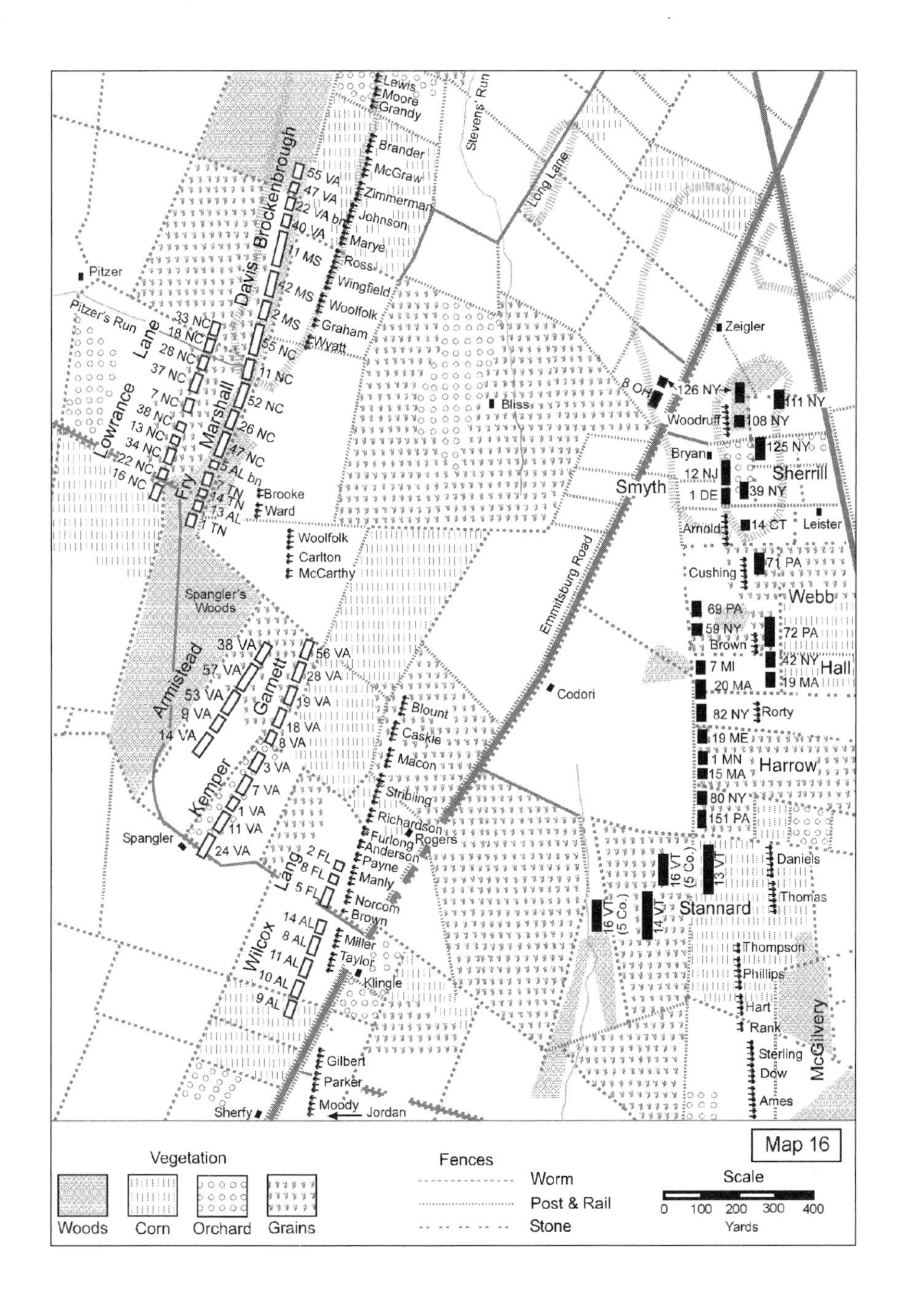

POSITIONS PRIOR TO THE
GREAT CANNONADE ON JULY 3

On Seminary Ridge, Confederate Maj. William T. Poague was surprised to hear five guns under Capt. James W. Wyatt open fire sometime between 7:00 a.m. and 8:00 a.m. The Federal guns responded as he rushed over, and it turned into a mismatch. "In a few minutes, the fire of several of their batteries was concentrated on these five guns, and seeing that the contest was a very unequal one . . . I directed the firing to cease."[50] While Wyatt's guns scored a direct hit on a Federal caisson, he lost eight of his best horses. This counterartillery fire on Wyatt's guns was ordered by 2nd Corps artillery commander Hazard. Cannoneer John Rhodes of T. Frederick Brown's battery noted, "The rebel fire succeeded in exploding one of our ammunition wagons and several of the limber chests along our line, in retaliation we performed the same service for them, which was acknowledged by both parties with continued shouts and cheers."[51] The greatest damage was to Cushing's battery. A shell exploded under the number-two limber, igniting the powder in number one and number three as well, blowing all three sky high. The lids were apparently left open, exposing 150 rounds of ammunition and almost a thousand pounds of powder.[52] "When the explosions occurred it frightened our horses and for a time it looks as though we were going to have a panic," recalled Christopher Smith.[53] The explosions were so powerful that a driver from Arnold's battery was knocked off his horse and a team stampeded away.[54] The gunners were finally able to calm the horses, however, and quickly returned the fire. According to Fuger, "when the Confederates saw this explosion they immediately jumped up and gave an immense yell; we, however, replied to their fire and within a few minutes after an explosion took place in their line, when our men jumped up and returned the compliment."[55] Poague later learned that Gen. A. P. Hill initially had ordered the pieces to open fire. Hill's report does not mention this action, so his motives are not known. Historian Elwood Christ surmised that Wyatt's guns were ordered to respond to Cushing's battery, which was shelling the Bliss farm.[56]

Hill again ignored orders and told his artillery to open fire later in the morning. Alexander, who was still positioning his guns to the south, was at first surprised then distraught. He wrote, "About 11, some of Hill's skirmishers and the enemy's began fighting over a barn [Bliss] between the lines, and gradually his artillery and the enemy's took part, until over a hundred were engaged, and a tremendous roar was kept up for some time."[57] Quiet then fell again on the battlefield, but not before more than sixty of

Hill's guns had expended precious ammunition in a pointless thirty-minute barrage.[58] Alexander noted, "The lack of which was much felt in the subsequent fighting."[59] He proudly added, "I never let one of my guns fire a shot [during that early barrage] & very few were fired at us."[60]

Although Alexander does not mention it in his voluminous postwar writings, some of the batteries of the Washington Artillery and James Dearing's battalion also opened fire on skirmishers earlier in the day who were trying to gain a ravine in front of them. Since there were apparently no Confederate skirmishers present, the artillery was forced to scatter the enemy.[61] "The enemy's batteries replied," wrote Maj. Benjamin Eshleman of the Washington Artillery, "but I paid little attention to them, seldom answering their fire . . . to save my ammunition for the grand attack."[62] A section of Edward S. McCarthy's battery and another from an unidentified battery were also advanced during the morning, where they fired twenty rounds, probably toward the Bliss farm, and succeeded in driving back a Federal skirmish line.[63]

Eshleman's men also had another exciting experience when they brought in an abandoned Federal gun, probably from James Thompson's battery in the no-man's-land between the two lines, about three hundred yards in front of their skirmishers. The Federal skirmishers immediately opened fire, but they were unable to stop the Confederate gunners from retrieving the piece and putting it in position along the line. The Southerners were disappointed that the limber contained only about fifty rounds of ammunition.[64]

The battlefield fell silent by about noon. John Rhodes of T. Frederick Brown's battery recalled, "As the forenoon wore on, there came a lull, a stillness even of death. A feeling of oppression weighed upon all hearts, the silence was ominous and portentous of coming evil. It was the calm which precedes the storm."[65] The cannoneers spent the time eating, resting, and checking their ammunition and equipment.[66]

Seeing the buildup of McGilvery's artillery on Cemetery Ridge, Alexander ordered several of Mathis W. Henry's batteries advanced toward Emmitsburg Road to reinforce Frank Huger's right flank. Leaving James Reilly's and William K. Bachman's batteries behind, Henry advanced Capt. Alexander C. Latham's Bravely Artillery (one 6-pounder, one 12-pounder howitzer, and three Napoleons) and Capt. Hugh R. Garden's Palmetto Light Artillery (two Napoleons and two 10-pounder Parrotts) to the Peach Orchard and placed the guns under the command of Maj. John Cheves Haskell.[67]

Alexander and Longstreet

James Longstreet, as commander of the assault on the Federal center, was to personally order the start of the cannonade. The signal was two shots fired in quick succession by the Washington Artillery near the Peach Orchard. Alexander recalled Gen. William Pendleton telling him earlier that the plan was to "give the enemy the most effective cannonade possible. It was not meant simply to make noise, but to try to cripple him—to tear him limbless, as it were, if possible."[68]

Taking his place near the left of his guns at about noon, Alexander received an unexpected communication from Longstreet: "Colonel: If the Artillery fire does not have the effect to drive off the enemy or greatly demoralize him, so as to make our efforts pretty certain, I would prefer that you should not advise General Pickett to make the charge. I shall rely a great deal on your good judgment to determine the matter, and shall expect you to let General Pickett know when the moment offers."[69]

The message startled Alexander, as he had not realized that the responsibility for launching the risky attack would be placed on his shoulders. He quickly responded:

> General: I will only be able to judge the effect of our fire on the enemy by his return fire, for his infantry is but little exposed to view and the smoke will obscure the whole field. If, as I infer from your note, there is any alternative to this attack, it should be carefully considered before opening our fire, for it will take all the artillery ammunition we have left to test this one thoroughly, and, if the result is unfavorable, we will have none left for another effort. And even if this is entirely successful, it can only be so at a very bloody cost.[70]

Alexander did not have long to wait for a reply from Longstreet: "Colonel: The intention is to advance the infantry if the artillery has the desired effect of driving the enemy's off, or having other effect such as to warrant us in making the attack. When the moment arrives advise General Pickett, and of course advance such artillery as you can use in aiding the attack."[71]

Perplexed about what to do next, Alexander visited Brig. Gen. Ambrose Wright, whose brigade had pierced the Federal line at the copse of trees the day before. He asked Wright, "What do you think of it? Is it as hard to get there as it looks?"

Wright answered, "The trouble is not in going there. I went there with my brigade yesterday. . . . The trouble is to stay there after you get there, for the whole Yankee army is there in a bunch."[72]

Alexander next found Maj. Gen. George E. Pickett, who was in a very positive and excited frame of mind. After a short consultation, Alexander sent a last communication to Longstreet: "General: When our fire is at its best, I will advise General Pickett to advance."[73]

All that Alexander could do now was to await the signal from Longstreet to open fire on the Federal line. According to two historians, he did not know that "Meade's and Hunt's line was rock-solid and as close to impregnable as any line constructed by the Army of the Potomac during the war."[74]

In the meantime, Alexander sent a courier to bring up John B. Richardson's nine howitzers. The aide returned with devastating news: the guns were gone and nowhere to be found. Unbeknownst to Alexander, Pendleton decided to move four of the guns, and Richardson had moved the others to a more secure area during Hill's earlier artillery action.

Alexander was still frustrated by these actions more than a decade after the war, writing, "God knows where he [Richardson] went to but it was where he could not be found."[75] He lamented the fact that they had full ammunition chests and could have done some damage to the enemy.[76] "It would not have made any difference in the result of the battle . . . but I feel bitterly about that to this day [1876]."[77]

9

JULY 3

THE GREAT CANNONADE

The Cannonade Begins

The men of Washington Artillery watched as a courier galloped toward them. Hastily dismounting, the aide made his way to Col. James B. Walton and handed him a message torn from a memorandum book. It read: "Colonel: Let the batteries open: order great care and precision in firing. If the batteries at the peach-orchard cannot be used against the point we intend attacking, let them open on the rocky hill [Little Round Top]."[1]

Walton briefly pondered this message from Longstreet then, turning to Maj. Benjamin Eshleman, he ordered the two signal guns fired. The first gun immediately fired, but then there was a pause, for the friction primer of the second failed to detonate. Another was quickly inserted, and the second gun fired. Almost immediately, more than 140 Confederate guns opened fire.[2]

Although the Federal troops had seen the Confederate artillery preparing for the cannonade, they could not have predicted its magnitude. One of the initial shots scored a direct hit on Lt. Alonzo H. Cushing's mess, overturning the coffeepot and sending pots and pans flying. The men were up in an instant, running toward their guns.[3] Gen. Henry J. Hunt was on Little Round Top, conferring with Lt. Benjamin F. Rittenhouse, when the cannonade began. He actually underestimated the number of enemy guns that opened fire, believing it was about 120. Hunt estimated that he had but 80 to respond.[4] He actually had 139 guns between Little Round Top and Cemetery

Hill and another 95 waiting in reserve. These 227 guns were arranged in three semidistinct groupings: the Cemetery Hill group under Thomas W. Osborn's command; the Union center group of the 2nd Corps Artillery Brigade; and the Union left group, composed of Freeman McGilvery's guns and those on Little Round Top. Arranged in this manner, the gunners could throw an enfilading fire into any Confederate attack column.[5]

According to Maj. James Dearing, his guns began the cannonade by firing "slowly and deliberately." To ensure greater accuracy and to conserve ammunition, he fired by battery. As a result, he considered the accuracy of his guns to be "very good, and most of the shell and shrapnel burst well."[6] His gunners aimed at the Federal batteries on Cemetery Ridge, and before long, three caissons exploded.

One of the Federal gunners recalled the initial bombardment: "There appeared to be but one flash, and those simultaneous reports pealed out deafening salvos, and were grand and impressive beyond description . . . the heavens had opened, and the Union soldiers found themselves in a pitiless storm of shot and shell which burst and tore up the ground in all directions, dealing out death and destruction on every side. So terrific was the cannonade . . . [that] the air was darkened by the heavy clouds of smoke which overhung the sky."[7]

Division commander Brig. Gen. John Gibbon, an old artilleryman, also watched the initial bombardment with a mixture of horror and fascination: "It was impossible to count the shots; and, along with these reports came every kind of bustle, whirr, whistle, and shriek that man has heard or can imagine; the most terrific of all proceeding from some elongated missile, which ceasing to revolve around its axis, dashed 'promiscuously' through the air, becoming visible on such event. The twelve-pound shots were also to be seen as they came; and the worst of it was that every shot seemed to be coming straight to hit you between the eyes."[8]

The Federal Response

The Federal gunners held their fire, except those on the heights at either end of the line, which numbered more than fifty guns. Another sixty or so from the 2nd Corps and Freeman McGilvery's line remained silent because Henry Hunt had given orders for these batteries not to fire but to conserve their ammunition.[9] According to Hunt, "It was of the first importance to subject the enemy's infantry, from the first moment of their advance, to such a cross-fire of our artillery as would break their formation, check their impulse, and drive them back, or at least bring to our lines in such condition as to make them easy prey."[10] The Federal gunners could only watch as the Confederates, "using every description of missiles and field artillery," banged away at their positions.[11]

Realizing that his artillery was not firing, 2nd Corps commander Maj. Gen. Winfield Hancock rode over to his artillery chief, Capt. John G. Hazard, and ordered him to respond to the enemy. Hazard informed him of Hunt's orders, but Hancock would hear none of it.[12] Given the massive explosions around him, Hancock emphatically believed that "his troops would not stand unless the reply was made."[13] Hazard wrote in his official report that "the batteries did not at first reply, till the fire of the enemy becoming too terrible, they returned it till all their ammunition, excepting canister, had been expended."[14] John Rhodes of T. Frederick Brown's battery, now under Lt. Walter S. Perrin, estimated that the gunners waited ten to fifteen minutes before opening fire. Because of the heavy losses, the battery was consolidated into four guns. The initial firing was slow and methodical but increased in intensity with time. Thick smoke soon enveloped the area, so Hazard's men directed their fire at the red flashes of the Confederate pieces as they fired.[15]

Hancock next rode south to McGilvery and ordered him to open fire as well. McGilvery was polite but firm—*his* commander had given him orders, and he intended to follow them.[16] Hancock later recalled that McGilvery looked him squarely in the eye and told him to "go to hell!" Politically savvy McGilvery did not divulge Hancock's identity in his battle report, writing instead, "Some general commanding the infantry line ordered three of the batteries to return the fire."[17] The battery commanders of Charles A. Phillips's and Patrick Hart's batteries all noted in their reports that the "culprit" was Hancock, and Hart noted after the war that the 2nd Corps commander used language that was "profane and Blasphemous such as a drunken Ruffian would use."[18] Hancock's language may have been appropriate, for Hart recalled his reaction when ordered to open fire: "I informed him that I had received my orders from General Hunt Chief of Arty and I would obey them. He ordered me to open fire that I was in his line. I replyed that should he give me a written order that I would open fire under protest."[19] Capt. James Thompson reported after the war that he received orders from Hancock as well.[20] Hunt had wanted his cannoneers to conserve their ammunition and lie down to protect themselves until the enemy guns stopped firing. McGilvery had wisely ordered the battery commanders to throw up a "slight earthwork" in front of each gun and then each man took refuge behind it during the cannonade.[21]

Hancock was not to be denied, however. Thompson was actually relieved when he received the orders as "it is much easier to fight than lay idle under such a storm of shot, shell and missiles."[22] Phillips noted after the war that the "rebels were not doing us any harm, and if they wanted to throw away their ammunition I do not see why we should prevent them."[23] Satisfied with his work, Hancock left the sector, and according to McGilvery, "After the discharge of a few rounds, I ordered the fire to cease and the men to be covered."[24]

Because they were not working their guns, and therefore not engulfed in smoke, McGilvery's cannoneers could see the effect of the Confederate cannonade. McGilvery called the fire "very rapid and inaccurate, most of the projectiles passing from 20 to 100 feet over our lines."[25] Phillips noted that "beyond the noise which was made no great harm was done."[26] Because of McGilvery's protected position, his batteries sustained virtually no casualties during the cannonade. It was a different story with the 2nd Corps batteries in the center of the Federal line, which drew the Confederates' fire.[27]

Hunt had intended for his batteries to reply fifteen or twenty minutes into the cannonade. He informed his battery commanders to first ascertain which Confederate batteries were doing the most harm and then, as Hunt put it, "concentrate our fire on that point, firing slowly, deliberately and making target practice of it."[28] Hunt understood that the goal of his artillery was not to disable the Confederate guns but to pulverize the enemy's infantry during the assault. He would do this "from the first moment of their advance and whilst beyond musketry range to a heavy concentrated cross fire of artillery in order to break their formation, check their impulse and bring them in as disordered a condition and with as much loss as possible."[29]

Pvt. Felix Galloway of James H. Lane's Confederate battalion on Seminary Ridge estimated that his cannon fired about forty times before the batteries in the Federal center replied. "All at once the entire front of Cemetery Ridge seemed to light up in a blaze. I well knew what was coming and bowed my head to the inevitable. It came like a fierce hailstorm," he wrote after the war.[30] Pvt. W. H. Routt of Osmond B. Taylor's battery wrote home that the air was filled with sulfurous clouds of smoke while "the shrieks of the wounded and dying rendered the scene more appalling & heart rendering the like I never saw the like I never want to see again."[31] A member of James Dearing's battalion wrote, "The very earth seemed to be rolling and tossing under us and the heavens shook above us . . . shot and shells screamed and bursted over and around us."[32]

No man on the battlefield had ever experienced such an intense cannonade. Lt. John Marye of the Fredericksburg battery (Willie Pegram's battalion) described the scene: "Round shot whistled by and plowed the ground. The air was alive with screaming, bursting shells and flying fragments. The simultaneous explosion of fifty, a hundred guns shook the earth, which rocked as if in the throes of an earthquake. Cannoneers with jackets off and perspiration streaming down their faces, blackened with powder, kept the guns cool by plunging the spongeheads in buckets of water, and as fast as a man fell another took his place; guns were dismounted, limbers and caissons blown up and horses ripped open and disemboweled. It was enough to try the stoutest heart."[33]

The Cannonade Pounds the Center of the Federal Line

The battery's battalion commander, Willie Pegram, waved his hat in the air and shouted with glee when one of his guns scored a direct hit on a Federal caisson, blowing it sky high. His exuberance was short-lived, for a Federal shell found one of his own caissons, having the same effect.[34] A member of William Mahone's brigade noted how badly the battalion suffered: it "presented a sad spectacle of war's destructive work."[35]

A soldier in the 69th Pennsylvania of the Philadelphia Brigade, lying in front of the copse of trees near Alonzo H. Cushing's battery, described the maelstrom after the war: "The air is filling with the whirling, shrieking, hissing sound of the solid shot and bursting shell; all threw themselves flat upon the ground, behind the little stone wall; nearly 150 guns belched forth messengers of destruction, sometimes in volleys, again in irregular, but continual sounds, traveling through the air, high above us, or striking the ground in front and ricocheting over us, to be imbedded in some object to the rear." Joseph McKeever, of the same regiment, noted, "After the cannonading began, we were all hugging the earth and we would have liked to get into it if we could."[36] The entire area was filled with dense sulphurous smoke.

Union Gen. John Gibbon watched the projectiles flying in the direction of the copse of trees. "The whole air," he noted, "seemed filled with rushing screaming and bursting shells. The larger round shells could be seen plainly as in their nearly completed course they curve in their fall toward Taneytown Road, but the long rifled shells came with a rush and a scream and could only be seen in their rapid flight when they 'upset' and went tumbling through the air, creating the uncomfortable impression that, no matter whether you were in front of the gun from which they came or not, you were liable to be hit."[37]

A shell scored a direct hit on one of Cushing's guns within fifteen minutes of the start of its firing, tearing away a wheel and sending the dazed survivors bolting for the rear. Pulling out his pistol, Cushing aimed it at the gun's commander and said, "Sergt. Watson, come back to your post. The first man who leaves his gun again, I'll blow his brains out."[38] The men replaced the damaged wheel, and the gun was quickly back in action. The battery would lose several more wheels during the cannonade.

Pudgy 5-foot-9, 170-pound Alonzo H. Cushing was a West Point graduate who quickly gained the respect of his men. One of his corporals noted that he "looked more like a schoolgirl than a warrior, but he was the best fighting man I ever saw."[39] While Cushing actively pinpointed each gun's target, he also paid special attention to the offending gun crew.[40] He was a special inspiration to his men. According to one of his gunners, "He was as cool and calm as I ever saw him, talking to the boys between shots with the glass constantly to his eyes, watching the effect of our shots."[41]

Gibbon watched another shell explode under an open limber box while a gunner retrieved some ammunition: "The poor gunner went hopping to the rear on one leg, the shreds of the other dangling about as he went."[42] Another cannoneer was in immense pain from a grave wound in his abdomen. Begging his comrades to shoot him to end his misery, he pulled a pistol out of his belt and blew his brains out.

A Confederate shell hit one of T. Frederick Brown's guns on the opposite side of the copse of trees (south) in the process of being loaded, leaving a distinct dent on the top of its barrel. One gunner was decapitated and another, whose arm was torn cleanly off of his body, lived long enough to exclaim, "Glory to God! I am happy! Hallelujah!"[43] The remaining gunners tried to load the gun, but the round stuck hard in the barrel. They pushed at the rammer and even took an ax to drive it home, but all their efforts were futile. Some of the men claimed that thirty-nine bullets and three Confederate shells hit the beleaguered cannon. They named it the "Gettysburg Gun," and it was later displayed at the Rhode Island capitol.[44]

Ernest Waitt of the 19th Massachusetts described the conditions in the center of the Federal line: "Fragments of bursting shell were flying everywhere. There seemed to be no place where they did not strike and no spot from whence they did not come."[45] At least they had the ability to seek shelter, unlike the cannoneers. Frederick Oesterle of the 7th Michigan wrote, "Oh how we hugged the ground and how we stuck our heads and as much as possible of our body out of the way of these merciless, approaching and bounding shells."[46]

The infantry supporting the artillery was in a most precarious position. According to Lt. L. A. Smith of the 136th New York (Orland Smith's brigade, Adolph von Steinwehr's division, 11th Corps):

> It is a terrible experience to support batteries when located in their front. . . . If you laid down on the ground and put your fingers in your ears you got, in addition to the crash in the air, the full effect of the earth's tremor and its additional force as a conductor. One of our men found afterwards that his teeth were loose and within a few days nearly all of them dropped out. If you rolled over on your back and looked up into the heavens fairly black with missiles exploding continually and sending their broken fragments in every direction, the situation is not more assuring. If you sat down with your back to the stone wall and looked over into the cemetery, you saw long fiery tongues leading toward you, and thick clouds of sulfurous smoke settle down around you. . . . If you turned around and looked over the wall toward the enemy each cannon ball seemed directed toward that particular spot.[47]

While most of the Confederate shells overshot the batteries, a large number plowed into the 2nd Corps batteries and created havoc. James M. Rorty's battery was among the hardest hit. A shell hit the battery's right caisson early in the cannonade, killing its horses and drivers.[48] Rorty, who had been in command for only a short time, had yet to be accepted by his men because Lt. Albert S. Sheldon, their temporary commander, had been passed over. One gunner wrote that the battery had become "a dumping ground for men with political and military pulls; and that merit within our ranks had been systematically overlooked to make room for favorites."[49] Another wrote of

Rorty, "Not a man likes him. All look sad." The battery performed magnificently during the cannonade, despite the fact that most of the men had gotten less than three hours sleep.[50] According to one soldier, "The men, begrimed with powder and smoke, loaded with precision and speed, sighting and firing their guns as if the fate of the nation depended upon their exertions. . . . With guns dismounted, caissons blown up, and rapidly losing men and horses, the intrepid commander moved from gun to gun as coolly as if at a West Point review. While bringing up ammunition, some of the men, to lessen their exposure, dismounted before reaching the battery; but the stern disciplinarian would not permit, and ordered them to remount and ride into position."[51]

Two of Rorty's guns were disabled by direct hits. Soon only one gun was in action, but it too fell silent when only three men were left to man it. The battery had entered the fray with sixty-five effectives, and the losses were so great that Rorty grabbed a rammer and assisted in getting the piece firing again. Seeing the 19th Massachusetts nearby, Rorty yelled to its commander, "For God's sake, Colonel, let me have twelve men to work my gun!"[52] Six were sent to help carry ammunition, and then another twenty were dispatched to serve the guns.[53] One of the privates yelled to his comrades, "Let's go and help. . . . We might as well get killed there as here."[54] Before long, two guns were back in action. A fragment from one of his exploding limbers mortally wounded Rorty soon after, so Sheldon again took command of the battery. Nearby, Evan Thomas's battery was also wrecked by the intense cannon fire, forcing it to withdraw.[55]

Although not being pounded as much as the other 2nd Corps batteries, George A. Woodruff's was exposed to the disconcerting effect of the British Whitworth rifles, which had been moved to Oak Hill from Seminary Ridge. According to Maj. Thomas W. Osborn, "The long steel bolt used for a solid shot . . . is readily recognized and distinguished from the reports of all other guns."[56] Exploding shells were not the only danger for Woodruff's battery in Ziegler's Woods; large branches constantly rained down on them, forcing the gunners to frequently halt their firing to clear the debris. At least one caisson was destroyed when a large limb fell squarely on it. "Limbs of trees, splinters of rails, gravel and dirt, pieces of stone from the stone walls filled the air and wounded many," recalled Col. Clinton MacDougall of the 111th New York (George L. Willard's brigade).[57] The men of the 108th New York suffered psychologically during the cannonade. "Our regiment came in for its full share of shot and shell aimed at Woodruff's guns," recalled Francis Wafer.[58] The men were called upon at times to help move the guns. So severe was the cannonade that wounded infantrymen opted to remain behind their breastworks and risk bleeding to death rather than move to the rear and risk the Confederate artillery fire that was blasting that area. It also forced any would-be skulker to remain in place.[59] Woodruff's gunners were not so complimentary. One noted, "If their artillery had been as good as their infantry, our loss would have been much greater."[60]

The cannoneers could not admire their work because they were engulfed in dense smoke for most of the cannonade.[61] Even the bright sun was obliterated. According to Christopher Smith of Cushing's battery, "We could occasionally get a glimpse of the green wheat field, and the very fury of the cannonading seemed to send waves across it like gusts of wind."[62] It was now fruitless to try to find targets, so the cannoneers "gave them all the proper elevation and kept on loading and firing as rapidly as possible."[63]

While the initial Confederate rounds were deadly accurate in pounding the Federal front line, the dense smoke precluded the gunners from seeing when their shots were overshooting their targets. Instead, the shells started crashing into the east slope of Cemetery Ridge or beyond. The inaccurate fire also could have been caused by the gunners not regularly resetting their cannon's elevating screws and not correctly cutting the fuses, which ignited the charges too late.[64] Henry Hunt wrote after the war that the "air was filled with projectiles there being scarcely an instant but that several were seen bursting at once. No irregularity of ground offered protection, and the plain in the rear of the line of battle was soon swept of everything movable."[65] The maelstrom was particularly great near George G. Meade's headquarters at the Leister farm, approximately fourteen hundred yards behind the front line. Samuel Wilkeson, a correspondent for the *New York Times,* wrote after the war:

> A shell screamed over the house, instantly followed by another and another, and in a moment the air was full of the most complete artillery prelude to an infantry battle that was ever exhibited. Every size and form of shell known to British and to American gunnery shrieked, whirled, moaned, whistled, and wrathfully fluttered over our ground. As many as six a second, constantly two in a second, bursting and screaming over and around the headquarters, made a very hell of fire that amazed the oldest officers. . . . The soldiers in Federal blue were torn to pieces in the road and died with the peculiar yells

> that blend the extorted cry of pain with horror and despair. Not an orderly, not an ambulance, not a straggler was to be seen upon the plain swept by this tempest of orchestral death thirty minutes after it commenced.[66]

Riding to the artillery reserve in the rear, Hunt was astonished to find it gone. He could understand why when he saw "the remains of a dozen exploded caissons, which had been placed under cover of a hill, but which the shells had managed to search out. In fact, the fire was more dangerous behind the ridge than on its crest."[67] Although only about 25 percent of the Confederate shells were exploding, the missiles created havoc among the mules. Ordnance officer Lt. Cornelius Gillett believed that it was just a matter of time before the mules stampeded, so he quickly moved the wagons to safety about a thousand yards to the rear.[68] Lt. Gulian V. Weir, whose battery was in this area, recalled after the war, "Men, horses, mules and wagons were moving everywhere, at top speed. Enemy shells plowed the earth and exploded to the point that nowhere was safe."[69] Weir quickly moved his guns toward the Jacob Hummelbaugh house.

Hunt was very active during this period, constantly riding along the Federal line, ascertaining the condition of his batteries, and replacing those that were badly battered. At one time or another, eighty-seven Federal cannon fired from Cemetery Ridge.[70]

Cemetery Hill

The Confederate batteries also shredded Thomas W. Osborn's artillery on Cemetery Hill. "The enemy's line of artillery was practically in a semi-circle around us. My batteries on the hill were then raked from every side, except the rear and the direct left," Osborn noted.[71] Because of the small space on Cemetery Hill, the Union batteries were crammed together, making them even more vulnerable. Ten rifled guns from Confederate Lt. Col. Thomas H. Carter's 2nd Corps artillery battalion opened fire from west of town.[72] Their goal was to "divert the fire of the enemy's guns from Hill's and Pickett's troops in their charge across the valley, and also to divert their fire from three batteries of the First Virginia Artillery, under Captain [Willis J.] Dance, and temporarily in my command," noted Carter.[73] William White of the 3rd Richmond Howitzers of Dance's battalion recalled that his orders were to fire on a specific Federal battery until it was silenced, then cease firing when the infantry attack was launched.[74] Carter believed that his guns were effective: "The effect of this concentrated fire on that part of the line was obvious to all. Their [the Federal] fire slackened and finally ceased."[75]

After the long days of stress during the campaign, some of the Federal gunners on Cemetery Hill broke down during the cannonade. Such was the case of George W. Norton's battery, on the left of the Federal line. Its men became bewildered when hit from frontal and enfilade fire. Capt. James F. Huntington, commanding the 3rd Volunteer Brigade of the artillery reserve, noted, "Most of the fire came from the left, and my battery was then faced to the front; we had to change the position of the guns under that fire. For the first time the men of Battery H hesitated to obey my orders; but they were almost exhausted by fatigue and the intense heat; and it was only for a moment that they forgot their duty. Soon [they] had our guns around, and opened one section to the left, another to the front, and the third I could not work for want of men able to stand on their feet."[76]

Osborn disagreed, noting that while the Confederates' aim was accurate, their range was too great, and most of the shells flew about twenty feet over the heads of the Federal gunners.[77] He did not mention that many of these shells landed among the batteries of the 12th Corps. Lt. Edward D. Muhlenberg noted, however, "The direction of their lines of fire was such that almost every projectile passing over Cemetery Hill found its bed within the battery line of these two batteries [Sylvanus T. Rugg's and David H. Kinzie's]."[78] Muhlenberg proudly wrote that the men withstood this fire without bolting. Because of the infantry in front of Dance's battalion and the defective nature of some of the ammunition, Confederate gunners were forced to use solid shot.[79] Dance nevertheless believed the firing to be "accurate and effective."[80]

The Federal guns on Cemetery Hill opened a rapid return fire—so fast that the caissons were constantly sent to the rear to replenish their ammunition.[81] Meade soon appeared and in a very agitated tone yelled to Osborn, "What are you drawing ammunition from the train for? . . . Don't you know that it is in violation of general orders and the army regulations to use up all your ammunition in a battle?"[82]

Just as the day before, Osborn's guns opened an effective crossfire on the nine guns on Benner's Hill, creating a living hell for the Confederate gunners. Horses and men went down, forcing the Confederate artillery to vacate the hill.[83] It was not quite so hot around Dance's battalion to the east of Cemetery Hill, so these veterans were able to hold their ground.[84]

Not all of the Federal artillery heroically weathered the storm. Osborn was told that an Ohio battery on the left of his line was dumping ammunition on the ground. By so doing, the gunners could say they were out of ammunition and had an excuse to remove the battery to the safety of the rear. Despite giving orders to the captain to maintain his position, Osborn looked over and saw the battery galloping down Baltimore Pike.[85] Capt. William H. McCartney of the 1st Massachusetts Light Artillery Battery A noted that his men collected "48 rounds of 3-inch projectiles, perfect; 22 rounds having been found near the position which had been occupied by one limber."[86] Osborn cited the battery as George W. Norton's 1st Ohio, Battery H. Originally commanded by Capt. James F. Huntington, this battery had had several unfounded charges lodged against it prior to Gettysburg. A recent analysis found that the offending battery was actually Capt. Wallace Hill's West Virginia Light Artillery, Battery C.[87]

Effects on the Infantry

The infantry of both armies, some lying under cover, some exposed, was also hit by the artillery fire. Gen. James L. Kemper's brigade of Pickett's division, occupying an exposed area, was the hardest hit Confederate unit. According to David Johnston of the 7th Virginia (Kemper's brigade):

> The very atmosphere seemed broken by the rush and crash of projectiles, solid shot, shrieking, bursting shells. The sun, but a moment before so brilliant, was now almost darkened by smoke and mist enveloping and shadowing the earth, and through which came hissing and shrieking, fiery fuses and messengers of death, sweeping, plunging, cutting, ploughing through our ranks, carrying mutilation, destruction, pain, suffering and death in every direction. . . . At almost every moment of time, guns, swords, haversacks, human flesh and bones, flying and dangling in the air, or bouncing above the earth, which now trembled beneath us as if shaken by an earthquake.[88]

Johnston noted that a Federal shell took out eight infantrymen lying prone on the ground. Kemper's brigade may have lost as many as 250 men, or 15 percent, of its effectives during the cannonade. James J. Archer's and Joseph R. Davis's brigades of James J. Pettigrew's division also sustained some losses, although fewer than Kemper's. For example, Davis's brigade lost about 23 men.[89]

The men of Cadmus M. Wilcox's brigade lying behind Alexander's guns near the Peach Orchard seldom raised their heads to see the commotion around them. When they did, they saw "shells . . . screeching & bursting over our heads, round shot were whizzing through the air," recalled Lt. Col. Hilary Herbert of the 8th Alabama.[90] Capt. George Clark of the 11th Alabama remembered that "men could be seen, especially among the artillery, bleeding at both ears from the effect of concussion and the wreck of the world seems to be upon us."[91]

Capt. J. Turney of the 1st Tennessee of Archer's brigade recalled that the "limbs and trunks of trees were torn to pieces and sent crashing to the earth to add to the havoc among the gallant boys who waited anxiously an order to charge."[92] One of the wounded was Col. Birkett D. Fry, now commanding Archer's brigade. Although hit by a shell fragment in the shoulder, he refused to relinquish command. He noted that "after lying inactive under that deadly storm of hissing and exploding shells, it seemed a relief to go forward to the desperate assault."[93]

Union Brig. Gen. George J. Stannard's brigade, just south of James M. Rorty's battery, was also exposed to the intense Confederate artillery fire. George Benedict of the 12th Vermont recalled that "shells whizzed and popped and fluttered on every side; spherical case shot exploded over our heads, and rained iron bullets upon us; solid shot tore the ground around us, and grape hurtled in an iron storm against the low breastworks."[94] Losses mounted in Stannard's exposed brigade. For example, the 14th Vermont lost about sixty men.[95]

Many Federal units lying in front of the copse of trees, the target of the impending attack, were largely immune to the tremendous cannonade. For example, Maj. Sylvanus Curtis of the 7th Michigan reported, "Nearly all the shot and shell struck in front and ricocheted over us, or passed over us and burst in our rear."[96]

According to Gen. John Gibbon, "Horses were the greatest sufferers here, for the men lay down and escaped; but the poor brutes had to take it standing."[97] A Confederate soldier in William G. Crenshaw's battery on Seminary Ridge noted that the horses were "ripped open and disemboweled." They subsequently screamed out in pain, causing him to write that it "was enough to try the stoutest heart."[98]

As the cannonade was about to end, Brig. Gen. Alexander Webb, commander of the Philadelphia Brigade, strode up to Lt. Alonzo H. Cushing and said, "It is my opinion that the Confederate infantry will now advance and attack our position." Cushing replied, "I had better run my guns right up to the stone fence and bring all my canister alongside each piece." To this, Webb replied, "All right, do so."[99]

Capt. Andrew Cowan's 1st New York Independent Battery (six ordnance rifles) was pulled out of the artillery reserve park at this time and added to the center of the line, where it was assigned a position near Stannard's brigade.[100] The battery galloped into position under a rain of missiles and shell fragments. Dropping trail to the left (south) of Rorty's battery, the six guns opened fire.[101] Cowan noted after the war, "We fired very slowly, but there was a tremendous waste of ammunition going on."[102] This was difficult for Cowan, for he had been trained to "make every shot tell if possible and to never get excited."[103]

Two other batteries were also called up from the artillery reserve. Capt. Robert H. Fitzhugh's 1st New York Light Artillery, Battery K (six ordnance rifles) and Lt. Augustin N. Parsons's 1st New Jersey Light Artillery (six 10-pounder Parrotts) galloped toward the copse of trees, halting about seventy-five yards south of it. Fitzhugh recalled, "There seemed to be a good deal of confusion."[104] This was an understatement, to say the least. Shells from Confederate batteries were flying everywhere. Fitzhugh's battery opened fire, and before long, the enemy's fire began to slacken.[105] This may have been because of ammunition depletion more than because of the effectiveness of the Federals' return fire. Other batteries also arrived in the vicinity, including Gulian V. Weir's, which had been so badly manhandled the day before.[106]

Col. Porter Alexander and the Cannonade

One of the major mistakes the Confederates made during the cannonade was not deploying all of their guns to neutralize the enemy's. According to Col. E. Porter Alexander, eighty guns from the 2nd and 3rd Corps were deployed, but the remaining fifty-six stood idle. "It was a phenomenal oversight not to place these guns, and many beside, in and near the town to enfilade the 'shank of the fish-hook' and cross fire with the guns from the west," noted Alexander.[107] Only William Pendleton could have coordinated the artillery of the three corps. Alexander only knew "his own ground" and "had but the vaguest notion of where [Richard S.] Ewell's corps was. And Ewell's chief doubtless had as vague ideas of my own situation & necessities. . . . Gen. Lee's chief [Pendleton] should have known & given every possible energy to improve the rear & great chance to the very utmost." Alexander concluded that the failure to use Ewell's batteries "was a serious loss. Every map of the field cries out about it."[108]

Those 2nd Corps guns that opened fire caused great devastation. According to Jennings Wise, when John Milledge Jr.'s battery of William Nelson's battalion opened fire on Cemetery Hill from the northeast, its results were devastating.[109] Maj. Thomas W. Osborn of the Union 11th Corps reported that these guns "opened directly on the right flank of my line of batteries. The gunners got our range at almost the first shot. Passing low over [Col. Charles S.] Wainwright's guns, they caught us square in the flank and with the elevation perfect. It was admirable shooting. They raked the whole line of batteries, killed and wounded the men and horses, and blew up the caissons rapidly."[110] Osborn soon neutralized this threat when he swung the big 20-pounder Parrotts of Elijah D. Taft's battery around and opened a withering fire, causing Milledge's battery to cease firing.[111]

Alexander initially believed the intense artillery barrage would take no longer than thirty minutes to neutralize the enemy's guns. But after twenty more minutes, additional Federal guns had joined the fray, and the two-mile line was "blazing like a volcano."[112] Alexander considered it "madness to order a column in the middle of a hot July day to undertake an advance of three-fourths of a mile over open ground against the centre of that line."[113]

Realizing that he could wait no longer, because his ammunition was almost depleted and the men were exhausted from their hard exertions in the high heat and humidity, Alexander sent the following message to Pickett: "General: If you are to advance at all, you must come at once or we will not be able to support you as we ought. But the enemy's fire has not slackened materially and there are still 18 guns firing from the cemetery."[114]

No sooner had Alexander sent this note than he noticed a distinct reduction in the intensity of the Federal artillery and could see several batteries withdrawing. He knew that Confederate artillery often exercised this ploy to save ammunition, but he had never seen it performed by the Federals. Watching intensely through his field glasses for the arrival of fresh batteries and seeing none, Alexander sent another message to Pickett: "For

God's sake come quick. The 18 guns have gone. Come quick or my ammunition will not let me support you properly."[115]

After conferring briefly with Longstreet, Pickett launched his attack. What Alexander saw was the effect of Henry J. Hunt's orders to gradually reduce the firing of his batteries. Hunt ordered the cessation, beginning with the batteries on Cemetery Hill and extending to Benjamin F. Rittenhouse's battery on the left. He did this for three reasons. First, to "induce the enemy to believe he had silenced us and to precipitate his assault."[116] Lee purportedly rode over to Capt. William J. Reese's four-gun battery and complimented the cannoneers for silencing nineteen guns on Cemetery Hill. Equally important was the fact that Federal resources of long-range ammunition were reaching dangerously low levels, and Hunt felt that it was too hazardous to bring up additional supplies. He wanted to conserve some long-range ammunition for the infantry attack he knew must surely come. Finally, Hunt also pulled out battered batteries and replaced them with fresh ones. He estimated that the cessation occurred at about 2:30 p.m.[117] Not comfortable with reducing the counterbattery fire without Meade's knowledge and permission, Hunt rode over to the commander's headquarters only to find it abandoned. He next rode toward Cemetery Hill, where Meade was supposed to be. He encountered Gen. Oliver O. Howard, who agreed that he was doing the right thing. Orders from Meade soon reached Hunt, ordering him to do just what he had done on his own.[118]

After reluctantly allowing Pickett to launch his division, Longstreet rode over to Alexander's position. His subordinate was not shy about voicing his concerns about his dwindling ammunition supply and about the attack's ultimate success.

Longstreet blurted out, "Go & halt Pickett right where he is, and replenish your ammunition."

Alexander replied, "We can't do that, sir. We nearly emptied the trains last night. Even if we had it, it would take an hour or two, & meanwhile the enemy would recover from the pressure he is now under. Our only chance is to follow it up now—to strike while the iron is hot."[119]

It did not help that Gen. William Pendleton, worried about the potential destructive Federal artillery fire, moved his precious ammunition wagons. He reported that because "frequent shell[ing] endangering the First Corps ordnance train in the convenient locality I had assigned it, it had been removed farther back. This necessitated longer time for refilling caissons."[120] Pendleton did not mention that he failed to inform his battalion commanders of his actions, forcing the caissons to gallop about aimlessly in search of the wagons.[121]

The Cannonade Ends

Because of the range of conflicting first-person accounts, the actual duration of the cannonade has been debated since the battle ended. For example, Alexander estimated that the barrage lasted at least forty-five minutes and possibly as long as an hour.[122] According to a careful analysis by historian Thomas Elmore, the Confederate gunners fired for about ninety to ninety-five minutes; their Federal counterparts were active for about seventy-five minutes. Earl Hess believed that the bombardment essentially ended at 2:00 p.m., or after about an hour. Elmore's estimate is probably the most accurate. Although, the Confederate gunners' overall aim was as poor as the quality of their ammunition, the batteries succeeded in essentially taking out most of the 2nd Corps guns at the point of attack. The same could not be said about the Federal infantry in the Federal center—the point of attack—where the enemy was full of fight. Elmore estimated that only 5 percent of the Federal infantry in this sector were put out of action by the Confederate artillery barrage; Earl Hess estimated 6 percent, or about 350 men. Both doubted that, given the heat and the gunners' exhaustion, the Confederate barrage could have gone on much longer.[123] Confederate gunner Felix Galloway wrote after the war about his increasing exhaustion as the cannonade continued, "covered with wet powder, standing half bent to dodge the balls."[124]

A second reason for the cessation of the cannonade was the growing depletion of the Confederate ammunition supply. Alexander noticed that the rate of firing of some Confederate batteries was noticeably slower as the cannonade proceeded, and some had even stopped firing entirely. These battery commanders apparently decided to ignore Alexander's orders and tried to husband at least a few rounds for the future when they might be needed. Others, like those in James Dearing's battalion, continued firing until they ran out of ammunition. Capt. Robert M. Stribling reported that his battery's ammunition was depleted approximately half an hour prior to the Pickett-Pettigrew-Trimble Charge.[125]

A third reason was the activities of the Federal artillery toward the end of the cannonade. Conferring with Howard and Osborn on Cemetery Hill,

Hunt noted that "General Meade had expressed a hope that the enemy would attack, and he had no fear of the result." After further discussion, Hunt was encouraged to order a cease-fire as a way of deceiving the Confederates into believing that their barrage had achieved its desired effect. Rather than seek permission of Meade, Hunt took matters into his own hands and ordered his cannoneers to stand down.[126]

While the Federal 2nd Corps artillery were fairly well beaten up by the cannonade, the Confederate infantry making the charge would soon learn firsthand of their artillery's ineffectiveness in neutralizing enough of the enemy forces in front of them and the artillery on either flank. After the war, 3rd Corps battalion commander David McIntosh noted:

> The impression that any very serious effect had been produced upon the enemy's lines by the artillery fire proved to be a delusion; the aim of the Confederate gunners was accurate, and they did their work as well as could be, but the distance was too great to produce the results which they sanguinely hoped for. Previous experience should have taught them better. It was not a little surprising that General Lee should have reckoned so largely upon the result. Both sides had been pretty well taught that sheltered lines of infantry cannot be shattered or dislodged when behind breastworks, by field artillery, at the distance of one thousand yards and upwards. The soldier who has been taught by experience to hug tight to his breastworks, and who knows that it is more dangerous to run than to lie still, comes to regard with stoical indifference the bursting missiles which are mostly above or behind him.[127]

William Pendleton explained, "With the enemy, there was advantage of elevation and protection from earthworks; but his fire was unavoidably more or less divergent, while ours was convergent. His troops were massed, ours diffused. We, therefore, suffered apparently much less. Great commotion was produced in his ranks, and his batteries were to such extent driven off or silenced as to have insured his defeat but for the extraordinary strength of his position."[128]

Pendleton could have added more important items to his list. At the top would have been his own ineffectiveness in coordinating the artillery of all three corps. Added to this was the very nature of the era's cannon. Recoiling after each round, the guns had to be repositioned and re-aimed prior to yanking the lanyard for the next round. Yet the dense smoke precluded the gunners from clearly seeing the Federal positions, and most shells landed in the rear of the Federal line. One officer estimated that nine out of every ten shots passed over his men's heads, and those that did explode usually did so when high above the ground, showering the men with hail-like hunks of metal, which, while discomforting, was rarely fatal. The quality of the Confederate fuses was a major problem. Another, seldom noted attribute, was the shape of Cemetery Ridge, which according to historian Peter Carmichael resembles a "narrow spinal column." He predicted that a "broad, flat plateau atop Cemetery Ridge would have significantly increased the Confederate chances of striking their target."[129]

A Union soldier probably summed up the grand cannonade best when he wrote, "Viewed as a display of fire-works, the rebel practice was entirely successful, but as a military demonstration, it was the biggest humbug of the season."[130]

The memory of the ineffectiveness of the Confederate artillery barrage remained with Henry J. Hunt throughout the war. At Appomattox he sought out an old associate, Armistead Long, by then a Confederate staff officer, and noted that the fire, "instead of being concentrated on the point of attack, as it ought to have been, and as I expected it would be, was scattered over the field." Long just smiled and replied, "When the fire became so scattered, [I] wondered what you would think about it!"[131]

James Longstreet's ambivalence about the effort to break the Federal center also contributed to the Confederates' ultimate defeat. Although in charge of the endeavor, he made little effort to coordinate any of the units, and by attempting to place major responsibilities on Alexander's shoulders, he removed the latter's ability to spend the time needed to effect a coordination with Col. R. Lindsay Walker, commander of the 3rd Corps artillery. The result was a disjointed and ineffective artillery barrage.[132] Brig. Gen. Fitzhugh Lee commented after the war that "the responsibility and fate of a great battle should be passed over to a lieutenant colonel of artillery, however meritorious he might be, is, and always will be, a subject of grave comment.[133]

Some historians have saved their greatest criticisms for Lee. According to noted Southern historian Peter Carmichael, Lee had about nine hours to study the Federal depositions prior to initiating his artillery cannonade and to make appropriate adjustments to his own units. Lee should have realized that Longstreet's guns occupied unfavorable ground, and although he roamed the field, he made no attempt to facilitate communication between the artillery of the three corps. Carmichael, who called Pendleton "stupendously incompetent," believed that this was the time for Lee to finally admit his subordinate's deficiencies as the nominal commander of the Southern artillery, and to intervene or appoint someone to assist. Pendleton's orders to move the ammunition wagons, without telling anyone, and then not providing ways to easily refill the depleted caissons, is a good example of his incompetence and negative impact on the battle. [134] Removal of commanders for incompetence had already been done at the battalion and corps levels when both Lt. Col. John Garnett and Col. James B. Walton were removed from their commands. Lee chose not to do what needed to be done and lost the battle as a result.

10

JULY 3

THE PICKETT-PETTIGREW-TRIMBLE CHARGE

PICKETT'S CHARGE

Their lines dressed and ready to go, Brig. Gen. Richard B. Garnett's brigade on the left and Brig. Gen. James L. Kemper's on its right stepped off near Seminary Ridge. Pickett's last brigade, under Brig. Gen. Lewis A. Armistead, followed in a second line. As they broke through the artillery line, James Dearing's gunners raised their hats and cheered. The youthful artillery officer yelled, "For God's sake wait till I get some ammunition and I will drive every Yankee from the heights." But waiting was out of the question, leaving Dearing to stomp his feet in disgust.[1]

Henry J. Hunt saw the long line of Confederate infantry forming near Seminary Ridge at about 3:00 p.m. Christopher Smith of Alonzo H. Cushing's battery described the sight: "All at once, as the smoke cleared away a little we saw a solid column of rebels. . . . It seemed almost as though they had sprung out of the ground."[2] Hunt could see two more lines advancing to his right (north) of Pickett's line. The first line was composed of Maj. Gen. Henry Heth's division, now under Brig. Gen. James J. Pettigrew. Behind these four brigades marched two additional ones from Maj. Gen. William Dorsey Pender's division, now under Maj. Gen. Isaac R. Trimble.

Galloping over to Lt. Col. Freeman McGilvery's batteries near Plum Run, Hunt directed them to pound Pickett's right flank as it advanced. Hunt was most upset about the activities of the 2nd Corps artillery prior to the charge. He bitterly wrote after the war, "Had their orders [the batteries of the 2nd Corps] not been interfered with [by Hancock], . . . the enemy would not have reached our lines in condition to make the vigorous attack they did . . . and caused so great a loss of life."[3] This was a direct slap at Winfield S. Hancock, who had ordered the 2nd Corps gunners to continue firing rather than conserve their ammunition for the infantry charge that was sure to follow. Hunt further related that Meade himself had sent a messenger to him for the batteries to hold their fire. The controversy continued to rage after the war.[4]

Pickett's men initially felt that they had a chance to succeed. No enemy fire raked their lines as they stepped off, and when the Federal artillery did open fire, it was largely ineffective. That soon changed, however. "The shells flew far over us at first, but this lasted but a moment. They soon obtained the range, and then Death commenced his work of destruction," noted James Walker of the 9th Virginia.[5] The Federal batteries now began wreaking havoc on Pickett's line. "Shot, shell, spherical case, shrapnel and cannister—thousands of deadly missiles racing through the air to thin our ranks!" recalled Randolph Shotwell of Kemper's brigade. "Whole regiments stoop like men running in a violent storm."[6] According to John W. Lewis of Armistead's brigade: "The crash of shell and solid shot, as they came howling and whistling through our lines, seemed to make no impression on the men. There was not a waver, but all was as steady as if on parade."[7] It was not unusual for eight, ten, or even fourteen men to fall with the explosion of a single shell. The grass even caught fire in some places. Sgt. William Robertson of the 14th Virginia noted, "Now and then a man's hand or arm or leg would fly like feathers before the wind."[8] The command of "Close up!" was constantly heard. Maj. Nathaniel Wilson of the 28th Virginia screamed to his men, "Now boys, put your trust in God and follow me." Wilson was mortally wounded within a matter of seconds.[9]

One Federal soldier marveled, "I can see no end to the right nor left to the line that is coming. . . . Men are being mowed down with every step. And men are stepping into their places. There is no dismay, no discouragement, no wavering."[10]

Alexander noted that while the Federal artillery fire was deadly, it did not halt Pickett's men. He estimated that there were only about half as many guns firing as during the cannonade. Most deadly were Charles E. Hazlett's guns, now commanded by Benjamin F. Rittenhouse, atop Little Round Top.

Capt. Basil C. Manly's battery opened fire on these guns, and its battalion commander, Col. Henry C. Cabell, wrote that it was "with effect."[11] Capt. Hugh R. Garden's battery of Mathis W. Henry's battalion also opened fire on Little Round Top, but it had little effect on Rittenhouse's devastation of the attacking Confederate infantry.[12]

Rittenhouse recalled that he opened fire with solid shot and shell as soon as Pickett's men emerged from Seminary Ridge. Then he changed to case shot, then to canister, and finally to double canister as the distance closed. "I watched Pickett's men advance, and opened on them with an oblique fire, and ended with a terrible enfilading fire," he recorded after the war.[13] As the distance lessened, Rittenhouse could only bring his right two pieces to bear on Kemper's brigade, forming Pickett's right flank. However, these guns fired so rapidly that it seemed as though the entire battery was still in action. "Many times a single percussion shell would cut out several files, and then explode in their ranks; several times almost a company would disappear, as the shell would rip from right to the left among them," Rittenhouse recalled.[14] It was almost impossible for any shell to miss its target at this point.

Pickett's men apparently initially did not see McGilvery's thirty-nine guns to their right. Because he had followed Hunt's orders not to respond to the Confederate barrage, McGilvery's men were fresh, and their ammunition chests were amply supplied with long-range munitions when Pickett's men stepped off.[15] "On they came whooping and yelling on double quick time their artillery playing on us all the time. When within 300 yards of us we got the word to fire. . . . We mowed them down. . . . They could not advance under our galling fire and began to recoil," noted a cannoneer in James Thompson's battery.[16] It was a surprise that attackers do not relish. James H. Cooper's beaten-up 1st Corps battery joined McGilvery's line at this time. The unit galloped into position amid the firestorm and took position between W. D. Rank's and John W. Sterling's batteries.[17] This was the type of situation that cannoneers could only wish for and infantry dread. McGilvery reported, "By training the whole line of guns obliquely to the right, we had a raking fire through all three of these lines." He further noted that the "execution of the fire must have been terrible, as it was over a level plain, and the effect was plain to be seen. In a few minutes, instead of a well-ordered line of battle, there were broken and confused masses, and fugitives fleeing in every direction."[18]

Capt. Patrick Hart noted that while his gunners briefly used solid shot at the Confederate artillery during the cannonade, they switched to shell and shrapnel when Pickett's men appeared. They then changed to deadly canister when the Virginians were within five hundred yards. "His [Pickett's] second line appeared to be coming direct for my battery. I turned all my guns on this line, every piece loaded with two canisters. I continued this dreadful fire on this line until there was not a man of them to be seen," wrote Hart.[19] Capt. Charles A. Phillips believed that the results were even more decisive than at the battle of Malvern Hill, when wave after wave of Confederate infantry attacked Federal artillery on the heights the year before, and Robert E. Lee halted Maj. Gen. George B. McClellan's Peninsula campaign against Richmond.[20]

As Pickett's men approached Emmitsburg Road, they were ordered to make a difficult "left oblique" to connect with Pettigrew's division on their left. While this took them out of the range of most of McGilvery's guns, it exposed their right flank to Robert H. Fitzhugh's, Jabez Daniel's, Augustin N. Parsons's, and Evan Thomas's batteries on Cemetery Ridge to the left (south) of the copse of trees. Parsons's battery opened fire with case shot, firing about 120 rounds into the enemy's flank "with good effect"[21]

Having expended all of their long-range ammunition during the artillery duel, Hazard's 2nd Corps batteries could only watch Pickett's men advance slowly toward them.[22] These batteries were beaten up and should have been pulled out and replaced with fresh units. Newly appointed battery commander James M. Rorty was dead, and his second in command, Lt. Albert S. Sheldon, was severely wounded. Lt. Robert E. Rogers assumed command of the all-but-destroyed battery that had only two guns ready for action. If it were not for nearby infantry, the guns would not have been in a position to fire again.[23]

Hunt later expressed his chagrin with these batteries: "For I never saw a finer opportunity to display the power of the arm, Hazard's guns were silent and the heavy cross fire relied upon to drive the enemy back, or throw his troops in disorder and so deliver them a comparatively easy prey to our infantry, was not obtained."[24]

The men of Capt. Andrew Cowan's battery, to the right of Fitzhugh's, saw an officer approaching and heard the order, "Cease firing, hold your fire for the infantry."[25] Cowan did not know what that meant, but he followed the order. When the smoke cleared, he finally understood. Each gun had fired about forty-five rounds at the Confederate positions on Seminary Ridge.[26] Looking up, Cowan saw another officer, who said, "Report to General Webb on your right."[27] This caused Cowan to hesitate, for his battery belonged to the 6th Corps and had been assigned to the 1st Corps. Now it was being ordered to support the 2nd Corps. Not knowing the officer, Cowan pondered his options. Gazing toward the copse of trees, Cowan saw an officer desperately waving his hat at him. This was Gen. Alexander S. Webb. Seeing T. Frederick Brown's battery withdrawing after being badly battered and almost out of ammunition, and glancing at the direction of the Confederate attack, Cowan made a quick decision.[28]

"Limber to the right, forward," he yelled and, the teams galloped toward the copse of trees.[29] His leading gun raced past the trees and dropped trail behind Alonzo H. Cushing's guns. Cowan deployed the remaining five guns in Brown's old position to the left (south) of the copse of trees. As the men went about preparing their guns, Cowan glanced toward Seminary Ridge and saw the Confederates stepping off.[30]

Realizing that he needed increased firepower, Hunt sprang into action, ordering several additional batteries to the center of the Federal line. Lt. William Wheeler's 13th New York Battery, which had seen extensive service on the July 1, galloped into position to the left of the copse of trees, taking position to the left of Rorty's beaten-up battery. Kemper's brigade of Pickett's division was about four hundred yards away and closing fast when Wheeler screamed for his men to open an oblique fire. He noted in his report that "this gave me a fine opportunity to enfilade their column with canister, which threw them in great disorder, and brought them to a halt three times."[31]

With the help of nearby infantry, the remaining two guns of Cushing's battery were rolled down to the stone wall just prior to the Confederate advance. The gunners piled canister—all the ammunition the battery had left—near the rear of the second gun. They also brought up additional handspikes. The men expected close-in fighting, and these large pieces of timber would work well under these conditions. The guns opened fire when the enemy came within four hundred yards, and the effect was devastating.[32] Christopher Smith recalled that Garnett's men "bowed their heads as men do in walking against a hail storm."[33] He estimated that the initial shots opened a fifty-foot-wide swath in the Confederates' ranks.[34]

The men of the 69th Pennsylvania of the Philadelphia Brigade behind the stone wall were not enthusiastic about Cushing's repositioning of his guns. Lt. Anthony McDermott wrote, "These pieces done more harm in that position to us than they did the enemy, as they only fired two or three rounds when their ammunition gave out, and one of those rounds blew the heads off two privates of the company, who were on one knee, at the time, besides these pieces drew upon us more than our share of fire from the battery that followed Pickett from the woods opposite to us."[35]

As Pickett's men approached the two fences along Emmitsburg Road, Federal infantry and 2nd Corps artillery finally opened fire. One Confederate soldier wrote, "A new storm of missiles—canister, shrapnel, and rifle shot belched forth from the Federal position, and even more men fell."[36] Whether a soldier survived the ordeal at Emmitsburg Road was a matter of timing and luck. He had to "climb up to the top of the fence, tumble over it, and fall flat into the bed of the road. All the while the bullets continued to bury themselves into the bodies of the victims and the sturdy chestnut rails."[37]

Watching the devastating effects of the Federal fire on the Confederate infantry at Emmitsburg Road, Hunt wrote that it "occasioned disorder, but still they advanced gallantly until they reached the stone wall behind which our troops lay. Here ensued a desperate conflict."[38] Charles Hazard wrote that the canister "was thrown with terrible effect into their ranks."[39] The effectiveness of the guns was directly related to the number of experienced gunners manning the piece. Those that had infantry serving the guns, or were short-staffed, had a noticeable reduction in the rate of firing. This was especially true of James M. Rorty's two guns, now under Robert E. Rogers, which slowly fired canister as Kemper's men approached.[40]

Cowan's guns, which opened fire when Pickett's men were within two hundred yards, blew apart the Confederate ranks. Cowan wrote soon after the battle that the "effect was greater than I could have anticipated."[41] He later elaborated on his observations: "They came on in splendid order, closing on their left as the shot and shell ploughed gaps though their ranks, and keeping their regular formation. . . . It was a wonderful sight!"[42] Hunt appeared next to Cowan's guns and excitedly yelled, "See 'em! See 'em!" as he emptied his pistol into the rapidly approaching line of battle. Just then, Hunt's horse went crashing to the ground, throwing its rider off in the process. Cowan found a mount for Hunt, who yelled above the din as he rode away, "Look out, or you will kill your own men," referring to the Federal infanry at the wall.[43] Kemper's infantry opened fire as they approached the stone wall, felling several gunners.

As the Confederates continued their approach, Cushing was almost immediately wounded in the shoulder and then in the groin. Although his wounds were intensely painful, he refused to be taken to the rear. "I will stay right here and fight it out or die in the attempt," he responded to a subordinate who suggested he head to the rear for aid.[44] As the Confederate infantry closed to within two hundred yards, the battery changed to double canister, causing "immense gaps" in their lines. The Number 3 man of the guns went down, and Cushing replaced him. Wearing a leather thumbstall, this man covered the vent hole with his thumb to prevent premature firing when the gun was loaded. Cushing did not have a thumbstall, so he was probably scalded to the bone as the gun continued firing. When the Confederates were at a hundred yards, Cushing ordered triple canister. He was hit a third time, this time near his mouth, killing him instantly.[45] The guns recoiled so violently with their massive loads that they almost flipped over.[46]

Still the Confederates surged forward. Sgt. Frederick Fuger recalled, "Owing to the dense smoke I could not see very far to the front, but to my utter astonishment I saw General Armistead leap over the stone fence with quite a number of his men (landing right in the midst of our battery), but my devoted cannoneers and drivers stood their ground, fighting hand-to-hand with pistols, sabers, handspikes and rammers." Many gallant cannoneers went down. Seeing the futility of further resistance, Fuger, now in command of the battery, yelled, "Men, run for your lives."[47]

The Confederates now began their ascent up the ridge. They had not gone far when brigade commander Armistead fell mortally wounded beside a wheel of one of Cushing's cannon, and soon after, the 72nd Pennsylvania of the Philadelphia Brigade and other units counterattacked and drove the enemy from Cushing's pieces. Fuger was rewarded for his heroics with a Medal of Honor.[48]

Over to the left, Cowan's guns fell silent as he felt he could not open fire on the rapidly approaching enemy as long as the infantry occupied the wall in front of him. Some of the Federal infantry, probably from the 59th New York (Norman J. Hall's brigade) suddenly began breaking for the rear. This infuriated some of the cannoneers. According to Cowan, "One was a captain, with his sword tucked under his arm, running like a turkey."[49] Cpl. James Plunkett was so incensed by what he considered to be cowardly behavior that he hit some of the infantry with his fists as they sprinted past. Picking up a large tin coffeepot, he smashed it over the head of a private. The bottom broke, causing it to slide down and over the unfortunate man's face.[50] "I can still see that fellow running with the tin pot well down over his ears," Cowan wrote after the war.[51]

With his front uncovered, Cowan's guns were able to open fire with double canister. After firing about three rounds, a group of Confederates who had sought shelter behind a slight elevation covered with bushes suddenly jumped to their feet and rushed for the guns. A young Confederate major leaped over the wall and yelled, "Take the gun!" Cowan yelled, "Fire!" and 220 lead balls plowed into the last semblance of organized attackers. When the smoke cleared, the enemy troops were gone. The men later buried the valiant young Confederate officer with honors.[52]

Out of ammunition, Cowan ordered his men to pull the guns to the crest of the ridge, where they opened again using percussion shells. By this time, whatever enemy soldiers who had not been hit or captured were now making their way back to Seminary Ridge, so his guns opened fire on the Confederate batteries that had advanced behind the infantry.[53]

As Kemper's men approached the stone wall, Rorty's two guns opened with double canister, blowing a hole in the line. One gun, loaded with a triple charge, flipped over when fired, crushing one of the gunners. Several Confederate soldiers were able to breech the line to the right (north) of Rorty's single gun and quickly approached the Parrott. The gun was quickly wheeled around, and it fired at least one load of canister before the Confederates threw themselves on the guns. They were met by the infantry serving the guns and cannoneers armed with handspikes and rammers. At least one Confederate soldier was killed with a handspike.[54] The fighting was short and intense, but the Union gunners were forced to either flee or surrender.

Jim Decker of Rorty's battery wrote to his sister three days after the fight, "They suffered terribly getting to us but they marched up as tigers keeping perfect line until they could almost put their hands on our guns but few of them lived to get back."[55]

The guns did not remain in the Confederates' possession for long, as Federal infantry converged on the area, forcing the enemy soldiers to either surrender or make their way back to Seminary Ridge.[56]

Pettigrew's and Trimble's Charge

While Pickett's division marched toward the copse of trees, six brigades in two lines advanced against the Federal line to the north. The Federal defense in this sector included Brig. Gen. Alexander Hays's division and William A. Arnold's and George A. Woodruff's batteries of the 2nd Corps, as well as the batteries on Cemetery Hill. As soon as the infantry broke from the cover of Seminary Ridge, the rifled guns on Cemetery Hill opened fire. "From the very first minute our guns created sad havoc in that line," Maj. Thomas W. Osborn wrote.[57] His guns initially used two types of ammunition, particularly on John M. Brockenbrough's brigade on the left of the line. Osborn reported that the "artillerymen endeavored to roll the solid shot through the ranks and explode the percussion shells in front of the line."[58]

It appears that the Confederate 2nd Corps artillery continued firing on the Federal artillery on Cemetery Hill, apparently hoping to divert some of the killing firepower away from their advancing infantry.[59] It would have worked—had Osborn permitted the gunners to do as they pleased. He later wrote:

> Their [Confederate] fire was exceedingly harassing and did us much damage. Several times my men swung their guns around to answer the fire of the enemy's batteries, which were annoying them so severely. I was often compelled to order them to turn their fire back upon the enemy's infantry, where their work would tell most effectively in deciding the fate of the day. It was hard for my men to have this fire concentrated on them from half the enemy's line, while they were not permitted to reply . . . but [I] ordered [them] to direct their fire upon a body of troops in no way annoying them at the time.[60]

The shells initially flew harmlessly over the line, causing William Peel of the 11th Mississippi (Joseph R. Davis's brigade) to write that it was "a storm of screaming, howling shells, across the field, that burst & tore the timber behind us in frightful manner."[61] The artillery soon found the advancing line of battle, and "we were met by a heavy fire of grape, canister, and shell, which told sadly upon our ranks," wrote General Davis.[62] Shells typically took out four or five men and often more. To William Peel, the sound of the artillery became louder than an ongoing thunderclap. "Shells, screaming & bursting around us, scattered their fragments & projectiles in every direction," he wrote.[63] The officers yelled, "Steady boys" or "Don't break yourself down by running." Davis noted that "under this destructive fire, which commanded our front and left with fatal effect, the troops displayed great coolness, were well in hand, and moved steadily forward, regularly closing up the gaps made in their ranks."[64]

The Federal troops were mesmerized by the Confederate charge. "I think the grandest sight I ever witnessed unfolded itself to our view, as the different lines came marching toward us, their bayonets glistening in the sun, from right to left, as far as the eye could reach; but on they come, their officers mounted, riding up and down their lines, apparently keeping them in proper formation. The lines looked to be as straight as a line could be, and at an equal distance apart," noted Capt. Az. Stratton of the 12th New Jersey (Thomas A. Smyth's brigade, Alexander Hays's division, 2nd Corps).[65] Maj. Theodore Ellis of the 14th Connecticut, also of Smyth's brigade, agreed, writing in his official report that "the spectacle was magnificent. They advanced in perfect order, the line of skirmishers firing."[66] Charles Page of the same regiment noted "gay war flags fluttering in the gentle summer breeze, while their sabers and bayonets flashed. . . . The advance seems as resistless as the incoming tide."[67]

"From the southwest, west, north, and northeast poured the missiles of death . . . the air all over the wide field was fierce and heavy with the iron hail. The greater portion of the field was swept by the fiery shot. Horses and men dropped, crushed and dead," wrote Ezra Simons of the 125th New York (George L. Willard's brigade, Alexander Hays's division).[68] The long straight lines advancing across the open, almost flat terrain were an artillerist's dream. When the line had progressed about a third of the distance to Cemetery Ridge, Osborn recalled that the Confederates "halted and closed their ranks from the right and left on the center and dressed their lines, which were materially shortened. This was done under a fearful artillery fire which was cutting them down by the hundreds every minute. They then moved forward as before, but the nearer they approached the more severe was their loss from our guns and the more seriously were the lines thinned. Still there was no hesitation or irregularity in the movement."[69]

John M. Brockenbrough's Virginia brigade on the far left of the Confederate line sustained the brunt of Osborn's pounding. It is difficult to imagine what it felt like to march toward a waiting line of Federal infantry, but the fear was compounded by the projectiles raining down on them, periodically taking out scores of men. Osborn wrote that the "havoc produced upon their ranks was truly surprising."[70] He noted after the war that while a round shot would take out about two men, shells exploding in front of the line "cut out four, six, eight, or even more men, making a wide gap in their line."[71] Capt. Frederick M. Edgell wrote that his battery's oblique fire had a "destructive effect."[72] Lt. Col. Franklin Sawyer of the 8th Ohio (Samuel S. Carroll's brigade, Alexander Hays's division) was much more graphic about the Virginians' destruction: "Arms, heads, blankets, guns, and knapsacks were thrown and tossed into the clear air. Their track, as they advanced, was strewn with dead and wounded. A moan went up from the field, distinctly to be heard amid the storm of battle, but on they went, too much enveloped in smoke and dust now to permit us to distinguish their lines or movements, for the mass appeared more like a cloud of moving smoke and dust than a column of troops. Still it advanced amid the now deafening roar of artillery and storm of battle."[73]

Reaching a swale beyond the recently burned Bliss barn, about five-hundred yards from the Federal line on Cemetery Ridge, Brockenbrough's men halted to reform their lines. It was here that many of the men realized that to continue was suicide, and they either remained here or began their retreat toward Seminary Ridge. The remainder, looking more like a skirmish line at this point, continued toward Cemetery Ridge, while the shells again tore their ranks. The 8th Ohio suddenly materialized on Brockenbrough's exposed left flank and fired into it. This, together with the murderous artillery fire, convinced the remainder of the Virginians to return to Seminary Ridge or surrender. Major Osborn did not exaggerate when he wrote in his report that "the fire from the hill was one of the main auxiliaries in breaking the force in this great charge."[74] Davis's brigade now became the left flank of the Confederate thrust and was unmercifully pounded.[75]

Pettigrew's brigade (now under Col. John Marshall) on the opposite side of the division was also being pounded. Capt. Albert Haynes (11th North Carolina) noted that the "storm of lead which now met us is beyond description. Grape and canister intermingled with minies and buckshot. The smoke was dense, and at times I could scarcely distinguish my own men."[76]

The losses mounted with each step as the line of battle approached the Union position. A gunner from William A. Arnold's battery wrote that "solid shot tore through the rebel ranks; shells were bursting under their feet, over their heads and in their faces. Men, or fragments of men, were being thrown in the air every moment, but, closing up the gaps and leaving swaths of dead and dying in their tracks."[77] A Federal infantryman noted that the number of killed and injured "could not be estimated by numbers, but must be measured by yards."[78]

Low on long-range ammunition, George A. Woodruff's battery of six Napoleons could only wait until the Confederate infantry came closer. The gunners in the left section ran their guns closer to the stone wall. When the Confederate infantry was within five hundred yards, Woodruff screamed for his men to open fire. "The slaughter was dreadful," noted Lt. Tully McCrae. "Never was there such a splendid target for Light Artillery."[79] Firing at a rate of two rounds per minute, McCrae noted that the "slaughter was fearful, and great gaps were made in the mass of the enemy upon each discharge."[80] The deadly canister blew apart the fences along Emmitsburg Road, and the bodies of dead and dying Confederates littered those sections of fencing still standing.

Lt. Col. Charles Morgan, the 2nd Corps chief of staff, was most impressed with Woodruff's fire: "I have never seen [it] surpassed, and when to it was added the still more destructive fire of Hays advanced regiments, posted behind the stone-wall, the enemy could not withstand it."[81]

Concerned that the Confederates were gaining a lane to his left (to the south of the Abraham Bryan house), Woodruff ordered Lt. John Egan to move his section in that direction and fire on any enemy in sight. Woodruff was wounded soon after issuing this order. From the nature of the wound, the men knew that it was but a matter of time before he died, which occurred the following day.[82]

Two guns from Lt. John G. Turnbull's battery and another section from Capt. John Bigelow's (now under Lt. Richard S. Milton) galloped up and dropped trail just south of the Bryan house and immediately opened fire on the approaching Confederates. Because Federal infantry occupied the wall in front of him, Turnbull ordered his gunners to use spherical case without fuses so they would not explode prematurely. Instead, the shells acted like solid shot, bounding along the Confederate line, tearing great gaps in their formations.[83]

Lt. Gulian V. Weir's battery, which had been overrun the day before, like Turnbull's and Bigelow's, was ordered north from the artillery reserve park. "Where shall I go in?" Weir yelled to an aide. He was told to "Go right up there [pointing up Taneytown Road], someone will show you where to go in."[84] Seeing a section of Turnbull's battery north of the Bloody Angle, Weir rode to its lieutenant and shouted, "I was sent here, where shall I go in?" The young officer pointed farther up Cemetery Ridge and yelled, "Go in over there, you'll come in on their flank and mow 'em down."[85]

Out of ammunition, Capt. William A. Arnold ordered all but two of his guns back over the top of the ridge to safety, "leaving the dead, some of the wounded, and two caissons behind," noted Theodore Reichardt.[86] They had followed Hancock's orders and responded to the Confederate artillery challenge and had little to show for it except disabled guns, blown-up limbers and canisters, and swarms of dead and wounded men and horses. The movement both disrupted Turnbull's battery and created a seventy-five-yard gap in the Federal line that was partially filled in by infantry. The remaining two cannon were left on the front line to help deal with the enemy, but only one was operational.[87]

Into this maelstrom galloped Weir's battery. Weir noted after the war, "I saw before me a small open plain to our front, our men on either side. There were several guns lying to the left, and open space to my front [Arnold's old position], and beyond this gap, a dense body of the enemy."[88] The gunners immediately came under fire from the infantrymen of Pettigrew's division, which were rapidly approaching the stone wall, and many horses and cannoneers fell. "I opened at once with canister. In a few minutes our infantry charged, and the enemy were driven back," Weir reported.[89] It was a terrible time for even the most experienced cannoneer. Lt. Homer Baldwin wrote home to his father on July 7, 1863, that the "recoil of our guns would send them over the dead and wounded, and flashes of our pieces would scorch and set fire to the clothes of those that lay in front of us."[90] Weir noted, "The artillery fire of the enemy burst all around and about us, but with little effect. The musketry fire was low, and I can distinctly remember hearing the shots striking the bodies of the wounded and dead which lay about me."[91] The battery fired a hundred rounds of canister within a few minutes.[92]

Pettigrew's men were joined by Trimble's, but they too were mowed down and the charge was finally repulsed. The Confederate survivors either surrendered or made their way back to Seminary Ridge. Maj. Thomas W. Osborn did not permit the latter to be uncontested, ordering his batteries to again open fire when knots of men came into view. "While to them this last fire was doubtless very demoralizing, yet I do not think the damage done could have been very great," he wrote after the war.[93]

The Wilcox-Lang Charge

Even though Pickett's, Pettigrew's, and Trimble's divisions were repulsed, two additional brigades under Brig. Gen. Cadmus Wilcox and Col. David Lang were ordered to launch a charge just to the right (south) of Pickett's path. Col. E. Porter Alexander wrote after the war, "Just then, Wilcox's Brigade passed by us, moving to Pickett's support. There was no longer anything to support, and with the keenest pity at the useless waste of life, I saw them advance. The men, as they passed us, looked bewildered, as if they wondered what they were expected to do, or why they were there."[94]

Their route brought them closer to Freeman McGilvery's guns than Pickett's men, and the Union cannoneers took full advantage of the opportunity.[95] Capt. Nelson Ames on the left of McGilvery's line did not wait for orders to open fire, instead yelling to his men, "Now it is your turn, up and give them the best you've got."[96] They used solid shot on Wilcox's men with "disastrous effects." The gaps closed and the line moved on. As many as fifty-nine Federal guns fired on the two small and beleaguered brigades that probably numbered fewer than fifteen hundred men.[97]

Wilcox reported that "all of the enemy's terrible artillery that could bear on them was concentrated upon them from both flanks and directly in front, and more than on the evening previous."[98] Col. Hilary Herbert of the 8th Alabama recounted that the Federal artillery at this time had no other targets than Wilcox's and Lang's brigades: "Shells bursting in the ranks made great gaps in the regt. These at the command of 'guide centre' were closed up as if on drill & we continued forward."[99] James Wentworth of the 5th Florida (Lang's brigade) noted that his men charged "amid a terrible shower of shell grape cannister and bullets. It was the hottest work I ever saw. My men falling all around me with brains blown out, arms off and wounded in every direction."[100]

Battery commander Capt. James H. Cooper wrote, "Into this line this battery, in connection with the adjacent batteries, fired case shot until they reached canister range, when a few charges were fired into them, completely routing them, without any infantry assistance."[101] Cooper was only partially correct. While his guns did dissuade Wilcox's men from continuing their attack, infantry from George J. Stannard's brigade barreled into the left flank of Lang's brigade. Raymond Reid of the 2nd Florida wrote home, "As we raised the Yell they poured a deadly fire of grape and canister upon us. On we rushed. Our men fell fast and thick. At last we were flanked."[102]

Watching this turn of events, Wilcox was confident that the Confederate artillery would quickly disperse this threat. Riding rapidly back to the batteries along Emmitsburg Road, he was horrified to learn that the Southern gunners were out of ammunition. "Not getting any artillery to fire upon the enemy's infantry that were on my left flank, and seeing none of the troops that I was ordered to support, and knowing that my small force could do nothing save to make a useless sacrifice of themselves, I ordered them back," he wrote.[103] Wilcox's brigade lost more than two hundred men during the aborted charge, all probably caused by McGilvery's unforgiving guns. Cooper estimated that his battery fired about 150 rounds during this brief engagement.[104]

Once again, Lt. Col. Freeman McGilvery's actions had helped to win the battle. Lt. Edwin B. Dow wrote in his after-battle report of McGilvery, "He was ever present, riding up and down the line in the thickest of fire, encouraging the men by his words and dashing example, his horse receiving eight wounds, of which he has since died, the gallant major [lieutenant colonel] himself receiving only a few scratches."[105]

Lee's Artillery During the Charge

Lee's plan for his artillery was not supposed to have ended with the initial cannonade. "The batteries were directed to be pushed forward as the infantry progressed, protect their flanks, and support their attacks closely," he wrote in his report.[106] Having launched the infantry, E. Porter Alexander now rode from battery to battery to ascertain which had sufficient ammunition to advance. Anywhere from twelve to nineteen guns answered the call of having at least fifteen rounds, and Alexander rode forward with them. The remaining guns were ordered to continue firing over the heads of Pickett's men.[107] The advancing guns encountered ground littered with the broken remains of a once proud division. Alexander ordered Henry A. Battles's section of the 4th Company of the Washington Artillery and C. H. C. Brown's section of the 1st Company to report to Maj. James Dearing to the right, where they were ordered to support Pickett's flank.[108]

With the charge at high tide, Alexander's cannon pounded the Federal reinforcements rushing toward the copse of trees to end Pickett's threat to the center of the line. According to Edmund Rice of the 19th Massachusetts, "A confederate battery near the Peach Orchard commenced firing. A cannon-shot tore a horrible passage through the dense crowd of men in blue, who were gathering outside the trees. Instantly another shot followed and fairly cut a road through the mass."[109]

Seeing Stannard's brigade attacking Pickett's vulnerable right flank, Alexander ordered the guns to drop trail and open fire when the Federals were near the Joseph Sherfy barn.[110] Lt. George Benedict heard a thud and then terrifying screams from the attacking 13th Vermont of Stannard's brigade. Looking around, he saw a gaping hole in the line and three prone men. The shell had not exploded, but when it did, two more men went down.[111] After Alexander's artillery scattered some of Stannard's men, they unmasked several of McGilvery's batteries, which immediately opened fire. The Confederates returned the fire, but many shots flew high over the Federal guns, creating havoc among the caissons in the rear. The Federal gunners' aim was much better, and "in a very few minutes these guns had disabled several of mine, killing and wounding quite a number of men and horses," noted Alexander. Despite the Federal losses, Alexander's guns were not able to blunt Stannard's charge.[112]

At least Alexander's men had tried. On the opposite side of the attacking line, the 8th Ohio attacked Col. John M. Brockenbrough's brigade and forced it to the rear. There was no assistance provided by R. Lindsay Walker's 3rd Corps artillery because they had exhausted their ammunition, leaving the infantry vulnerable.[113] Lee reported that because the 3rd Corps artillery remained stationary, "The enemy was enabled to throw a strong force of infantry against our left, already wavering under a concentrated fire of artillery from the ridge in front, and from Cemetery Hill, on the left."[114]

The slow but steady depletion of ammunition was a growing concern of the battery commanders. James Dearing's ammunition was almost completely exhausted, and what was left was used on Stannard's brigade. He sent aides back to find additional ammunition, but none was forthcoming, so his men stood idly by their guns for about an hour. Realizing the futility of the situation, Dearing finally pulled his guns back to Seminary Ridge. Capt. Osmond B. Taylor's battery of Alexander's (Huger's) battalion also pulled back for the same reason. [115]

Five guns from Mathis W. Henry's battalion, three Napoleons from Alexander C. Latham's battery, and a section of Parrotts from Hugh R. Garden's battery also moved forward to support Pickett's beleaguered division. Dropping trail near the Peter Rogers house, the five guns opened fire on McGilvery's line. It did not take long for Patrick Hart's, Charles Phillip's, and James Thompson's batteries, making up part of McGilvery's line, to blast these exposed Confederate artillery pieces, and the effect was devastating. Phillips wrote that all of the batteries focused on these guns, which was done with such good effect that the Rebel cannoneers were driven from their posts "almost immediately."[116] Lt. Benjamin F. Rittenhouse on Little Round Top said to his gunners, "Now boys, turn your pieces to the front and blow them out of there."[117] The area around the Confederate guns became a whirlwind of shells, flying debris, and clouds of dust. Limbers exploded, and men and horses were killed and wounded in rapid succession. Both of Henry A. Battles's guns were disabled, and C. H. C. Brown's section of Merrit B. Miller's battery was silenced and its commander severely wounded.[118]

One gun in Garden's battery under Lt. William McQueen was pounded into submission by at least eight Federal batteries, a total of more than forty guns. According to Capt. Hugh R. Garden, the ground around the gun was "ploughed and torn and great clouds of dirt and debris thrown up everywhere; both earth and air were rended by the detonations and mighty rush of iron hail intermingled with the pelting of leaden bullets; man after man went down."[119] Every horse and man were hit, and the survivors were told to lie down. Seeing these events unfold, Garden, who was barely out of his teens, collected his courage and a group of volunteers to reclaim the gun. He and his men were driven away at first by explosions of Federal metal all around them, but they were successful in their second attempt. He later wrote, "After two attempts we succeeded, under the same concentrated fire, made more terrible by the explosion of caissons and the fire overhead of our friends in the rear. It was [a] . . . carnival of hell."[120]

Capt. Andrew Cowan reported that after he had repelled the last of the Confederate infantry, he turned his attention to several smoothbore guns that had advanced and opened fire on the area. Targeting a single battery, he wrote that he hit "four of its limbers in rapid succession, driving it from the field."[121]

Confederate battalion commander Col. Henry C. Cabell advanced Edward S. McCarthy and Henry H. Carlton's (now under Lt. C. W. Motes) batteries from their positions on Seminary Ridge as Pickett's broken division retreated. They were joined by a gun from John C. Fraser's battery. Lt. William J. Furlong, who was now commanding the battery, must have questioned the order; his limber contained only three rounds of canister. Enemy skirmishers harassed the gunners in their exposed positions between the two ridges. Looking behind them, the gunners could see Richard S. Anderson's division far in the rear, about four hundred yards away. All were concerned about the dearth of ammunition and their vulnerable positions. When it seemed that Federal infantry were about to advance, Cabell's gunners opened fire, dampening their ardor. The guns remained here until dark, when they pulled back to a safer position on Seminary Ridge.[122]

Maj. William T. Poague also received orders to prepare to advance when he began seeing knots of Pickett's men returning. Spying Pickett, Poague approached and told him of his orders and his concerns. Pickett looked straight ahead and did not answer. Poague directly asked if he should advance, to which Pickett answered, "I think you had better save your guns."[123] Lee soon appeared and asked Poague about the status of his ammunition. He replied that his chests were about a fourth full, but his howitzers were fully stocked as they did not have the range to participate in the cannonade. "Ah! That's well; we may need them," noted Lee, anticipating a Federal counterattack.[124]

According to historian Earl Hess, the Confederate artillery was ineffectual in supporting Pickett's infantry. He attributed their poor performance at this stage of the battle to gunner exhaustion, scarcity of ammunition, the superiority of the Federal artillery positions, and a poorly constructed tactical plan.[125]

Seeing the broken remnants of Pickett's division returning, Alexander ordered his guns to cease fire. He knew that a Federal counterattack was a distinct possibility, and he wanted to husband his remaining ammunition. It was a critical time for the Confederates as they were in no shape to beat off a determined Federal thrust.[126]

Federal artillery commander Henry J. Hunt realized that a well-planned and well-executed Confederate artillery barrage could have had the necessary effect in causing the infantry to find success in their charge. It did not, as Hunt wrote in his report: "Their artillery fire was too much dispersed, and failed to produce the intended effect."[127] After the war, he expanded his thoughts, writing that "most of the enemy's projectiles passed overhead, the effect being to sweep all the open ground in our rear, which was of little benefit to the Confederates—a mere waste of ammunition, for everything here could seek shelter."[128]

Lee took responsibility for the charge's failure. He expected his massive artillery to silence the enemy's at the point of attack. His report noted that the Federal batteries opened again as soon as the Confederate infantry advanced. "Our [artillery] having nearly exhausted their ammunition in the protracted cannonade . . . were unable to reply, or render the necessary support to the attacking party. Owing to this fact, which was unknown to me when the assault took place, the enemy was enabled to throw a strong force of infantry against our left, already wavering under a concentrated fire of artillery from the ridge in front, and from Cemetery Hill on the left."[129]

For the attack to succeed, it also required coordination between the various units involved: Pickett's, Pettigrew's, and Trimble's divisions as well as the different corps artilleries' with each other and with the infantry. None of this occurred, primarily because of Lee's hands-off approach. E. Porter Alexander bluntly wrote, "It must be remembered that the preparations for this charge were made deliberately, & under the observation of Gen. Lee himself, & of all of his staff. . . . If there was one thing they might be supposed to take an interest in, it would be in seeing that the troops which were to support the charge were in position to do it."[130] James Longstreet concurred, stating that Lee "gave no orders or suggestions after his early designation of the point for which the column should march."[131] After issuing vague orders to his commanders, Lee essentially left them to their own devices. Most troublesome was the apparent uncertainty about whether the artillery was primarily to break the Federal line prior to the charge or to closely follow the charge, providing needed support. To accomplish the latter, the cannoneers had to conserve their ammunition during the former.[132]

11

JULY 3

CAVALRY ACTIONS AND AFTERMATH

THE CAVALRY FIGHT NORTHEAST OF GETTYSBURG

While both armies' attention was riveted on the Pickett-Pettigrew-Trimble Charge in the afternoon, two other actions involving artillery bracketed the attack. Both involved cavalry, and both merit but a footnote for most historians of the campaign.

The first action, northeast of Gettysburg, involved the remnants of Jeb Stuart's four cavalry brigades—about three thousand men—attempting to slide around Brig. Gen. David Gregg's division, plus Brig. Gen. George A. Custer's brigade from H. Judson Kilpatrick's division. The Federal horsemen equaled the Southerners in numbers. Any surprise Stuart hoped to gain ended when he ordered his cannon to fire four shots to inform Lee that he was in position and ready to attack the Federal rear. According to the plan, the Confederates would send out a strong and aggressive skirmish line from Cress Ridge to preoccupy the Federal horsemen, then they would swing around their flank.[1]

Capt. Thomas Jackson's Charlottesville Horse Battery (two 12-pounder howitzers and two ordnance rifles) dropped trail on Cress Ridge and opened an effective fire on Custer's men, scattering some of the units. Custer quickly responded by ordering up Lt. Alexander Pennington's Battery M, 2nd U.S. (six ordnance rifles) and then Capt. Alanson Randol's Batteries E & G, 1st

U.S. (four ordnance rifles). Captain Randol dismissively wrote, "As a rule their Horse Art'y was so badly handled in battle that we Art'y officers paid little attention to it." But the Federals did pay attention this afternoon, and the superior guns of the Union batteries quickly made their presence felt, accurately dropping shells around Lt. Micajah Wood's section, wounding four cannoneers and killing half the horses. Jackson's guns were unable to reach Pennington's, so it was a mismatch, but the Confederates held their ground until shooting off all their ammunition at the Federal horsemen.[2]

Two sections of Pennington's battery deployed just to the northeast of the Abraham Reever house, north of Hanover Road. The third section dropped trail to the southeast, on the opposite side of Hanover Road, near the J. Spangler house. Randol's two sections also fought separately, about five hundred yards apart, both deployed north of the G. Howard house, just west of Low Dutch Road and north of Hanover Road.[3]

The Federal artillery also pounded Stuart's troopers, forcing them to change position from time to time. Gregg noted that "never was there more accurate and effective fire delivered by Artillery than by the guns of Randol and Pennington."[4]

Capt. Charles A. Green's Louisiana Guard Artillery (two ordnance rifles and two 10-pounder Parrotts) arrived and dropped trail at Jackson's exposed position on the ridge, and it too was raked by the eight Federal guns. Green's gunners also ignored the Federal pounding, concentrating instead on the aggressive Union cavalry. Additional Confederate batteries arrived, including Capt. James Breathed's Virginia Battery (four ordnance rifles). Breathed also focused on the Federal cavalry charging against their Confederate counterparts in the open fields, and the shells devastated the ranks. When the horsemen engaged in close contact, the Confederate guns fell silent. During one of the latter charges of the day, the Federal cannon wreaked havoc on Gen. Wade Hampton's troops as they galloped across the open plain until they closed with the enemy horsemen.[5]

When Federal cavalry targets were not available, Breathed's cannoneers turned their attention to their old adversary: Pennington's battery. H. H. Matthews of Breathed's battery explained, "The guns became so hot that I was afraid the old 2nd gun would have a premature explosion and blow No. 1 into eternity. My thumb on my left hand became burnt from thumbing the vent so that it became very painful to use it. We stayed with Pennington until dark, expending every round of ammunition. Both batteries

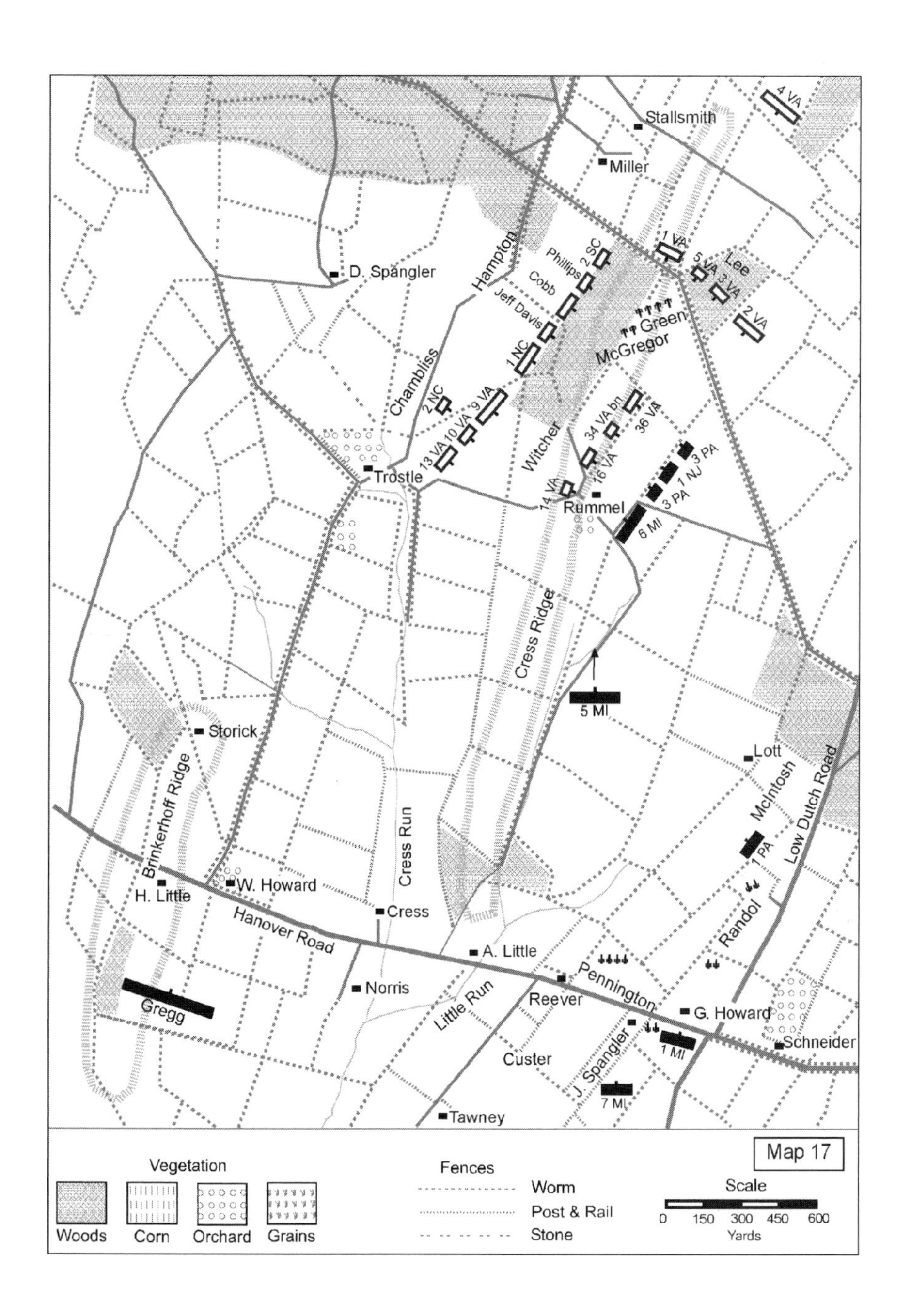
Stallsmith
4 VA
Miller
1 VA
Lee
5 VA
3 VA
2 VA
Phillips
2 SC
Cobb
Jeff Davis
Green
McGregor
D. Spangler
Hampton
1 NC
Chambliss
2 NC
9 VA
10 VA
13 VA
Trostle
Witcher
34 VA bn
36 VA
16 VA
14 VA
Rummel
3 PA
1 NJ
3 PA
6 MI
Cress Ridge
5 MI
Storick
Lott
McIntosh
Low Dutch Road
Brinkerhoff Ridge
Cress Run
1 PA
H. Little
W. Howard
Hanover Road
Cress
A. Little
Randol
Norris
Little Run
Reever
Pennington
G. Howard
Schneider
Gregg
Custer
J. Spangler
1 MI
7 MI
Tawney
Map 17
Vegetation
Woods
Corn
Orchard
Grains
Fences
Worm
Post & Rail
Stone
Scale
0 150 300 450 600
Yards

suffered heavily." Pennington's report was much less expressive, writing that he engaged an enemy battery and silenced it.[6]

The heroics of the Federal cavalry and artillery blunted the Confederate drive toward the Federal rear, and Stuart again failed Lee. The guns were in action for about four hours, expending a large quantity of ammunition. Randol's four guns fired more than 380 rounds; Pennington's six fired more than 240. An appreciative cavalry division commander wrote, "The fire of the artillery during this engagement was the most accurate I have ever witnessed."[7]

At the opposite end of the Federal line, another failed mission was about to be played out.

Actions on the South Field

On the south side of the battlefield Brig. Gen. H. Judson Kilpatrick stewed that his cavalry division had not participated in the defeat of Lee's army. In response to Gen. Alfred Pleasonton's vague orders to press the enemy and attack him at every opportunity, Kilpatrick decided to launch Brig. Gen. Elon Farnsworth's brigade on a charge against the Confederate right, between Emmitsburg Road and the Round Tops. Lt. Samuel Elder's Battery E, 4th U.S. Artillery (four ordnance rifles) dropped trail on Bushman's Knoll at about 1:00 p.m. to support the cavalry in this sector. Parts of three infantry brigades of Evander M. Law's (A. P. Hood's) division were deployed here with several batteries, including William K. Bachman's Palmetto Light Artillery (four Napoleons), Capt. James Reilly's Rowan Artillery (two 10-pounder Parrotts, two Napoleons, and two ordnance Rifles), and Capt. James Hart's Horse Artillery. The latter was badly battered at the June 21, 1863, battle of Upperville, so only two Blakely rifles traveled with the slower-moving infantry. Bachman's battery and four guns of Reilly's battery were deployed along the Bushman farm lane, facing southeast and Farnsworth's brigade. Reilly's remaining section was positioned west of Emmitsburg Road, in support of the 1st Texas. Hart's battery accompanied the 1st South Carolina Cavalry on the east side of Emmitsburg Road.[8]

Brig. Gen. Wesley Merritt's reserve cavalry brigade (John Buford's division), supporting Capt. William Graham's Battery K, 1st U.S., composed of six ordnance rifles, faced Hart's battery along Emmitsburg Road. Although two Federal cavalry brigades occupied this sector, they did not coordinate their actions and ultimately both failed in breaching the Confederate defenses. The same is not true for Graham's battery. After pummeling Hart's battery and its 1st South Carolina Cavalry supports, Graham turned his attention to Reilly's and Bachman's batteries and the portion of the 1st Texas infantry facing Farnsworth's men to the east.[9]

At the same time, a dismounted skirmish line of the 1st Vermont Cavalry inched forward toward the eight guns of Bachman's and Reilly's batteries and then opened on them with their carbines. "We annoyed them some as they favored us with several charges of canister," wrote Sgt. Horace Ide. Seeing the futility of these actions, the Federal skirmish line drifted backward.[10]

The action began with an artillery duel as Elder attempted to develop the enemy's force in the fields south of the Bushman house. The trees on Bushman's Knoll caused Elder's men to elevate the tubes of their guns, resulting in their overshooting the Confederate batteries. Elder fired about 150 rounds to try to knock out the Confederate defenses prior to the subsequent charges. He failed miserably. The trees, however, screened the battery from the Confederates, so losses were minimal: one man slightly wounded and four horses killed.[11]

Spying the depleted 1st Texas occupying a long half-mile front in essentially a skirmish line, Kilpatrick ordered a charge. Farnsworth objected, but Kilpatrick would not reverse his orders, so the 1st West Virginia Cavalry galloped from the Federal line near Bushman Hill toward the Texans. Despite canister blasts from the Confederate cannon, the horsemen charged. The artillery fire created dread in both friend and foe. A Texan wrote, "A fusillade of shot & shell from Rileys Battery passed a couple of feet above our heads." When the Union cavalry was only fifty yards away, the Texans opened fire, blasting the Mountaineers from their mounts. Still the charge continued, and hand-to-hand fighting erupted. A cannoneer from Reilly's battery unseated two horsemen with his rammer. The two detached guns from Reilly's battery quickly rejoined the main unit as the Federal threat deepened.[12]

Battered and beaten, the West Virginians fell back to regroup and lick their wounds. Kilpatrick next spied the 18th Pennsylvania Cavalry and yelled to its commander, "Why in hell and damnation don't you move those troops out?" They charged, but within minutes the beaten Pennsylvanians returned. Watching Reilly's battery's destructiveness, Kilpatrick suggested that Farnsworth's last regiment—the 1st Vermont Cavalry—drive forward to capture it. No amount of protests could dissuade Kilpatrick, so Farnsworth rode

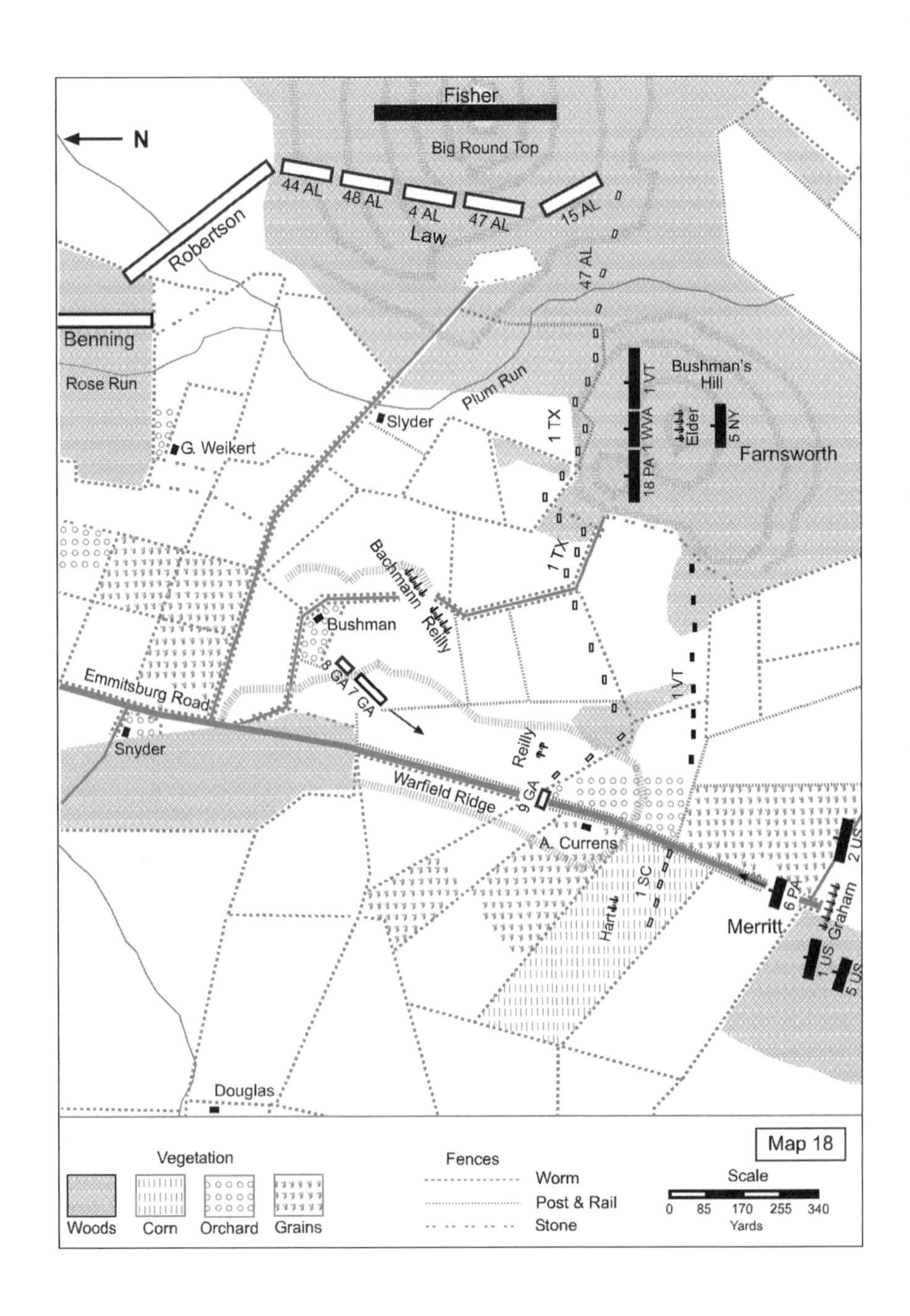
N
Fisher
Big Round Top
44 AL
48 AL
4 AL
47 AL
15 AL
Law
Robertson
47 AL
Benning
Rose Run
Plum Run
Slyder
G. Weikert
1 TX
1 VT
1 WVA
18 PA
Bushman's Hill
Elder
5 NY
Farnsworth
Bachmann
Reilly
1 TX
Bushman
8 GA
7 GA
Emmitsburg Road
Snyder
1 VT
Reilly
9 GA
Warfield Ridge
A. Currens
1 SC
Hart
6 PA
2 US
Graham
Merritt
1 US
5 US
Douglas
Map 18
Vegetation
Woods
Corn
Orchard
Grains
Fences
Worm
Post & Rail
Stone
Scale
0
85
170
255
340
Yards

to the front of the regiment, unsheathed his sword, and ordered his men forward in three separate battalions.[13]

Evander M. Law, in command of this sector, meanwhile had ordered several infantry regiments to the aid of the 1st Texas, and soon Alabamians and Georgians were sprinting in their direction. William K. Bachman's guns shifted their orientation ninety degrees to the east, while James Reilly's remained facing south. Both opened fire on the fast-moving Vermont horseman, bringing down a man now and then. Turning, the horsemen of one battalion rode straight for Reilly's battery, which now shifted to double canister loads. Infantry from Law's old infantry brigade was now on the scene, throwing volleys into the Federal horsemen. While many of the men were able to make their way to safety, Farnsworth was trapped and shot several times, tumbling dead from his saddle.[14]

This ended the last artillery-related action at Gettysburg.

Aftermath

Late in the afternoon of July 3, E. Porter Alexander pulled all of his guns back from their forward locations to reoccupy their July 2 positions on Seminary Ridge.[15] The battle essentially over, the battery commanders tallied their losses. The Federal 2nd Corps batteries were especially hard hit during the cannonade and the charge, losing 25 percent of their men. Three of the five commanders (Alonzo H. Cushing, James M. Rorty, and George A. Woodruff) were killed or mortally wounded, and all of the batteries were pummmeled. William A. Arnold's and T. Frederick Brown's batteries were consolidated, as were Cushing's and Woodruff's, to form two serviceable batteries.[16] In all, five Federal batteries (John Bigelow's, John K. Bucklyn's, Cushing's, James Stewart's, and Malbone F. Watson's) lost more than 25 percent of their effectives. Many of the Confederate batteries had also taken a beating. For example, S. Capers Gilbert's battery lost forty-two of its seventy men, twenty-five horses, two caissons destroyed, and two of the four howitzers were dismounted.[17] Two of Joe Norcom's guns were knocked out of action during the cannonade, as were others from Edward S. McCarthy's and John B. Richardson's batteries.[18] Six Confederate batteries (Thomas A. Brander's, John C. Carpenter's, John C. Fraser's, Richard C. M. Page's, and Pichegru Woolfolk Jr.'s) lost more than 25 percent of their men. At the battalion level, Frank Huger's suffered the highest casualties—24 percent.

Artillery played a major role at Gettysburg. Henry J. Hunt's use of his batteries was far superior to William Pendleton's. Being on the defensive, Hunt was everywhere on the battlefield, skillfully deploying his batteries to maximize the damage to the enemy, but it came at a high cost. The Federal artillery lost 737 men, about 10 percent of its strength. The Confederates lost between 580 and 610, which was approximately the same ratio as the Federal losses. Equally concerning to Hunt was the loss of 881 horses. Federal gunners expended some 32,781 rounds of ammunition during the battle (or 270 rounds per gun), about a third of the available 97,740 rounds. The Confederates fired 22,000 rounds (150 per gun) of their 40,800 total rounds, more than half. The Confederates' lower number of rounds fired per gun probably was related to the relatively large number of artillery pieces, particularly from the 2nd Corps, that were not in action.[19]

While eight Confederate cannon were severely damaged, none were captured. The Federal army lost several guns, particularly on July 2, but most were recaptured. Only seven guns remained in Confederate hands by the end of the battle: one gun from Gilbert H. Reynolds's battery (1st Corps), three guns from James E. Smith's (3rd Corps), two guns from Lewis Heckman's (11th Corps), and one gun from James Thompson's (Reserve).[20]

Ammunition proved to be the bane of the Confederates, and many battery commanders complained bitterly about it. A major complaint was the confusion between 3-inch and 2.9-inch ammunition. Lt. John Gregory, the 2nd Corps chief of ordnance, reported that two guns were put out of service because they used the wrong caliber of ammunition. Gregory also complained that the friction primers were too often defective because of improper filling and because their tops were not properly closed. Much of the ammunition that had been captured could not be used because of an inherrent incompatibility with the Confederate-made fuses. Finally, he complained that the Richmond-made ammunition was packed in "miserably weak boxes" that were always bursting, making the ammunition within unusable.[21]

Few battery commanders left records of the ammunition they expended during the battle, and most were units active on July 1. The data for two batteries using Napoleons are very revealing:[22]

Battery	Solid Shot	Spherical	Case Shell	Canister	Total
Wilkeson-Bancroft	616 (45%)	588 (43%)	158 (11%)	18 (1%)	1380
Stevens-Whittier	384 (39%)	380 (39%)	112 (11%)	103 (11%)	979

Bayard Wilkeson's–Eugene A. Bancroft's battery dropped trail on Blocher's Knoll on July 1, then briefly took position near the town in an attempt to stem the Confederate tide later that afternoon. For the rest of the battle, the battery was deployed in the middle of the Cemetery Hill artillery line, facing north. This battery—which had to contend with a fast-moving Confederate attack on July 1 and counterbattery fire during the rest of the battle—primarily used nonexploding ammunition that is best against cannon, limbers, and caissons. Greenlead T. Stevens's–Edward N. Whittier's battery was positioned on Seminary Ridge on July 1, where it attempted to stave off an attack by W. Dorsey Pender's division's. During the rest of the battle the battery was deployed on what was later called Stevens Knoll, which was between Cemetery and Culp's hills. As a result of engaging Confederate infantry here, the battery shot off a higher percentage of antipersonnel ammunition: canister. Thus, 22 percent of the ammunition fired by this battery consisted of case shell and canister, compared with only 12 percent for Wilkeson's–Bancroft's battery.

Both sides lost significant numbers of horses, and this was a worse hardship on the Confederates as they had great difficulty replacing them. After praising his men and mourning his losses in his official report of the campaign, Col. R. Lindsay Walker of the 3rd Corps devoted considerable space to the want of shoes for his artillery horses. "On the first day I was placed in command of this corps, I applied to the Ordnance Department for horseshoes and nails. I repeated this application. . . . I am satisfied that most of the horses lost on the march were lost . . . because of their lameness in traveling over turnpikes . . . without shoes. . . . The value of horses abandoned from this cause during the march, was . . . $75,000."[23]

The fortunes of war waxed and waned for some of the Federal artillerymen after Gettysburg. Henry J. Hunt—who wrote his wife, "I have never been in so much and so long continued peril as on yesterday [July 3] and the day before"—continued his effective leadership of the Union artillery through the Appomattox campaign. He spent much of his postwar years fretting that he did not receive the credit he deserved for helping to defeat Lee at Gettysburg. He also had a running fight with Winfield S. Hancock about the artillery's role in the great cannonade.[24]

Some of the Federal artillery brigade commanders continued to find success after Gettysburg. Col. Charles S. Wainwright, who performed so ably at Gettysburg, was assigned to head the 5th Corps artillery when the army was reorganized during March 1864. He received the coveted brigadier general star later that year and served through the end of the war. Capt. John G. Hazard continued serving as the 2nd Corps artillery chief and was also promoted to brigadier general in 1864. The same is true of Capt. Charles H. Tompkins, commander of the 6th Corps artillery brigade. Although the brigade did not see much action at Gettysburg, he continued to command the brigade through the remainder of the war, gaining promotion to brigadier general in 1864.[25]

Maj. Thomas W. Osborn of the 11th Corps was not destined to receive a general's star but did see considerable action after Gettysburg. Transferred to the Army of the Tennessee after the battle of Gettysburg with the rest of the 11th and 12th Corps, Osborn commanded his brigade until the two corps were merged. He followed his friend and mentor Maj. Gen. Oliver Howard to the 4th Corps and commanded its artillery. Later he again followed Howard, this time to command the artillery in the Department of the Tennessee. After the war, he worked with Howard at the newly formed Freedmen's Bureau.[26]

Brig. Gen. Robert O. Tyler's post-Gettysburg career took an interesting twist when he was ordered to bring a division of artillery regiments into the field as infantry. He fought his new division at Spotsylvania and at Cold Harbor, where he was wounded. The wound ended his career in the Army of the Potomac.[27]

Other artillery commanders did not fare as well. Such was the case of Capt. George E. Randolph of the 3rd Corps, who resigned his commission and left the army in January 1864, probably because the army's reorganization would leave him without a command.[28]

Lt. Col. Freeman McGilvery would not join in these postwar debates. After receiving much-deserved praise for his performance after the battle, he was promoted to colonel and given command of the ammunition train. He became artillery chief of the Tenth Corps in August 1864, but he was wounded in the hand three days later. Amputation was attempted on September 2, 1864, but he died after receiving an improper dose of chloroform.[29]

It was status quo for many of Lee's artillerists. Although many deemed William Pendleton incompetent, Lee continued defending him, and he ended the war as nominal chief of the Army of Northern Virginia's artillery. Similarly, Col. Henry C. Cabell retained command of his battalion through the war, despite an attempt to transfer him to an artillery battalion defending

Richmond during the late fall of 1863. Although R. Lindsay Walker continued to command the 3rd Corps artillery through the remainder of the war, he was one of the few artillerists to gain the rank of brigadier general. Maj. Benjamin F. Eshleman, commander of the Washington Artillery Battalion, remained with his command through the war, rising to the rank of lieutenant colonel. The same is true of William Nelson, who commanded a 2nd Corps reserve battalion at Gettysburg. He was promoted to the rank of colonel just two months before the end of the war. Maj. William T. Poague of the 3rd Corps also served with his battalion through the rest of the war. He helped save Lee's army from disaster on May 6, 1864, when his battalion stood alone against a relentless Federal attack.[30]

E. Porter Alexander was given permanent command of the 1st Corps artillery and fought with distinction until Appomattox. Through two books, his thoughts about the battle of Gettysburg continue to be heard. Maj. Frank Huger, who assumed temporary command of Alexander's battalion at Gettysburg, received permanent command of the unit and fought with it through the end of the war. He was promoted to colonel about two months prior to Appomattox Court House.[31]

A number of artillery commanders saw their stars continue to rise after Gettysburg. Lt. Col. Hilary P. Jones, commander of the battalion linked to Jubal Early's division, was promoted to colonel in February 1864 and temporarily commanded the 3rd Corps artillery. He commanded two artillery battalions in North Carolina the following May and later led Richard H. Anderson's corps artillery. He surrendered at Appomattox. Col. Thomas H. Carter assumed command of the 2nd Corps artillery in 1864 and led it through the end of the war. Likewise, Maj. David G. McIntosh, commander of a reserve battalion in the 3rd Corps, rose to command its artillery by the end of the war at the rank of colonel.[32]

At the opposite extreme, Lt. Col. John Garnett—who was relieved of his duties as a battalion commander attached to Henry Heth's division—was officially suspended by Lee on February 18, 1864, and on April 1, 1864, he was relieved of all duties associated with the Army of Northern Virginia. Unwilling to leave the military, Garnett commanded an outpost at Hicksford, Virginia, until November 1864, when he was appointed inspector of artillery of the Army of Tennessee.[33]

Col. James B. Walton, who was unceremoniously removed from command of the 1st Corps artillery reserve during the battle, was reassigned as inspector general of field artillery. He resigned before assuming this post, partly because he was outraged that Alexander was made the permanent commander of his old command. Maj. Mathis W. Henry left the army in a happier state of mind. He was promoted to colonel and transferred to the Trans-Mississippi Department in the early months of 1864. This ended the command confusion with Maj. John Cheves Haskell, who assumed sole oversight of the 1st Corps battalion.[34]

Some of Lee's battalion commanders did not see the end of the war. Col. J. Thompson Brown, who as 2nd Corps artillery chief had so much difficulty positioning his artillery at Gettysburg, was killed by a sniper's bullet at the battle of the Wilderness as he positioned his battalions. Promoted to full colonel, William J. "Willie" Pegram fought with his battalion until his death at the battle of Five Forks, less than a couple of weeks before Appomattox. Impatient with his lack of promotion, 1st Corps battalion commander James Dearing sought a cavalry command, receiving it in April 1864 and eventually earning a general's star. Dearing has the distinction of being the last general in Lee's army to die of his wounds (April 23, 1865).[35]

The Confederacy's high tide had come and passed. The battle of Gettysburg was won for several reasons, but among the most important was the magnificent role the Federal artillery played in the engagement.

APPENDIXES

APPENDIX A: Order of Battle, Strengths, and Losses of the Artillery at Gettysburg

APPENDIX B: Characteristics of Civil War Cannon Used at Gettysburg

NOTES

BIBLIOGRAPHY

INDEX

APPENDIX A

Order of Battle, Strengths, and Losses of the Artillery at Gettysburg

Unit	Strength	Losses (k:w:m)	Percent
ARMY OF THE POTOMAC			
1st Corps Artillery Brigade: Col. Charles S. Wainwright	**596**	**106 (9:86:11)**	**16.8**
Maine Light, 2nd Battery (B): Capt. James A. Hall 6 ordnance rifles	117	18 (0:18:0)	15.4
Maine Light, 5th Battery (E): Capt. Greenleaf T. Stevens 6 Napoleons	119	23 (3:13:7)	19.3
1st New York Light, Batteries L & E: Capt. Gilbert H. Reynolds Lt. George Breck 6 ordnance rifles	124	17 (1:15:1)	13.7
1st Pennsylvania Light, Battery B: Capt. James H. Cooper 4 ordnance rifles	106	12 (3:9:0)	11.3
4th U.S., Battery B: Lt. James Stewart 6 Napoleons	123	36 (2:31:3)	29.3
2nd Corps Artillery Brigade: Capt. John G. Hazard	**605**	**149 (27:119:3)**	**24.6**
1st New York Light, Battery B: Capt. James M. Rorty Lt. Albert S. Sheldon Lt. Robert E. Rogers 4 10-pounder Parrotts	117	26 (10:16:0)	22.2
1st Rhode Island, Battery A: Capt. William A. Arnold 6 ordnance rifles	117	32 (3:28:1)	27.4
1st Rhode Island, Battery B: Lt. T. Frederick Brown 6 Napoleons	129	28 (7:19:2)	21.7
1st U.S., Battery I: Lt. George A. Woodruff Lt. Tully McCrea 6 Napoleons	112	25 (1:24:0)	22.3
4th U.S., Battery A: Lt. Alonzo H. Cushing Sgt. Frederick Fuger 6 ordnance rifles	126	38 (6:32:0)	30.2

Unit	Strength	Losses (k:w:m)	Percent
3rd Corps Artillery Brigade: Capt. George E. Randolph Capt. A. Judson Clark	**596**	**106 (8:81:17)**	**17.8**
1st New Jersey Light, 2nd Battery (B): Capt. A. Judson Lt. Robert Sims 6 10-pounder Parrotts	131	20 (1:16:3)	15.3
1st New York, Battery D: Capt. George B. Winslow 6 Napoleons	116	18 (0:10:8)	15.5
New York Light, 4th Battery: Capt. James E. Smith 6 10-pounder Parrotts	126	13 (2:10:1)	10.3
1st Rhode Island Light, Battery E: Lt. John K. Bucklyn Lt. Benjamin Freeborn 6 Napoleons	108	30 (3:26:1)	27.8
4th U.S., Battery K: Lt. Francis W. Seeley Lt. Robert James 6 Napoleons	113	25 (2:19:4)	22.1
5th Corps Artillery Brigade: Capt. Augustus P. Martin	**432**	**43 (8:33;2)**	**10.0**
Massachusetts Light, 3rd Battery (C): Lt. Aaron F. Walcott 6 Napoleons	115	6 (0:6:0)	5.2
1st New York Light, Battery C: Capt. Almont Barnes 4 ordnance rifles	62	0	0
1st Ohio Light, Battery L: Capt. Frank C. Gibbs 6 Napoleons	113	2 (0:2:0)	1.8
5th U.S., Battery D: Lt. Charles E. Hazlett Lt. Benjamin F. Rittenhouse 6 10-pounder Parrotts	68	13 (7:6:0)	19.1
5th U.S., Battery I: Lt. Malbone F. Watson Lt. Charles C. MacConnell 4 ordnance rifles	71	22 (1:19:2)	31.0
6th Corps Artillery Brigade: Col. Charles H. Tompkins	**937**	**12 (4:8:0)**	**1.3**
Massachusetts Light, 1st Battery (A): Capt. William H. McCartney 6 Napoleons	135	0	0
New York Light, 1st Battery: Capt. Andrew Cowan 6 Ordnance Rifles	103	12 (4:8:0)	11.6
New York Light, 3rd Battery: Capt. William A. Harn 6 10-pounder Parrotts	111	0	0
1st Rhode Island Light, Battery C: Capt. Richard Waterman 6 ordnance rifles	116	0	0

Unit	Strength	Losses (k:w:m)	Percent
1st Rhode Island Light, Battery G: Capt. George A. Adams 6 10-pounder Parrotts	126	0	0
2nd U.S., Battery D: Lt. Edward B. Williston 6 Napoleons	126	0	0
2nd U.S., Battery G: Lt. John H. Butler 6 Napoleons	101	0	0
5th U.S., Battery F: Lt. Leonard Martin 6 10-pounder Parrotts	116	0	0
11th Corps Artillery Brigade: Maj. Thomas W. Osborn	**604**	**69 (7:53:9)**	**11.4**
1st New York Light, Battery I: Capt. Michael Weidrich 6 ordnance rifles	141	13 (3:10:0)	9.2
New York Light, 13th Battery: Lt. William Wheeler 4 ordnance rifles	110	11 (0:8:3)	10.0
1st Ohio Light, Battery I: Capt. Hubert Dilger 6 Napoleons	127	13 (0:13:0)	10.2
1st Ohio Light, Battery K: Capt. Lewis Heckman 4 Napoleons	110	15 (2:11:2)	13.6
4th U.S., Battery G: Lt. Bayard Wilkeson Lt. Eugene A. Bancroft 6 Napoleons	115	17 (2:11:4)	14.8
12th Corps Artillery Brigade: Lt. Edward D. Muhlenberg	**391**	**9 (0:9:0)**	**2.3**
1st New York Light, Battery M: Lt. Charles E. Winegar 4 10-pounder Parrotts	90	0	0
Pennsylvania Light, Battery E: Lt. Charles A. Atwell 6 10-pounder Parrotts	139	3 (0:3:0)	2.2
4th U.S., Battery F: Lt. Sylvanus T. Rugg 6 Napoleons	89	1 (0:1:0)	1.1
5th U.S., Battery K: Lt. David H. Kinzie 4 Napoleons	72	5 (0:5:0)	6.9
CAVALRY CORPS HORSE ARTILLERY			
1st Brigade: Capt. James M. Robertson	**493**	**8 (2:6:0)**	**1.6**
9th Michigan Battery: Capt. Jabez J. Daniels 6 ordnance rifles	111	5 (1:4:0)	4.5
6th New York Battery: Capt. Joseph W. Martin 6 ordnance rifles	103	1 (0:1:0)	.9
2nd U.S., Batteries B & L: Lt. Edward Heaton 6 ordnance rifles	99	0	0

Unit	Strength	Losses (k:w:m)	Percent
2nd U.S., Battery M: Lt. A. C. M. Pennington Jr. 6 ordnance rifles	117	1 (0:1:0)	.9
4th U.S., Battery E: Lt. Samuel S. Elder 4 ordnance rifles	61	1 (1:0:0)	1.6
2nd Brigade: Capt. John C. Tidball	**276**	**15 (2:13:0)**	**5.4**
1st U.S., Batteries E & G: Capt. Alanson M. Randol 4 ordnance rifles	85	0	0
1st U.S., Battery K: Capt. William M. Graham 6 ordnance rifles	114	3 (2:1:0)	2.6
2nd U.S., Battery A: Lt. John H. Calef 6 ordnance rifles	75	12 (0:12:0)	16.0

ARTILLERY RESERVE: BRIG. GEN. ROBERT O. TYLER

Capt. James M. Robertson

Unit	Strength	Losses (k:w:m)	Percent
1st Brigade (Regular): Capt. Dunbar R. Ransom	**445**	**68 (13:53:2)**	**15.3**
1st U.S., Battery H: Lt. Chandler P. Eakin Lt. Philip D. Mason 6 Napoleons	129	10 (1:8:1)	7.8
3rd U.S., Batteries F & K: Lt. John G. Turnbull 6 Napoleons	115	24 (9:14:1)	20.9
4th U.S., Battery C: Lt. Evan Thomas 6 Napoleons	95	18 (1:17:0)	18.9
5th U.S., Battery C: Lt. Gulian V. Weir 6 Napoleons	104	16 (2:14:0)	15.4
1st Volunteer Brigade: Lt. Col. Freeman McGilvery	**385**	**93 (17:71:5)**	**24.2**
Massachusetts Light, 5th Battery E: Capt. Charles A. Phillips 6 ordnance rifles	104	21 (4:17:0)	20.2
Massachusetts Light, 9th Battery: Capt. John Bigelow 6 Napoleons	104	28 (8:18:2)	26.9
New York Light, 15th Battery: Capt. Patrick Hart 4 Napoleons	70	16 (3:13:0)	22.9
Pennsylvania Light, Batteries C & F: Capt. James Thompson 6 ordnance rifles	105	28 (2:23:3)	26.7
2nd Volunteer Brigade: Capt. Elijah D. Taft	**241**	**8 (1:5:2)**	**3.3**
Connecticut Light, 2nd Battery: Capt. John W. Sterling 4 James rifles; 2 12-pounder howitzers	93	5 (0:3:2)	5.4

Unit	Strength	Losses (k:w:m)	Percent
New York Light, 5th Battery: Capt. Elijah D. Taft 6 20-pounder Parrotts	146	3 (1:2:0)	2.1
3rd Volunteer Brigade: Capt. James F. Huntington	**431**	**37 (10:24:3)**	**8.6**
New Hampshire Light, 1st Battery: Capt. Frederick M. Edgell 4 ordnance rifles	86	3 (0:3:0)	3.5
1st Ohio Light, Battery H: Lt. George W. Norton 6 ordnance rifles	99	7 (2:5:0)	7.1
1st Pennsylvania Light, Batteries F & G: Capt. R. Bruce Ricketts 6 ordnance rifles	144	23 (6:14:3)	16.0
West Virginia Light, Battery C: Capt. Wallace Hill 4 10-pounder Parrotts	100	4 (2:2:0)	4.0
4th Volunteer Brigade: Capt. Robert H. Fitzhugh	**499**	**36 (2:34:0)**	**7.2**
Maine Light, 6th Battery (F): Lt. Edwin B. Dow 4 Napoleons	87	13 (0:13:0)	14.9
Maryland Light, Battery A: Capt. James H. Rigby 6 ordnance rifles	106	0	0
1st New Jersey Light, Battery A: Lt. Augustin N. Parsons 6 10-pounder Parrotts	98	9 (2:7:0)	9.2
1st New York Light, Battery G: Capt. Nelson Ames 6 Napoleons	84	7 (0:7:0)	8.3
1st New York Light, Battery K/11th New York Battery: Capt. Robert H. Fitzhugh 6 ordnance rifles	122	7 (0:7:0)	5.7

ARMY OF NORTHERN VIRGINIA

1ST CORPS

Unit	Strength	Losses (k:w:m)	Percent
McLaws's Division Artillery: Col. Henry Coalter Cabell	**378**	**52 (15:37:0)**	**13.8**
1st North Carolina Artillery Battery A: Capt. Basil C. Manly 2 12-pounder howitzers; 2 10-pounder Parrotts	131	13 (3:10:0)	9.9
Pulaski (Georgia) Artillery: Capt. John C. Fraser Lt. William J. Furlong 2 ordnance rifles; 2 10-pounder Parrotts	63	19 (7:12:0)	30.2
1st Richmond Howitzers: Capt. Edward S. McCarthy 2 Napoleons; 2 ordnance rifles	90	13 (3:10:0)	14.4
Troup (Georgia) Artillery: Capt. Henry H. Carlton Lt. C. W. Motes 2 12-pounder howitzers; 2 10-pounder Parrotts	90	7 (2:5:0)	7.8

Unit	Strength	Losses (k:w:m)	Percent
Pickett's Division Artillery: Maj. James Dearing	**419**	**29 (9:16:4)**	**6.9**
Fauquier (Virginia) Artillery: Capt. Robert M. Stribling 4 Napoleons; 2 20-pounder Parrotts	134	5 (1:4:0)	3.7
Hampden (Virginia) Artillery: Capt. William H. Caskie 2 Napoleons; 1 ordnance rifle; 1 10-pounder Parrott	90	4 (0:3:1)	4.4
Richmond Fayette Artillery: Capt. Miles C. Macon 2 Napoleons; 2 10-pounder Parrotts	90	5 (3:1:1)	5.6
Lynchburg (Virginia) Artillery: Capt. Joseph G. Blount 4 Napoleons	96	10 (5:3:2)	10.4
Hood's Division Artillery: Maj. Mathis W. Henry	**403**	**27 (5:22:0)**	**6.7**
Branch (North Carolina) Artillery: Capt. Alexander C. Latham 3 Napoleons; 1 12-pounder howitzer; 1 6-pounder field gun	112	3 (1:2:0)	2.7
German (South Carolina) Artillery: Capt. William K. Bachman 4 Napoleons	71	—	—
Palmetto (South Carolina) Light Artillery: Capt. Hugh R. Garden 2 Napoleons; 2 10-pounder Parrotts	63	7 (2:5:0)	11.1
Rowan (North Carolina) Artillery: Capt. James Reilly 2 Napoleons; 2 ordnance rifles; 2 10-pounder Parrotts	148	6 (2:4:0)	4.1
Artillery Reserve: Col. James B. Walton Col. Edward P. Alexander			
Alexander's Battalion: Col. Edward P. Alexander Maj. Frank Huger	**576**	**139 (22:111:6)**	**24.1**
Ashland (Virginia) Artillery: Capt. Pichegru Woolfolk Jr. Lt. James Woolfolk 2 Napoleons; 2 20-pounder Parrotts	103	28 (3:24:1)	27.7
Bedford (Virginia) Artillery: Capt. Tyler C. Jordan 4 ordnance rifles	78	9 (1 7:1)	11.5
Brooks (South Carolina) Artillery: Lt. S. Capers Gilbert 4 12-pounder howitzers	71	36 (7:29:0)	50.7
Madison (Louisiana) Artillery: Capt. George V. Moody 4 24-pounder howitzer	135	33 (4:29:0)	24.4
Virginia (Richmond) Battery: Capt. William W. Parker 3 ordnance rifles; 1 10-pounder Parrott	90	18 (3:14:1)	20.0
Virginia (Bath) Battery: Capt. Osmond B. Taylor 4 Napoleons	90	13 (4:8:1)	14.4

Unit	Strength	Losses (k:w:m)	Percent
Washington (Louisiana) Artillery: Maj. Benjamin F. Eshleman	**338**	**30 (8:11:11)**	**8.9**
1st Company: Capt. Charles W. Squires 1 Napoleon	77	4 (1:0:3)	5.2
2nd Company: Capt. John B. Richardson 2 Napoleons; 1 12-pounder howitzer	80	4 (1:0:3)	7.5
3rd Company: Capt. Merritt B. Miller 3 Napoleons	92	10 (5:2:3)	10.9
4th Company: Capt. Joe Norcom Lt. H. A. Battles 2 Napoleons; 1 12-pounder howitzer	80	10 (0:6:4)	12.5
2nd Corps Artillery: Col. J. Thompson Brown			
Early's Division Artillery: Lt. Col. Hilary P. Jones	**290**	**12 (2:6:4)**	**4.1**
Charlottesville (Virginia) Artillery: Capt. James McD. Carrington 4 Napoleons	71	2 (0:0:2)	2.8
Courtney (Virginia) Artillery: Capt. William A. Tanner 4 ordnance rifles	90	2 (0:0:2)	2.2
Louisiana Guard Artillery: Capt. Charles A. Green 2 ordnance rifles; 2 10-pounder Parrotts	60	7 (2:5:0)	11.7
Staunton (Virginia) Artillery: Capt. Asher W. Garber 4 Napoleons	60	1 (0:1:0)	1.7
Rodes's Division Artillery: Lt. Col. Thomas H. Carter	**385**	**77 (14:25:34)**	**20.0**
Jeff Davis (Alabama) Artillery: Capt. William J. Reese 4 ordnance rifles	79	8 (0:0:8)	10.1
King William (Virginia) Artillery: Capt. William P. Carter 2 Napoleons; 2 10-pounder Parrotts	103	23 (7:4:12)	22.3
Morris (Virginia) Artillery: Capt. Richard C. M. Page 4 Napoleons	114	39 (7:25:7)	34.2
Orange (Virginia) Artillery: Capt. Charles W. Fry 2 ordnance rifles; 2 10-pounder Parrotts	80	7 (0:0:7)	8.8
Johnson's Division Artillery: Lt. Col. R. Snowden Andrews Maj. James W. Latimer Capt. Charles I. Raine	**356**	**51 (22:29:0)**	**14.3**
1st Maryland Battery: Capt. William F. Dement 4 Napoleons	90	5 (1:4:0)	5.6
Alleghany (Virginia) Artillery: Capt. John C. Carpenter 2 Napoleons; 2 ordnance rifles	91	24 (10:14:0)	26.4

Unit	Strength	Losses (k:w:m)	Percent
Chesapeake (Maryland) Artillery: Capt. William D. Brown 4 10-pounder Parrotts	76	17 (8:9:0)	22.4
Lee (Virginia) Battery Capt. Charles I. Raine Lt. William M. Hardwicke 1 ordnance rifle; 1 10-pounder Parrott; 2 20-pounder Parrotts	90	4 (2:2:0)	4.4
Artillery Reserve: Col. J. Thompson Brown			
1st Virginia Artillery Battalion: Capt. Willis J. Dance	**367**	**50 (3:21:26)**	**13.6**
2nd Richmond (Virginia) Howitzers: Capt. David Watson 4 10-pounder Parrotts	64	3 (2:1:0)	4.7
3rd Richmond (Virginia) Howitzers: Capt. Benjamin H. Smith Jr. 4 ordnance rifles	62	4 (1:1:2)	6.5
Powhatan (Virginia) Artillery: Lt. John M. Cunningham 4 ordnance rifles	78	15 (0:3:12)	19.2
Rockbridge (Virginia) Artillery: Capt. Archibald Graham 4 20-pounder Parrotts	85	21 (0:14:7)	19.2
Salem (Virginia) Artillery: Lt. Charles B. Griffin 2 Napoleons; 2 ordnance rifles	69	7 (0:2:5)	10.1
Nelson's Battalion: Lt. Col. William Nelson	**277**	**24 (0:1:23)**	**8.7**
Amherst (Virginia) Artillery: Capt. Thomas J. Kirkpatrick 3 Napoleons; 1 ordnance rifle	105	13 (0:0:13)	12.4
Fluvanna (Virginia) Artillery: Capt. John L. Massie 3 Napoleons; 1 ordnance rifle	90	11 (0:1:10)	12.2
Georgia Battery: Capt. John Milledge Jr. 2 ordnance rifles; 1 10-pounder Parrott	73		—
3rd Corps Artillery: Col. R. Lindsay Walker			
Anderson's Division Artillery (Sumter Battalion): Maj. John Lane	**384**	**42 (3:35:4)**	**10.9**
Company A: Capt. Hugh M. Ross 3 10-pounder Parrotts; 1 12-pounder howitzer; 1 Napoleon; 1 ordnance rifle	130	13 (1:11:1)	10.0
Company B: Capt. George M. Patterson 4 12-pounder howitzers; 2 Napoleons	124	9 (2:6:1)	7.3
Company C: Capt. John T. Wingfield 3 3-inch Navy rifles; 2 10-pounder Parrotts	121	20 (0:18:2)	16.5
Heth's Division Artillery: Lt. Col. John Garnett	**396**	**22 (0:5:17)**	**5.6**
Norfolk Light Artillery Blues: Capt. Charles R. Grandy 2 12-pounder howitzers; 2 ordnance rifles	106	2 (0:1:1)	1.9
Pittsylvania Artillery: Capt. John W. Lewis 2 ordnance rifles; 2 Napoleons	90	—	—

Unit	Strength	Losses (k:w:m)	Percent
Donaldsonville Artillery: Capt. Victor Maurin 2 ordnance rifles; 1 10-pounder Parrott	114	6 (0:2:4)	5.3
Virginia Battery: Capt. Joseph D. Moore 2 Napoleons; 1 10-pounder Parrott; 1 ordnance rifle	77	—	—
Pender's Division Artillery: Maj. William T. Poague	**377**	**34 (2:24:8)**	**9.0**
Albemarle (Virginia) Artillery: Capt. James W. Wyatt 2 ordnance rifles; 1 12-pounder howitzer; 1 10-pounder Parrott	94	13 (0:12:1)	13.8
Charlotte (North Carolina) Artillery: Capt. Joseph Graham 2 12-pounder howitzers; 2 Napoleons	125	5 (0:0:5)	4.0
Madison (Mississippi) Light Artillery: Capt. George Ward 3 Napoleons; 1 12-pounder howitzer	91	—	—
Virginia (Warrington) Battery: Capt. James V. Brooke 2 12-pounder howitzers; 2 Napoleons	58	5 (1:2:2)	8.6
Artillery Reserve: Col. R. Lindsay Walker			
Mcintosh's Battalion: Maj. D. G. Mcintosh	**357**	**48 (8:24:16)**	**13.4**
Danville (Virginia) Artillery: Capt. R. Sidney Rice 4 Napoleons	114	2 (0:1:1)	1.8
Hardaway (Alabama) Artillery: Capt. William B. Hurt 2 ordnance rifles; 2 Whitworth rifles	71	8 (0:4:4)	11.3
2nd Rockbridge (Virginia) Artillery: Lt. Samuel Wallace 2 ordnance rifles; 2 Napoleons	67	6 (3:3:0)	9.0
Virginia (Richmond) Battery: Capt. Marmaduke Johnson 4 ordnance rifles	96	10 (5:1:4)	10.4
Pegram's Battalion: Maj. William J. Pegram Capt. E. B. Brunson	**375**	**51 (12:36:3)**	**13.6**
Crenshaw (Virginia) Battery: Lt. Andrew B. Johnson 2 12-pounder howitzers; 2 ordnance rifles	76	15 (1:14:0)	19.7
Fredericksburg (Virginia) Artillery: Capt. Edward A. Marye 2 ordnance rifles; 2 Napoleons	71	2 (2:0:0)	2.8
Letcher (Virginia) Artillery: Capt. Thomas A. Brander 2 10-pounder Parrotts; 2 Napoleons	65	17 (3:11:3)	26.2
Pee Dee (South Carolina) Artillery: Lt. William E. Zimmerman 4 ordnance rifles	65		—
Purcell (Virginia) Artillery: Capt. Joseph McGraw 4 Napoleons	89	6 (1:5:0)	6.7

Unit	Strength	Losses (k:w:m)	Percent
Stuart's Division's Horse Artillery: Major Robert F. Beckham			
Breathed's (Virginia) Battery: Capt. James Breathed 4 ordnance rifles	106	1 (0:1:0)	0.1
2nd Baltimore Maryland Artillery: Capt. William H. Griffin 4 ordnance rifles	106	?	?
2nd Stuart Horse Artillery (Virginia): Capt. William M. McGregor 2 ordnance rifles; 2 Napoleons	106	2 (0:0:2)	1.9
Washington Artillery (South Carolina): Capt. James F. Hart 3 Blakely rifles	107	1 (1:0:0)	0.9
Ashby Horse Artillery (Virginia): Capt. R. Preston Chew 1 ordnance rifle; 1 12-pounder howitzer	99	?	?
Beauregard Rifles (Virginia): Capt. Marcellus N. Moorman 3 ordnance rifles; 1 Napoleon			
Imboden's Command			
Staunton Horse Battery (Virginia): Capt. John H. McClanahan 4 12-pounder howitzer; 1 ordnance rifle	64	?	?
Charlottesville Horse Battery (Virginia): Capt. Thomas Jackson 2 ordnance rifles; 2 12-pounder howitzer	75	?	?

APPENDIX B

CHARACTERISTICS OF CIVIL WAR CANNON USED AT GETTYSBURG

Type of Cannon	Bore Diam.	Type	Weight of Tube (lbs.)	Range at 50 Elevation (yds.)	CS	US
Napoleon	4.62"	Smooth	1,227	1,619	98	142
6-pounder field gun	3.67"	Smooth	884	1,523	1	—
12-pounder howitzer	4.62"	Smooth	788	1,072	33	2
24-pounder howitzer	5.82"	Smooth	1,318	1,322	4	—
3-inch ordnance rifle	3.0"	Rifled	816	1,835	78	142
10-pounder Parrott	2.9"	Rifled	890	2,000	44	60
20-pounder Parrott	3.67"	Rifled	1,750	2,100	10	6
James rifle	3.80"	Rifled	918	1,700	—	4
Whitworth rifle	2.75"	Rifled	1,092	2,800	2	—

NOTES

(Full bibliographical data can be found in the bibliography.)

Foreword

1. Cole, *Civil War Artillery at Gettysburg*, 34–38; Naisawald, *Grape and Canister*, 260–61; Murray, *Artillery Tactics of the Civil War*, 13; Longacre, *The Man Behind the Guns*, 125; U.S. War Department, *The War of the Rebellion*, ser. 1, vol. 27, pt. 1, 750–51 (hereinafter cited as *OR*).

2. Cole, *Civil War Artillery at Gettysburg*, 55–56, 65; Murray, *Artillery Tactics of the Civil War*, 9–10; Wert, *Gettysburg, Day Three*, 105; Barnett, "The Batteries Fired with Very Decided Effect," 200; James K. P. Scott, "The Artillery at Gettysburg."

3. Coco, *A Concise Guide to the Artillery at Gettysburg*, 75, 83; Cole, *Civil War Artillery at Gettysburg*, 56.

4. Alexander, "Confederate Military Service," 98–113.

5. Stewart, *Pickett's Charge*, 33.

6. Cole, *Civil War Artillery at Gettysburg*, 71, 80, 85, 85–99; Coddington, *The Gettysburg Campaign*, 89; Barnett, "The Batteries Fired with Very Decided Effect," 198, 200; Dew, *Ironmaker to the Confederacy*, 190.

7. Thomas, *Cannons*, 27, 28, 31, 33, 39, 43; Cole, *Civil War Artillery at Gettysburg*, 56–59; *OR*, ser. 1, vol. 25, pt. 2, 615.

8. Thomas, *Cannons*, 16–17; Cole, *Civil War Artillery at Gettysburg*, 122–36.

9. Gettysburg National Military Park (hereafter cited as GNMP).

10. Stewart, *Pickett's Charge*, 117–18.

11. Cole, *Civil War Artillery at Gettysburg*, 172–77.

12. Newton, *Silent Sentinels*, 54, 56; Cole, *Civil War Artillery at Gettysburg*, 228–29.

13. Pfanz, *Gettysburg—The Second Day*, 125.

14. Murray, *Artillery Tactics of the Civil War*, 2–5.

Chapter 1: *July 1: The Fight Along Chambersburg Pike*

1. Gottfried, *Roads to Gettysburg*, 6, 69.

2. Martin, *Gettysburg, July 1*, 60–61; Barnett, "The Batteries Fired with Very Decided Effect," 201.

3. Krick, *The Fredericksburg Artillery*, 58.

4. Marye, "The First Gun at Gettysburg," 1228.

5. Martin, *Gettysburg, July 1*, 60–61.

6. Busey and Martin, *Regimental Strengths and Losses at Gettysburg*, 99, 172; Pfanz, *Gettysburg, The First Day*, 59.

7. Downey, *The Guns at Gettysburg*, 10.

8. Ibid., 11, 14.

9. Gibbon, *The Artillerist's Manual*, 352–56; Shultz and Rollins, "The Most Accurate Fire Ever Witnessed," 55.

10. *OR*, vol. 27, pt. 1, 1031; Martin, *Gettysburg, July 1*, 60–61, 78; Calef, "Gettysburg Notes," 47–48; Naisawald, *Grape and Canister*, 267. Calef had initially massed his battery along Chambersburg Pike that morning (*OR*, vol. 27, pt. 1, 1030–31).

11. Marye, "The First Gun at Gettysburg," 1228–29; *OR*, vol. 27, pt. 2, 677; Fleet, "First Appearance of the Confederate States Flag with White Field," 240; Carmichael, *The Purcell, Crenshaw, and Letcher Artillery*, 184–85; Krick, *The Fredericksburg Artillery*, 58–59.

12. *OR*, vol. 27, pt. 2, 677. Marye recalled that only one gun was unlimbered in the road (Marye, "The First Gun at Gettysburg," 1229).

13. Fleet, "First Appearance of the Confederate States Flag with White Field," 240.

14. Marye, "The First Gun at Gettysburg," 1229.

15. Ibid., 31; Fleet, "First Appearance of the Confederate States Flag with White Field," 240; *OR*, vol. 27, pt. 2, 677.

16. Krick, *The Fredericksburg Artillery*, 59–60.

17. Pfanz, *Gettysburg, The First Day*, 60–61; Fleet, "First Appearance of the Confederate States Flag with White Field," 240; *OR*, vol. 27, pt. 2, 677–78; Krick, *The Fredericksburg Artillery*, 59–60; Goolsby, "Crenshaw's Battery, Pegram's Battalion, Confederate States Artillery," 336; Purifoy, "The Artillery at Gettysburg," 424. Several men from Johnston's Crenshaw battery insisted that it was their shrapnel that killed Union Gen. John F. Reynolds (*OR*, vol. 27, pt. 2, 677; Krick, *The Fredericksburg Artillery*, 59).

18. Fleet, "First Appearance of the Confederate States Flag with White Field," 242.

19. Downey, *The Guns at Gettysburg*, 19.

20. Ibid., 16–18; Freeman, *Lee's Lieutenants*, 3:179.

21. Calef, "Gettysburg Notes," 47–48; *OR*, vol. 27, pt. 1, 1031; Shultz and Rollins, "The Most Accurate Fire Ever Witnessed," 55.

22. *OR*, vol. 27, pt. 1, 1031.

23. *OR*, vol. 27, pt. 2, 678; Barnett, "The Batteries Fired with Very Decided Effect," 201; Hankins, *Simple Story of a Soldier*, 44.

24. Carmichael, *The Purcell, Crenshaw, and Letcher Artillery*, 100–101; *OR*, vol. 27, pt. 1, 1031; Naisawald, *Grape and Canister*, 269; Shultz and Rollins, "The Most

Accurate Fire Ever Witnessed," 57. According to Downey, Newman later returned to the dead horses to salvage their harnesses (*The Guns at Gettysburg*, 23).

25. Calef, "Gettysburg Notes," 48; *OR*, vol. 27, pt. 1, 1031.

26. *OR*, vol. 27, pt. 1, 927.

27. Smith, *History of the Seventy-Sixth Regiment*, 237.

28. New York Monuments Commission, *Final Report of the Battlefield of Gettysburg*, 3:990 (hereafter, *New York at Gettysburg*).

29. James A. Hall to John Bachelder, December 29, 1869, Bachelder Papers, New Hampshire Historical Society (hereafter cited as NHHS).

30. James A. Hall to John Bachelder, February 27, 1867, Bachelder Papers, NHHS.

31. *OR*, vol. 27, pt. 1, 359; Hall to Bachelder, December 29, 1869.

32. Downey, *The Guns at Gettysburg*, 27; Small, *The Road to Richmond*, 65.

33. *OR*, vol. 27, pt. 1, 359.

34. *New York at Gettysburg*, 3:991.

35. *OR*, vol. 27, pt. 1, 359; *Lewiston Journal*, copy in Brake Collection, U.S. Army Military History Institute (hereinafter cited as USAMHI); Hall to Bachelder, December 29, 1869.

36. Hall to Bachelder, December 29, 1869; *OR*, vol. 27, pt.1, 359.

37. Hall to Bachelder, December 29, 1869.

38. *OR*, vol. 27, pt. 1, 359; Hall to Bachelder, December 29, 1869. Several Union soldiers in that part of the field swore that Hall lost two pieces. In his official report, prepared on July 27, 1863, Lt. John Calef wrote, "This battery [Hall's], and leaving two pieces on the field, withdrew" (*OR*, vol. 27, pt. 1, 1031). This was confirmed after the war by Lt. J. Volnay Pierce, who commanded Company G, 147th New York, in a letter to John Bachelder (November 1, 1882): "I noticed Hall's battery limber up, and dash down the Chambersburg Pike with all but 2 guns."

39. *OR*, vol. 27, pt. 1, 1031.

40. *OR*, vol. 27, pt. 1, 1031; Calef, "Gettysburg Notes," 49; Shultz and Rollins, "The Most Accurate Fire Ever Witnessed," 60.

41. Nevins, *A Diary of Battle*, 233.

42. Martin, *Gettysburg, July 1*, 168; *New York at Gettysburg*, 3:1256.

43. Pfanz, *Gettysburg—Culp's Hill and Cemetery Hill*, 175.

44. Wise, *The Long Arm of Lee*, 618–19; Jorgensen, "Confederate Artillery at Gettysburg," 22; *OR*, vol. 27, pt. 2, 674–675; Martin, *Gettysburg, July 1*, 190; Pfanz, *Gettysburg, The First Day*, 119–20. There is some confusion about the position of Wallace's battery. Wise states that it was "just to the left of the pike," while the Bachelder maps and Jorgensen place it to the right of the line. Pfanz seems to support Wise, noting that the battery took position "somewhere near the pike."

45. *OR*, vol. 27, pt. 2, 674–75.

46. Thomas, *Cannons*, 43.

47. *OR*, vol. 27, pt. 2, 674; Sifakis, *Who Was Who in the Civil War*, 417. Historian Harry Pfanz believes that the Whitworths probably shelled Shultz's Woods on Seminary Ridge (*Gettysburg, The First Day*, 119–20).

48. Wise, *The Long Arm of Lee*, 616–17.

49. Wise, *The Long Arm of Lee*, 619; *OR*, vol. 27, pt. 2, 652; Garnett, *Gettysburg*, 18.

50. *OR*, vol. 27, pt. 2, 652; Wise, *The Long Arm of Lee*, 720.

51. Chamberlin, *History of the One Hundred and Fiftieth Regiment, Pennsylvania Volunteers*, 118.

52. Francis Bacon Jones, "Excerpt from Chronicles of Francis Bacon Jones," Brake Collection, USAMHI.

53. Matthews, *The 149th Pennsylvania Volunteer Infantry Unit in the Civil War*, 82.

54. Ibid., 82–83; Murray, *Artillery Tactics of the Civil War*, 50.

55. *OR*, vol. 27, pt. 1, 355–56.

56. Martin, *Gettysburg, July 1*, 191; *OR*, vol. 27, pt. 1, 360.

57. *OR*, vol. 27, pt. 1, 360.

58. Nevins, *A Diary of Battle*, 234.

59. *OR*, vol. 27, pt. 1, 355–56, 360; Nevins, *A Diary of Battle*, 234.

60. Herdegen and Beaudot, *In the Bloody Railroad Cut of Gettysburg*, 207.

61. Murray, *Artillery Tactics of the Civil War*, 29.

62. *OR*, vol. 27, pt. 1, 1031–32; Calef, "Gettysburg Notes," 50–51; McLean, *Cutler's Brigade at Gettysburg*, 128.

63. *OR*, vol. 27, pt. 1, 1031–32; Calef, "Gettysburg Notes," 51; Naisawald, *Grape and Canister*, 273; Coddington, *The Gettysburg Campaign*, 270, 687 (n. 58). Wadsworth was able to bring some of Cutler's regiments forward to support Calef's battery (Calef, "Gettysburg Notes," 50).

64. Charles S. Wainwright Journal, Huntington Library, San Marino, California.

65. *OR*, vol. 27, pt. 2, 678.

66. *OR*, vol. 27, pt. 1, 356, 362, 1032.

67. *OR*, vol. 27, pt. 1, 356.

68. *OR*, vol. 27, pt. 1, 1032; Pfanz, *Gettysburg, The First Day*, 157; Clark, *Histories of the Several Regiments and Battalions from North Carolina*, 2:255.

69. Jones, "Excerpt from the Chronicles of Francis Bacon Jones."

70. *OR*, vol. 27, pt. 1, 355–356; Bachelder map series; Nevins, *A Diary of Battle*, 234; Murray, *Artillery Tactics of the Civil War*, 32.

71. Cook, "Personal Reminiscences of Gettysburg," 324–25.

72. *OR*, vol. 27, pt. 1, 364; John A. Gardener letter (October 22, 1902), 1st Pennsylvania Battery B folder, GNMP; Martin, *Gettysburg, July 1*, 240; Nevins, *A Diary of Battle*, 235.

73. *OR*, vol. 27, pt. 1, 356; Felton, "The Iron Brigade Battery," 148–49; Stew-

art, "Battery B Fourth United States Artillery at Gettysburg," 1:368–69; McLean, *Cutler's Brigade at Gettysburg*, 140. The battery was probably on Seminary Ridge for about an hour before it was ordered across the road. According to Stewart's postwar account, the right wing, which was under cover of some woods, was about two hundred yards in advance of the left.

74. Martin, *Gettysburg, July 1*, 244.

75. *OR*, vol. 27, pt. 2, 675.

76. Fleet, "First Appearance of the Confederate States Flag with White Field," 241.

77. *OR*, vol. 27, pt. 1, 356; Pennsylvania Gettysburg Battle-field Commission, *Pennsylvania at Gettysburg*, 2:808; Nevins, *A Diary of Battle*, 235.

78. Nevins, *A Diary of Battle*, 235.

79. Maine Gettysburg Commission, *Maine at Gettysburg*, 84.

80. Clark, *North Carolina Regiments*, 3:89–90.

81. Moore, *The Danville, Eight Star New Market and Dixie Artillery*, 30.

82. *OR*, vol. 27, pt. 1, 356, 363.

83. Martin, *Gettysburg, July 1*, 347.

84. Nevins, *A Diary of Battle*, 235; *OR*, vol. 27, pt. 1, 356, 363.

85. William Shelton, "Autobiography," New York Historical Society, New York.

86. Nevins, *A Diary of Battle*, 235; OR, vol. 27, pt. 1, 356; Nevins, *A Diary of Battle*, 235; Maine Gettysburg Commission, *Maine at Gettysburg*, 83–84; Martin, *Gettysburg, July 1*, 397–98; Pfanz, *Gettysburg, The First Day*, 296, 297.

87. Maine Gettysburg Commission, *Maine at Gettysburg*, 85.

88. Ibid., 84–85; Brown, *A Colonel at Gettysburg and Spotsylvania*; Downey, *The Guns at Gettysburg*, 35.

89. *OR*, vol. 27, pt. 2, 670; Gottfried, "To Fail Twice: Brockenbrough's Brigade at Gettysburg," 71.

90. Buell, *Cannoneer*, 66.

91. Ibid., 67.

92. Tervis, *History of the Fighting Fourteenth*, 85.

93. Buell, *Cannoneer*, 67; Tervis, *History of the Fighting Fourteenth*, 85; Felton, "The Iron Brigade Battery," 148–49. Augustus Buell's account has generated much controversy, particularly because he was not at Gettysburg but wrote as though he was. This has led many historians to ignore his vivid descriptions of James Stewart's battery. However, a close study by Silas Felton showed that Buell's account was probably based on the observations of at least three eyewitnesses and therefore recommended that the account not be dismissed (Silas Felton, "Pursuing the Elusive 'Cannoneer,'" 38–39).

94. Downey, *The Guns at Gettysburg*, 35–36.

95. Ibid., 37.

96. Buell, *Cannoneer*, 70.

97. Hubler, "Just a Plain Unvarnished Story."

98. Buell, *Cannoneer*, 68.

99. Downey, *The Guns at Gettysburg*, 39–40.

100. Buell, *Cannoneer*, 70.

101. *OR*, vol. 27, pt. 2, 670.

102. *OR*, vol. 27, pt. 2, 670; Nevins, *A Diary of Battle*, 236.

103. Brown, "Some Recollections of Gettysburg," 53.

104. Caldwell, *A Brigade of South Carolinians*, 96; *OR*, vol. 27, pt. 2, 661–62; Brown, *A Colonel at Gettysburg and Spotsylvania*, 80, 84; Calef, "Gettysburg Notes," 51. The battle was finally over for Calef's battery. The battery pulled back toward the left of the army, just east of Emmitsburg Road, and it remained there until the following morning; it then rode to Westminster with the rest of Buford's division (Calef, "Gettysburg Notes," 51–52).

105. *OR*, vol. 27, pt. 2, 665.

106. Gottfried, *Brigades of Gettysburg*, 644.

107. Nevins, *A Diary of Battle*, 236; *OR*, vol. 27, pt. 1, 361.

108. Nevins, *A Diary of Battle*, 236; Hubler, "Just a Plain Unvarnished Story"; Buell, *Cannoneer*, 73. Harry Pfanz believed that the troops approaching Stevens's battery were from Scales's brigade (Pfanz, *Gettysburg, The First Day*, 313). According to the historian of Cooper's battery, it was an officer of the 142nd Pennsylvania who informed them that the enemy was approaching and they needed to retreat (*Pennsylvania at Gettysburg*, 2:909).

109. Perrin, "A Little More Light on Gettysburg," 522.

110. Krick, "Three Confederate Disasters on Oak Ridge," 87.

111. Nevins, *A Diary of Battle*, 236–37; *OR*, vol. 27, pt. 1, 357.

112. Maine Gettysburg Commission, *Maine at Gettysburg*, 86.

113. Martin, *Gettysburg, , July 1*, 429; Stewart, "Battery B Fourth Artillery at Gettysburg," 370–72; Buell, *Cannoneer*, 73, 74.

114. Buell, *Cannoneer*, 74; Barnett, "If Ever Men Stayed by Their Guns," 96.

115. Wise, *Long Arm of Lee*, 622.

116. Tervis, *The History of the Fighting Fourteenth*, 87.

117. *OR*, vol. 27, pt. 2, 349.

118. *OR*, vol. 27, pt. 2, 349; Martin, *Gettysburg, July 1*, 499.

119. Freeman, *Lee's Lieutenants*, 1:614; Downey, *The Guns at Gettysburg*, 44–45; Carmichael, "Every Map of the Field Cries Out About It," 271–72; Barnett, "The Batteries Fired with Very Decided Effect," 199.

120. *OR*, vol. 27, pt. 2, 349.

121. Ibid.

122. Ibid.

123. Wise, *Long Arm of Lee*, 623; McIntosh, "A Review of the Gettysburg Campaign by One Who Participated," 118–19.

124. *OR*, vol. 27, pt. 1, 357, 365; James A. Gardner Letter (March 2, 1923), 1st Pennsylvania Battery B Folder, GNMP.

125. Wise, *The Long Arm of Lee*, 616–17.

126. *OR*, vol. 27, pt. 2, 610, 675, 678.

127. McIntosh, "The Gettysburg Campaign"; McIntosh, "A Review of the Gettysburg Campaign by One Who Participated," 118–19.

128. Garnett, *Gettysburg*, 20.

Chapter 2: *July 1: The Battle North of Gettysburg*

1. *OR*, vol. 27, pt. 2, 602.

2. Krick, *Lee's Colonels*, 72; Sifakis, *Who Was Who in the Civil War*, 110.

3. *OR*, vol. 27, pt. 2, 602. The two Carters were half brothers (Barnett, "The Batteries Fired with Very Decided Effect," 205).

4. *OR*, vol. 27, pt. 2, 602. According to Pfanz, this opening fire served little purpose except to advertise Rodes's position on Oak Hill (Pfanz, *Gettysburg, The First Day*, 156).

5. Matthews, *History of the 149th Pennsylvania Infantry*, 82.

6. Ibid., 84. Stone forgot about the flags when he ordered his brigade back to Seminary Ridge, which resulted in their capture (ibid., 93–95).

7. Bachelder, "Movements of the 151st Penn. Vols. of Biddle's Brigade," 1:271; *OR*, vol. 27, pt. 1, 320.

8. *OR*, vol. 27, pt. 2, 603.

9. *New York at Gettysburg*, 1:378; Martin, *Gettysburg, Day 1*, 260.

10. Laboda, *From Selma to Appomattox*, 136; *New York at Gettysburg*, 1:378.

11. *OR*, vol. 27, pt. 1, 754; Kepf, "Dilger's Battery at Gettysburg," 56. Dilger accompanied Schurz's (Schimmelfennig's) division on its march to Gettysburg. According to Naisawald (*Grape and Canister*, 281), Schimmelfennig showed remarkable confidence in permitting Dilger to select his own position. But Schimmelfennig was a brigade commander who had no prior experience commanding a division, much less positioning a battery (*OR*, vol. 27, pt. 1, 754). Dilger mistakenly wrote in his report that he first took position between Taneytown Road and Baltimore Pike.

12. Kepf, "Dilger's Battery at Gettysburg," 49–50; Barnett, "If Ever Men Stayed by Their Guns," 88.

13. Applegate, *Reminiscences and Letters of George Arrowsmith of New Jersey*, 211–12. An infantryman confirmed this shot in a letter home soon after the battle (Marcus, *A New Canaan Private in the Civil War*, 42). Capt. James McD. Carrington

claimed that the shot plugged one of Asher W. Garber's battery's guns opposite Blocher's Knoll (Carrington, "First Day on the Left at Gettysburg," 330).

14. *OR*, vol. 27, pt. 1, 289–311; vol. 27, pt. 2, 552–53.

15. *New York at Gettysburg*, 1:379.

16. Barnett, "The Batteries Fired with Very Decided Effect," 206; *OR*, vol. 27, pt. 2, 603; T. H. Carter to D. H. Hill, July 1, 1885, Lee Family Papers, Virginia Historical Society; Pfanz, *Gettysburg, The First Day*, 168–69. Page's battery lost more men than any other Confederate artillery unit at Gettysburg (Martin, *Gettysburg, Day 1*, 262).

17. *OR*, vol. 27, pt. 2, 603; vol. 27, pt. 1, 747, 752, 754; Wheeler, *Letters of William Wheeler of the Class of 1855, Yale College*, 408; Naisawald, *Grape and Canister*, 281; Martin, *Gettysburg, Day 1*, 262. William Wheeler's battery had accompanied Adolph von Steinwehr's division during the last stage of the march to Gettysburg.

18. Pfanz, *Gettysburg—Culp's Hill and Cemetery Hill*, 175; Sifakis, *Who Was Who in the Civil War*, 480–81.

19. *OR*, vol. 27, pt. 1, 752, 754; Naisawald, *Grape and Canister*, 282; Kepf, "Dilger's Battery at Gettysburg," 58; Wheeler, *Letters of William Wheeler of the Class of 1855, Yale College*, 409–11.

20. *OR*, vol. 27, pt. 1, 754; Murray, *Artillery Tactics of the Civil War*, 38.

21. Martin, *Gettysburg, July 1*, 263; Wise, *The Long Arm of Lee*, 620.

22. *OR*, vol. 27, pt. 2, 603.

23. Kiefer, *History of the One Hundred and Fifty-Third Regiment Pennsylvania Volunteers Infantry*, 209, 210; Dodge, "Left Wounded on the Field," 321.

24. Wise, *The Long Arm of Lee*, 620; Barnett, "The Batteries Fired with Very Decided Effect," 209; *OR*, vol. 27, pt. 2, 583.

25. Ward, *Sketch of Battery "I" First Ohio Artillery*, 45–46; *OR*, vol. 27, pt. 1, 752.

26. Lee, "Reminiscences of Gettysburg Battle," 55.

27. *OR*, 28, 1, 756; Martin, *Gettysburg, July 1*, 281. According to Harry Pfanz, Wilkeson's battery arrived about an hour after Dilger's (Pfanz, *Gettysburg, The First Day*, 232).

28. Busey and Martin, *Regimental Strengths and Losses at Gettysburg*, 80, 157; Krick, *Lee's Colonels*, 349; Sifakis, *Who Was Who in the Civil War*, 183.

29. *OR* vol. 28, pt. 2, 495, 496, 497, 498; Wise, *The Long Arm of Lee*, 621; Carrington, "First Day on the Left at Gettysburg," 330.

30. *OR*, vol. 27, pt. 1, 717.

31. Wheeler, *Letters of William Wheeler of the Class of 1855, Yale College*, 409–10; *OR*, vol. 27, pt. 1, 753, 754.

32. Marcus, *A New Canaan Private*, 41.

33. Anonymous, 75th Ohio folder, GNMP.

34. *OR*, vol. 27, pt. 1, 748.

35. *OR*, vol. 27, pt. 1, 756; Osborn, *The Eleventh Corps Artillery at Gettysburg*, 10. Wilkeson quickly applied a tourniquet made from a handkerchief and then he amputated his dangling limb.

36. Kiefer, *History of the One Hundred and Fifty-Third Regiment Pennsylvania Volunteers Infantry*, 211.

37. *OR*, vol. 27, pt. 1, 757.

38. Ibid., 756.

39. Ibid., 757.

40. Ibid., 757.

41. Ibid., 756.

42. Ibid., 757.

43. Ibid., 495.

44. *OR*, vol. 27, pt. 2, 495, 496, 497; *OR*, vol. 27, pt. 1, 756; Murray, *Artillery Tactics of the Civil War*, 43. Bancroft wrote in his report that he opened fire with three guns. He did not indicate what happened to the fourth (*OR*, vol. 27, pt. 1, 756).

45. Lee, "Reminiscences of the Gettysburg Battle," 56.

46. *OR*, vol. 27, pt. 1, 753, 754. Capt. Alfred Lee of the 82nd Ohio insisted that most of the artillery fire came from Oak Hill during the attack (Lee to John Bachelder, February 16, 1888, Bachelder Papers, NHHS).

47. *OR*, vol. 27, pt. 1, 748.

48. *OR*, vol. 27, pt. 2, 479.

49. *OR*, vol. 27, pt. 1, 755.

50. Ibid., 748.

51. Ibid., 755.

52. Ibid., 748; vol. 27, pt. 2, 495.

53. *OR*, vol. 27, pt. 1, 754, Kepf, "Dilger's Battery at Gettysburg," 60.

54. *OR*, vol. 27, pt. 1, 753.

55. Ibid., 748.

56. *OR*, vol. 27, pt. 2, 554; Barnett, "The Batteries Fired with Very Decided Effect," 209.

57. *OR*, vol. 27, pt. 2, 603.

58. Ibid., 495, 498, 603; Nicholas and Servis, *Powhatan, Salem and Courtney Henrico Artillery*, 171; Carrington, "First Day on the Left at Gettysburg," 330, 332; Carrington, "First Day on the Left at Gettysburg," *Richmond Times Dispatch*; Daniel, "First Day on the Left at Gettysburg."

59. *OR*, vol. 27, pt. 1, 748, 751; *New York at Gettysburg*, 1:247; Martin, *Gettysburg, July 1*, 280–81; Callihan, "Captain Michael Wiedrich's Company I, First New York Light Artillery at Gettysburg," 85.

60. *OR*, vol. 27, pt. 1, 748, 751; *OR*, vol. 27, pt. 2, 484; Clark, *North Carolina Regiments*, 3:414; Samuel Eaton Diary, Southern Historical Collection, University of

North Carolina. Wiedrich's battery was placed under the command of Col. Charles S. Wainwright of the 1st Corps during this period.

61. Nevins, *A Diary of Battle*, 237–38; Martin, *Gettysburg, July 1*, 474–75.

62. *OR*, vol. 27, pt. 1, 233.

63. Ibid., 748.

64. Nevins, *A Diary of Battle*, 237.

65. Stewart, "Battery B Fourth Artillery at Gettysburg," 373–74.

66. Maine Gettysburg Commission, *Maine at Gettysburg*, 88–89.

67. Nevins, *A Diary of Battle*, 238.

68. Martin, *Gettysburg, July 1*, 476; *OR*, vol. 27, pt. 1, 357.

69. *OR*, vol. 27, pt. 1, 756, 757–58; *New York at Gettysburg*, 3:1247.

70. *OR*, vol. 27, pt. 1, 357, 756, 757–58. Ever the artilleryman, Dilger spent more space in his report discussing his defective ammunition than praising his men or bemoaning his losses: "In regard to the ammunition, I must say that I was completely dissatisfied with the results observed of the fuses for the 12-pounder shells and spherical case, on the explosion of which, by the most preparation, you cannot defend" (*OR*, vol. 27, pt. 1, 754–55).

71. *OR*, vol. 27, pt. 1, 357.

72. Nevins, *A Diary of Battle*, 238.

73. McKelvey, "George Breck's Civil War Letters from Reynolds' Battery," 131–32.

74. Osborn, *Trials and Triumphs*, 96.

75. Frassanito, *Gettysburg: A Journey in Time*, 118–19; Wheeler, *Witness to Gettysburg*, 156.

76. George Breck letter, 1st New York, Battery L folder, GNMP.

77. Pfanz, *Gettysburg, The First Day*, 333.

78. *OR*, vol. 27, pt. 2, 457, 495, 603, 675, 678. One of the guns in the Danville battery (McIntosh's battalion) fired prematurely. Infantryman John Prewett, who was pressed into service as the sponger for gun number 3, had neglected to sponge the gun after it fired, and the loading process continued. All were surprised when the gun prematurely discharged, severely wounding Prewett's right hand and arm (Moore, *Danville Battery*, 31).

79. *OR*, vol. 27, pt. 2, 495, 675, 678.

80. Marye, "The First Gun at Gettysburg," 32.

81. Shultz and Rollins, "A Combined and Concentrated Fire," 50.

Chapter 3: *July 2: The Two Sides Take Position*

1. *OR*, vol. 27, pt. 1, 232.

2. Stewart, *Pickett's Charge*, 71; Sifakis, *Who Was Who in the Civil War*, 326.

3. Longacre, *The Man Behind the Guns*, 158.

4. *OR*, vol. 27, pt. 1, 361. Edward Johnson's division was probably the target.

5. Stephen Wallace Diary, Pennsylvania State Archives.

6. Maine Gettysburg Commission, *Maine at Gettysburg*, 91–92.

7. *OR*, vol. 27, pt. 1, 232–33, 870. Hunt recalled that Meade's words were "This is your affair. Take the proper measures to provide against the attack, and make the line safe with artillery until it is properly occupied ("Account of Brig. Gen. Henry Hunt, Chief of Artillery, Army of the Potomac, *Bachelder Papers*, 1:426). Historian Philip Cole believed that this exchange empowered Hunt and led to his contentious interactions with Gen. Winfield Hancock on July 3 (Cole, *Civil War Artillery at Gettysburg*, 48–49).

8. Hunt indicated that he ordered Hall to reinforce Osborn on the left during the morning of July 2. Yet the three guns were in that position the evening before, and there is no indication that they returned to Wainwright before reassuming their position on the left (*OR*, vol. 27, pt. 1, 233).

9. *OR*, vol. 27, pt. 1, 233; Piatek, "Col. Charles Wainwright's Account of Cooper's Company B, 1st Pennsylvania Light Artillery on East Cemetery Hill," 95.

10. Sifakis, *Who Was Who in the Civil War*, 299.

11. *OR*, vol. 27, pt. 1, 478.

12. Pfanz, *Gettysburg—The Second Day*, 482; *OR*, vol. 27, pt. 1, 928.

13. *OR*, vol. 27, pt. 1, 581; Jorgensen, "Clark's Battery B, 1st New Jersey Artillery on July 2, 1863," 40.

14. *OR*, vol. 27, pt. 1, 581.

15. Ibid., 82.

16. Ibid., 582; Murray, "The Artillery Duel in the Peach Orchard July 2, 1863," 72. The actual position of Smith's guns has been questioned over the years. Historian Garry Adelman found evidence to support the claim that the four guns on Houck's Ridge were near the location of the 99th Pennsylvania memorial, and the two rear guns were seventy-five to one hundred yards behind them, not in the position currently ascribed on the battlefield (Adelman, "The Fight for and Location of the 4th New York Independent Battery at Gettysburg," 67–68).

17. *OR*, vol. 27, pt. 1, 234–35, 586, 900; Jorgensen, "Clark's Battery B, 1st New Jersey Artillery on July 2, 1863," 40; Hanifen, *History of Battery B, First New Jersey Artillery*, 68. According to Pfanz, Ames's battery took position at about 4:00 p.m. (Pfanz, *Gettysburg—The Second Day*, 134).

18. *OR*, vol. 27, pt. 1, 900.

19. Ibid.

20. Shultz, "Gulian V. Weir's 5th U.S. Artillery, Battery C," 82.

21. Hunt, "The Second Day at Gettysburg," 3:297–99.

22. Ibid.

23. Ibid., 297, 299–300; Longacre, *The Man Behind the Guns*, 159. The 3rd

Corps was not the only one to have problems with its ammunition supply. The 2nd Corps arrived on the battlefield with half of its ammunition wagons. The same would be true later on July 3, when the 6th Corps arrived on the battlefield. To Hunt, this was but another example of the corps commanders' insensitivity, as he believed they put too little a value on the artillery's ammunition in their haste to arrive on the battlefield (Longacre, *The Man Behind the Guns*, 159).

24. Sifakis, *Who Was Who in the Civil War*, 666.

25. Wise, *The Long Arm of Lee*, 629–30; Sifakis, *Who Was Who in the Civil War*, 686.

26. Pfanz, *Culp's Hill and Cemetery Hill*, 176. While Lt. Col. R. Snowden Andrews prepared his battalion's official report, he was not present at Gettysburg, because he had been wounded at Winchester. The same was true of Capt. Willis J. Dance, who led a battalion in the absence of its permanent commander, Lt. Col. Robert Hardaway. Dance returned to his battery, the Powhatan Artillery, after Hardaway's return after the battle. Carter's and Nelson's battalions probably deployed on the ridge that straddled Carlisle Road, just to the east of the Hagey farm.

27. Krick, *Lee's Colonels*, 61; Sifakis, *Who Was Who in the Civil War*, 80.

28. *OR*, vol. 27, pt. 2, 635, 652, 678; Pfanz, *Culp's Hill and Cemetery Hill*, 177–78. According to the battlefield tablet for Rice's battery, two of the guns were engaged and two were in reserve.

29. *OR*, vol. 27, pt. 2, 604.

30. Ibid., 605.

31. Ibid.

32. Krick, *Lee's Colonels*, 248.

33. *Pennsylvania at Gettysburg*, 2:900.

34. *OR*, vol. 27, pt. 2, 610.

35. *Pennsylvania at Gettysburg*, 2: 900.

36. Wise, *The Long Arm of Lee*, 72–74; Pfanz, *Gettysburg—The Second Day*, 117.

37. Krick, *Lee's Colonels*, 333.

38. Warner, *Generals in Gray*, 3; Sifakis, *Who Was Who in the Civil War*, 6.

39. Klein, *Edward Porter Alexander*, 76–77. Alexander was also junior in rank to Col. Henry Cabell, who commanded the battalion assigned to Lafayette McLaws's division. According to historian Jennings Wise (*The Long Arm of Lee*, 756, 851–52), both Walton and Cabell lacked the necessary assertiveness and aggressiveness needed on the field.

40. Alexander, *Fighting for the Confederacy*, 235.

41. Pfanz, *Gettysburg—The Second Day*, 117; Murray, *E. P. Alexander and the Artillery Action in the Peach Orchard*, 38–9; Alexander, *Military Memoirs of a Confederate*, 391; Barnett, "The Severest and Bloodiest Artillery Fight I Ever Saw," 69.

42. *OR*, vol. 27, pt. 2, 350.

43. Alexander, *Fighting for the Confederacy*, 236.

44. Ibid. Sgt. Henry Wentz, a member of Lt. Osmond B. Taylor's Virginia Battery (Alexander's battalion) grew up on his father's farm just north of the soon-to-be-famous Peach Orchard (Pfanz, *Gettysburg—The Second Day*, 118).

45. Alexander, *Fighting for the Confederacy*, 237.

46. Krick, *Lee's Colonels*, 162; Haskell, *The Haskell Memoirs*, 46–47; Gottfried, "The Story of Henry's Artillery Battalion at Gettysburg," 30–31.

47. Pfanz, *Gettysburg—The Second Day*, 160–61. Alexander C. Latham's battery contained the only 6-pounder smoothbore left in the Army of Northern Virginia (Barnett, "The Severest and Bloodiest Artillery Fight I Ever Saw," 69).

48. Wise, *The Long Arm of Lee*, 720; Sifakis, *Who Was Who in the Civil War*, 99; Krick, *Lee's Colonels*, 66.

49. Alexander, *Fighting for the Confederacy*, 238.

50. Pfanz, *Gettysburg—The Second Day*, 305–6; Jorgensen, "Confederate Artillery at Gettysburg," 28–29; Alexander, *Military Memoirs of a Confederate*, 394. The two Napoleons in McCarthy's battery were apparently kept in reserve.

51. Murray, *E. P. Alexander and the Artillery Action in the Peach Orchard*, 44.

52. Alexander, *Military Memoirs of a Confederate*, 395.

53. Pfanz, *Gettysburg—The Second Day*, 155. Alexander incorrectly wrote that he had eighteen guns deployed for action out of his twenty-six guns. His actual numbers were sixteen deployed out of a total of twenty-four. Two battery commanders in Alexander's battalion were feuding bitterly when the guns arrived in Gettysburg. A quarrel arose between Capts. Pichegru Woolfolk and George V. Moody when one's battery was slow in taking its position in the column during the march to Gettysburg and the other's took its place. So intense was the feud that the two agreed to a duel on July 2. The weapon of choice was infantry rifles at ten paces. The battle precluded the duel; the two never served together again afterward, as Woolfolk was later wounded and Moody was captured (Pfanz, *Gettysburg—The Second Day*, 155–56).

Chapter 4: *July 2: The Battle South of Wheatfield Road*

1. Alexander, *Fighting for the Confederacy*, 239. Although it appears that all of Henry Cabell's guns opened fire at about the same time, there were many discrepancies in the battery commanders' official reports filed after the battle. Capt. Basil C. Manly recalled that his guns opened fire at 2:30 p.m., Lt. C. W. Motes of Carlton's battery noted that it was at 3:00 p.m., and Capt. Edward S. McCarthy said it was at 4:00 p.m. (*OR*, vol. 27, pt. 2, 379, 380, 384).

2. *OR*, vol. 27, pt. 2, 380.

3. Ibid., 282.

4. *OR*, vol. 27, pt. 1, 582, 587.

5. Hanifen, *History of Battery B, First New Jersey Artillery*, 68.

6. Ibid., 68–69.

7. Ames, *History of Battery G, First Regiment, New York Light Artillery*, 64–65.

8. Ibid., 65.

9. Ibid.

10. *OR*, vol. 27, pt. 2, 382.

11. Ibid.

12. Alexander, *Fighting for the Confederacy*, 239.

13. Pfanz, *Gettysburg—The Second Day*, 303, 305; *OR*, vol. 27, pt. 2, 380–81; Krick, *Parker's Virginia Battery, C.S.A.*, 155; Hewett, *Supplement to the Official Records of the Union and Confederate Armies*, 5:367.

14. Pfanz, *Gettysburg—The Second Day*, 310.

15. Alexander, *Military Memories of a Confederate*, 398–99; Murray, "The Artillery Duel in the Peach Orchard," 75; Ames, *History of Battery G, First Regiment, New York Light Artillery*, 71.

16. Ames, *History of Battery G, First Regiment, New York Light Artillery*, 66.

17. Wise, *The Long Arm of Lee*, 646.

18. Callihan, "A Cool, Clear Headed Old Sailor: Freeman McGilvery at Gettysburg," 46–47.

19. Baker, *History of the Ninth Massachusetts Battery*, 57.

20. Bigelow, *The Peach Orchard*, 52.

21. Murray, *E. P. Alexander and the Artillery Action in the Peach Orchard*, 60.

22. Bigelow, *The Peach Orchard*, 52.

23. Baker, *The Ninth Massachusetts Battery*, 56–57; Campbell, "Baptism of Fire: The Ninth Massachusetts Battery at Gettysburg, July 2, 1863," 58. Bigelow spied Augustus Hessie dropping down to the ground after each shot. Thinking this to be a cowardly act, Bigelow rode over to him and learned that Hessie was trying to get under the smoke to ascertain the effectiveness of each shot (Baker, *The Ninth Massachusetts Battery*, 79).

24. Fred T. Waugh, "Account," 5th Massachusetts Battery Folder, GNMP.

25. Hunt, "The Second Day at Gettysburg," 305.

26. Bradley, "At Gettysburg."

27. Hunt, "The Second Day at Gettysburg," 305–6.

28. *OR*, vol. 27, pt. 1, 588; Pfanz, *Gettysburg—The Second Day*, 161.

29. Patrick Hart to John Bachelder, January 24, 1891, Bachelder Papers, NHHS.

30. *OR*, vol. 27, pt. 1, 235, 881, 887; Ladd and Ladd, *Bachelder Papers*, 1:166; Appleton, *History of the Fifth Massachusetts Battery*, 630; "Bigelow's Battery," Brake Collection, USAMHI. In a letter to John Bachelder, Hart indicated that Sickles ini-

tially told him where to place the battery, but Hunt countermanded these orders as the battery moved into position (Patrick Hart to John Bachelder, January 24, 1891, Bachelder Papers, NHHS).

31. *OR*, vol. 27, pt. 1, 583, 887; Appleton, *History of the Fifth Massachusetts Battery*, 626; Ladd and Ladd, *Bachelder Papers*, 1:173.

32. *OR*, vol. 27, pt. 1, 586.

33. Pfanz, *Gettysburg—The Second Day*, 309.

34. *OR*, vol. 27, pt. 2, 432.

35. Ladd and Ladd, *Bachelder Papers*, 1:167; 3:1632; Murray, "The Artillery Duel in the Peach Orchard," 80. Artillery guidons were about the same size as a regimental flag.

36. *OR*, vol. 27, pt. 1, 887; Murray, "The Artillery Duel in the Peach Orchard," 81.

37. Naisawald, *Grape and Canister*, 297.

38. *OR*, vol. 27, pt. 1, 881; Murray, *E. P. Alexander and the Artillery Action in the Peach Orchard*, 70, 72; Murray, "The Artillery Duel in the Peach Orchard," 83; Gibbon, *The Artillerists' Manual*, 358.

39. Appleton, *Fifth Massachusetts Battery*, 635,

40. A. C. Sims, "Recollections of A. C. Sims at the Battle of Gettysburg," Brake Collection, USAMHI; John A. Wilkerson, "Experiences of 'Seven Pines' at Gettysburg," 3rd Arkansas folder, GNMP; Barziza, *The Adventures of a Prisoner of War*, 44; Polley, *Hood's Texas Brigade: Its Marches, Its Battles, Its Achievements*, 167; Collier, *"They'll Do to Tie To!"—The Story of the Third Regiment Arkansas*, 138. The historian of Bigelow's battery believed that his unit's fire caused the loss of a third of the 50th Georgia of Paul J. Semmes's brigade while it waited to charge across the fields west of the Rose house. There is little evidence to substantiate this claim (Baker, *History of the Ninth Massachusetts Battery*, 59).

41. George Hillyer, "Battle of Gettysburg Address," *Walton Tribune*, 9th Georgia folder, GNMP.

42. J. C. Reid, "Diary," Alabama State Archives.

43. McLaws, "Gettysburg," 72–73.

44. Murray, "The Artillery Duel in the Peach Orchard," 75; McLaws, "Gettysburg," 73. Bert Barnett believed that the battery may have been S. Capers Gilbert's (Barnett, "The Severest and Bloodiest Artillery Fight I Ever Saw," 71).

45. William H. Hill, "Diary," Mississippi Department of Archives and History.

46. Leftwich, "The Carreer of a Veteran."

47. McNeily, "Barksdale's Mississippi Brigade at Gettysburg," 235–36.

48. Dyer, *The Gallant Hood*, 194; Pfanz, *Gettysburg—The Second Day*, 172.

49. Pfanz, *Gettysburg—The Second Day*, 333.

50. Haynes, *History of the Second New Hampshire: Its Camps, Marches, and Battles*, 171.

51. John Burrill, "Letter," Civil War Times Illustrated Collection, USAMHI.

52. Haynes, *History of the Second New Hampshire: Its Camps, Marches, and Battles*, 169; *OR*, vol. 27, pt. 1, 570.

53. Bloodgood, *Personal Reminiscences of the War*, 134–35; Bowen, "Collis' Zouaves: The 114th Pennsylvania at Gettysburg"; Bates, *History of Pennsylvania Volunteers, 1861–1865*, 4:676; 3:251; 7:7, 441.

54. Dowdey, *The Guns of Gettysburg*, 69.

55. J. S. Henley, "On the Way to Gettysburg," 17th Mississippi Folder, GNMP.

56. Alexander, "Artillery Fighting at Gettysburg," 3:360.

57. Adams, "The Fight at the Peach Orchard"; Loring, "Gettysburg." According to an account written by a member of the 68th Pennsylvania soon after the war, Gen. Charles K. Graham was wounded during this cannonade and carried to the rear. Command of the brigade devolved upon Col. Andrew H. Tippin. Graham purportedly returned to the field a little later in the engagement and was subsequently captured (Bates, *History of Pennsylvania Volunteers*, 4:676).

58. Murray, "The Artillery Duel in the Peach Orchard," 85.

59. Smith, *A Famous Battery and Its Campaigns*, 95–96.

60. *OR*, vol. 27, pt. 2, 375.

61. Hunt, "The Second Day at Gettysburg," 305.

62. Smith, *A Famous Battery and Its Campaigns*, 102–3.

63. Tucker, "Orange Blossoms: Services of the 124th New York at Gettysburg."

64. Smith, *A Famous Battery and Its Campaigns*, 101–2.

65. Ibid., 102.

66. Elijah Walker to John Bachelder, January 5, 1885, Bachelder Papers, NHHS.

67. Smith, *A Famous Battery and Its Campaigns*, 109.

68. Ibid., 103.

69. Oates, *The War Between the Union and Confederacy and Its Lost Opportunities*, 207; William Oates, letter, William Clements Library, University of Michigan; *OR*, vol. 27, pt. 2, 392, 393.

70. *OR*, vol. 27, pt. 2, 404.

71. Smith, *A Famous Battery and Its Campaigns*, 103.

72. Adelman and Smith, *Devil's Den: A History and Guide*, 31.

73. *OR*, vol. 27, pt. 1, 588; Smith, *A Famous Battery and Its Campaigns*, 103.

74. *OR*, vol. 27, pt. 1, 493; Weygant, *History of the One Hundred and Twenty-Fourth Regiment*, 175, 176.

75. *OR*, vol. 27, pt. 2, 415; "A Letter from the Army," *Savannah Republican*, July 22, 1863. Because some of the soldiers referred to Houck's Ridge as "the heights" or "peak," it is often confused with Little Round Top. Thus, in his report, Benning wrote, "When my line reached the foot of the peak, I found there a part of the First Texas, struggling to make the ascent." In reality, the 1st Texas was battling

the 124th New York in the Triangular Field near Devil's Den (*OR*, vol. 27, pt. 2, 415).

76. Tucker, "Orange Blossoms: Services of the 124th New York at Gettysburg."

77. Bradley, "At Gettysburg."

78. Pfanz, *Gettysburg—The Second Day*, 191.

79. Hanifen, *History of Battery B, First New Jersey*, 71.

80. *OR*, vol. 28, pt. 1, 589.

81. Ibid., 589; *OR*, vol. 27, pt. 2, 409, 421; Adelman, "The Fight for and Location of the 4th New York Independent Battery at Gettysburg," 62.

82. *OR*, vol. 27, pt. 2, 426–27; Lokey, "Wounded at Gettysburg," 400.

83. Smith, *A Famous Battery and Its Campaigns*, 104.

84. Ladd and Ladd, *Bachelder Papers*, 2:1094–95; Smith, *A Famous Battery and Its Campaigns*, 105; *OR*, vol. 27, pt. 1, 589.

85. *OR*, vol. 27, pt. 1, 589.

86. Smith, *A Famous Battery and Its Campaigns*, 106.

87. *OR*, vol. 27, pt. 1, 583.

88. *New York at Gettysburg*, 3:1292–93; Smith, *A Famous Battery and Its Campaigns*, 147. Some comic relief occurred when Smith appeared to be wounded. He tumbled to the ground when his horse was hit, then, borrowing another mount, he continued moving his remaining three guns to safety. One of his men yelled, "Captain, you're shot," pointing to Smith's boot. Smith could feel blood squishing around his foot and immediately felt a stab of pain. His boot was removed but revealed no wound; the blood was from his unfortunate horse, and the pain was imaginary.

89. *OR*, vol. 27, pt. 1, 582; Pfanz, *Gettysburg—The Second Day*, 246; Winslow, "On Little Round Top"; Gottfried, *The Maps of Gettysburg*, 167.

90. Dowdey, *The Guns at Gettysburg*, 69.

91. *OR*, vol. 27, pt. 1, 587.

92. Ibid.

93. Winslow, "On Little Round Top." Lt. George Verrill of the 17th Maine told a different story: "Of course many projectiles entering our ranks" (George Verrill to John Bachelder, February 11, 1884, Bachelder Papers, NHHS).

94. *OR*, vol. 27, pt. 1, 583.

95. Ibid., 587.

96. Winslow, "On Little Round Top."

97. *OR*, vol. 27, pt. 1, 587.

98. Winslow, "On Little Round Top."

99. *OR*, vol. 27, pt. 1, 587; Albert Ames to mother, July 9, 1863, Brake Collection, USAMHI; George B. Winslow to John Bachelder, May 17, 1878, Bachelder Papers, NHHS.

100. *OR*, vol. 27, pt. 1, 587–88.

101. Ibid., 583.

102. Ibid.

103. Gottfried, *Maps of Gettysburg*, 144–52.

104. *OR*, vol. 27, pt. 1, 659; Martin, "Little Round Top." The other two batteries in the column were Aaron F. Walcott's and Malbone F. Watson's.

105. Taylor, *Gouveneur Kemple Warren: The Life and Letters of an American Soldier*, 129.

106. Augustus P. Martin, "Artillery Brigade 5th Corps at the Battle of Gettysburg," Joshua Chamberlain Papers, Library of Congress.

107. Rittenhouse, "The Battle of Gettysburg as Seen from Little Round Top," 1:37.

108. Scott, "On Little Round Top."

109. O. W. Damon, "Civil War Diary," 5th U.S. Battery D folder, GNMP.

110. Farley, "Bloody Round Top."

111. Smith, "Account."

112. Martin, "Little Round Top."

113. According to Harry Pfanz, the actual location of Hazlett's battery is unknown (Pfanz, *Gettysburg—The Second Day*, 224).

114. Nash, *A History of the 44th New York Infantry*, 145.

115. Hazen, "Fighting the Good Fight: The 140th New York and Its Work on Little Round Top."

116. Taylor, *Gouveneur Kemple Warren*, 129.

117. Rittenhouse, "The Battle of Gettysburg as Seen from Little Round Top," 38.

118. Bennett, *Sons of Old Monroe: A Regimental History of Patrick O'Rorke's 140th New York Volunteers*, 219.

119. Scott, "On Little Round Top."

120. *OR*, vol. 27, pt. 1, 237; Naisawald, *Grape and Canister*, 262.

121. Scott, "On Little Round Top."

122. *OR*, vol. 27, pt. 1, 237.

Chapter 5: *July 2: The Battle of the Peach Orchard*

1. *OR*, vol. 27, pt. 1, 235. One of eight batteries that were consolidated after the battle of Chancellorsville, the gunners were not happy with the union (Murray, *E. P. Alexander and the Artillery Action in the Peach Orchard*, 67).

2. *OR*, vol. 27, pt. 1, 885.

3. Ibid., 900.

4. Ibid., 900–901.

5. Hanifen, *History of Battery B, First New Jersey Battery*, 73.

6. Ibid., 72.

7. Ibid., 73.

8. *OR*, vol. 27, pt. 1, 881.

9. Ibid.

10. J. C. Reid, "Diary," Alabama State Archives.

11. Hanifen, *History of Battery B, First New Jersey*, 70.

12. *OR*, vol. 27, pt. 1, 586.

13. *OR*, vol. 27, pt. 1, 901. Capt. Nelson Ames said that he received some ammunition from the battery to his left. This would have been James Thompson's, which was the most recent arrival to this sector (Ames, *History of Battery G, First Regiment, New York Light Artillery*, 72–73).

14. Ames, *History of Battery G, First Regiment, New York Light Artillery*, 69–70.

15. Ibid., 70. The lieutenant purportedly told others after the battle that "the last thing he would ever think of doing was to make a suggestion to fall back to that 'little bull dog' as long as he was in command of the battery" (ibid., 70–71).

16. Ibid., 71.

17. Ibid.

18. Appleton, *History of the Fifth Massachusetts Battery*, 636.

19. Ames, *History of Battery G, First Regiment, New York Light Artillery*, 74–75.

20. Ladd and Ladd, *Bachelder Papers*, 3:1632.

21. *OR*, vol. 27, pt. 1, 584.

22. Alexander, *Military Memoirs of a Confederate*, 397–98.

23. *OR*, vol. 27, pt. 1, 881–82.

24. *OR*, vol. 27, pt. 1, 586. A. Judson Clark's guns fired so rapidly that their vents were burned out and expanded from the normal .2 inch to the size of a large finger (.5 inch). The battery's Parrotts were replaced with Napoleons after the battle (George Bonnell to John Bachelder, March 24, 1882, Bachelder Papers, NHHS).

25. Hanifen, *History of Battery B, First New Jersey Battery*, 74–75.

26. Ibid., 75.

27. *OR*, vol. 27, pt. 1, 882.

28. Bigelow, *The Peach Orchard*, 53.

29. Ibid., 53–54.

30. *OR*, vol. 27, pt. 2, 368; Kershaw, "Kershaw's Brigade at Gettysburg," 3:335. While most of the south-facing cannon opened fire on Kershaw's men, it appears that only two of Nelson Ames's cannon were involved. The four others continued to engage Alexander's artillery on Seminary Ridge (Ames, *History of Battery G, First Regiment, New York Light Artillery*, 67; *OR*, vol. 27, pt. 1, 901).

31. Kershaw, "Kershaw's Brigade at Gettysburg," 3:335.

32. *OR*, vol. 27, pt. 2, 368.

33. Alex McNeill, "Letter," 2nd South Carolina folder, GNMP.

34. Coxe, "The Battle of Gettysburg," 433.

35. W. T. Shumate, "With Kershaw at Gettysburg."

36. W. A. Johnson, "The Battle of Gettysburg," 2nd South Carolina folder, GNMP.

37. Kershaw, "Kershaw's Brigade at Gettysburg," 3:335.

38. Smith, *One of the Most Daring of Men*, 83. The veterans' anger did not fade with time. John Coxe wrote, "To think of it makes my blood curdle even though nearly fifty years afterwards—the insane order was given by 'right flank'" (Coxe, "The Battle of Gettysburg," 434).

39. Campbell, "Baptism of Fire: The Ninth Massachusetts Battery at Gettysburg, July 2, 1863," 63.

40. Bandy and Freeland, *The Gettysburg Papers*, 2:735.

41. *OR*, vol. 27, pt. 1, 882.

42. Moran, "A Fire Zouave, Memoirs of a Member of the 73rd New York."

43. Gibbon, *The Artillerists' Manual*, 401.

44. *OR*, vol. 27, pt. 1, 882.

45. Hanifen, *History of Battery B, First New Jersey Battery*, 76.

46. *OR*, vol. 27, pt. 1, 882, 885, 887; Baker, *History of the Ninth Massachusetts Battery*, 59–60; Hanifen, *History of Battery B, First New Jersey Battery*, 77. James Thompson's battery was apparently one of the first to depart, to the chagrin of Capt. Patrick Hart (*OR*, vol. 27, pt. 1, 887).

47. Ladd and Ladd, *Bachelder Papers*, 1:168.

48. Phillips, *History of the Fifth Massachusetts Battery*, 632.

49. Appleton, *History of the Fifth Massachusetts Battery*, 638; Ladd and Ladd, *Bachelder Papers*, 1:168. Charles A. Phillips's battery abandoned a limber during the retreat but retrieved it that night. Phillips explained a prolonge after the war: "The long rope coiled on the trail of the gun is called the 'prolonge.' It is used when you want to retreat and to fire while you are retreating. To do this the order is given 'Fix prolonge to fire retiring!' Then, in the lucid language of the book, 'the limber inclines to the right, wheels to the left about, and halts 4 yards from the trail. N. 5 uncoils the prolonge and passes the toggle to the gunner, who fixes it in the trail by passing it upwards through the lunette, whilst he attaches the other end to the limber by passing the ring over the pintle and keying it. At the command 'Retire!' the cannoneers face about, all march on the left of the piece except Nos. 1 and 3. They keep their implements in their hands &c. &c. That is to say, they go on loading and firing, but the horses all the time dragging the gun away from the enemy" (Appleton, *History of the Fifth Massachusetts Battery*, 628).

50. Appleton, *History of the Fifth Massachusetts Battery*, 625–26.

51. Ames, *History of Battery G, First Regiment, New York Light Artillery*, 73–74.

52. *OR*, vol. 27, pt. 1, 901; Appleton, *History of the Fifth Massachusetts Battery*, 631–32.

53. *OR*, vol. 27, pt. 1, 887. Patrick Hart believed that Freeman McGilvery had

moved the caissons and went to his grave angry with his former commander (Patrick Hart to John Bachelder, February 23, 1891, Bachelder Papers, NHHS). Lt. Edward Knox won the Medal of Honor for his actions near the Peach Orchard. According to the story, Hart brought his guns back to the Peach Orchard, where they were overrun. Realizing that he could not get his guns to safety, Knox ordered his men to play dead until a Federal counterattack saved the pieces (Arrington, *The Medal of Honor at Gettysburg*, 18).

54. Appleton, *History of the Fifth Massachusetts Battery*, 627.

55. Bigelow, *The Peach Orchard*, 55.

56. Benjamin Humphreys to John Bachelder, May 1, 1876, Bachelder Papers, NHHS.

57. Baker, *History of the Ninth Massachusetts Battery*, 60.

58. "Bigelow's Battery," Brake Collection, USAMHI.

59. John Bigelow to John Bachelder, Bachelder Papers, NHHS.

60. Bigelow, *The Peach Orchard*, 55–56.

61. Naisawald, *Grape and Canister*, 308.

62. Bigelow, *The Peach Orchard*, 46.

63. Campbell, "Baptism of Fire: The Ninth Massachusetts Battery at Gettysburg, July 2, 1863," 65–66.

64. *OR*, vol. 27, pt. 1, 889; Woods, "Defending Watson's Battery," 46. Pvt. Casper Carlisle received the Medal of Honor for his actions in helping to save one of Thompson's guns (Arrington, *The Medal of Honor at Gettysburg*, 11).

65. Lt. John K. Bucklyn report, in Hewett, *Supplement to the Official Records of the Union and Confederate Armies*, 5:187–88; Dowdey, *The Guns of Gettysburg*, 69.

66. "Bucklyn's Battery E, 1st Rhode Island Artillery," RG 94, National Archives; John K. Bucklyn to John Bachelder, December 31, 1863, Bachelder Papers, NHHS; *OR*, vol. 27, pt. 1, 590. According to George Lewis, historian of Bucklyn's battery, the guns dueled with S. C. Gilbert's and George V. Moody's batteries in front of them and William W. Parker's and Osmond B. Taylor's to the right (Lewis, *History of Battery E, First Rhode Island Light Artillery*, 206–7). In actuality, Parker's and Taylor's were to Bucklyn's left, not right.

67. *Pennsylvania at Gettysburg*, 2:910.

68. Owen, *In Camp and Battle with the Washington Artillery*, 245.

69. McNeily, *Barksdale's Mississippi Brigade at Gettysburg*, 236.

70. Henley, "On the Way to Gettysburg."

71. Leftwich, "The Carreer of a Veteran."

72. McNeily, *Barksdale's Mississippi Brigade at Gettysburg*, 239.

73. Henley, "On the Way to Gettysburg."

74. Moran, "About Gettysburg."

75. *OR*, vol. 27, pt. 1, 590.

76. "Bucklyn's Battery E, 1st Rhode Island Artillery," RG 94, National Archives.

77. *OR*, vol. 27, pt. 1, 502.

78. Imhof, *Gettysburg: Day Two*, 139.

79. Lewis, *History of Battery E, First Rhode Island Light Artillery*, 208.

80. Ibid., 208–9.

81. John K. Bucklyn to John Bachelder, December 31, 1863, Bachelder Papers, NHHS.

Chapter 6: *July 2: Stemming the Confederate Tide*

1. *OR*, vol. 27, pt. 1, 636.

2. Adolphus Cavada, "Diary," Historical Society of Pennsylvania.

3. *OR*, vol. 27, pt. 1, 553; Marbaker, *History of the Eleventh New Jersey Volunteers*, 97; Blake, *Three Years in the Army of the Potomac*, 206–7.

4. *OR*, vol. 27, pt. 1, 559; Brown, *History of the 3d Regiment, Excelsior Brigade*, 104.

5. *OR*, vol. 27, pt. 1, 532, 591.

6. Francis W. Seeley to John Bachelder, May 23, 1878, Bachelder Papers, NHHS.

7. *OR*, vol. 27, pt. 1, 591.

8. Ibid., 584.

9. Ibid., 533.

10. Ibid., 534.

11. Francis W. Seeley to John Bachelder, May 23, 1878, Bachelder Papers, NHHS.

12. *OR*, vol. 27, pt. 1, 591.

13. Ibid., 532; Ladd and Ladd, *Bachelder Papers*, 1:230–31; "Report of Lieutenant John Graham Turnbull," Henry Jackson Hunt Papers, Library of Congress.

14. *OR*, vol. 27, pt. 1, 873.

15. John G. Turnbull to Henry Hunt, July 1863, Henry Jackson Hunt Papers, Library of Congress.

16. Ladd and Ladd, *Bachelder Papers*, 3:1976; "Report of Lieutenant John Graham Turnbull," Henry Jackson Hunt Papers, Library of Congress.

17. Naisawald, *Grape and Canister*, 299.

18. Alexander, *Fighting for the Confederacy*, 240.

19. Sifakis, *Who Was Who in the Civil War*, 175; Alexander, *Fighting for the Confederacy*, 523.

20. Colston, "Gettysburg as I Saw It," 552.

21. Wise, *The Long Arm of Lee*, 648.

22. Pfanz, *Gettysburg—The Second Day*, 361; Wise, *The Long Arm of Lee*, 646; A. Prince, "Recollections," South Carolina Department of Archives and History.

23. Wise, *The Long Arm of Lee*, 647.

24. According to Alexander, these two batteries were about to be called up to

reinforce the line on Seminary Ridge when they were ordered to follow the infantry (Alexander, *Military Memoirs of a Confederate*, 399).

25. Wise, *The Long Arm of Lee*, 648–49.

26. Alexander, "Gettysburg," Southern Historical Collection, University of North Carolina.

27. *Pennsylvania at Gettysburg*, 2:613.

28. Pfanz, *Gettysburg—The Second Day*, 337.

29. Alexander, "Gettysburg," Southern Historical Collection, University of North Carolina.

30. Alexander, "Gettysburg," Southern Historical Collection, University of North Carolina; Martin, *Confederate Monuments at Gettysburg*, 65.

31. *OR*, vol. 27, pt. 1, 890; Ladd and Ladd, *Bachelder Papers*, 1:173; Pfanz, *Gettysburg—The Second Day*, 339.

32. *OR*, vol. 27, pt. 1, 882.

33. Baker, *History of the Ninth Massachusetts Battery*, 60.

34. Campbell, "Baptism of Fire: The Ninth Massachusetts Battery at Gettysburg, July 2, 1863," 68.

35. Baker, *History of the Ninth Massachusetts Battery*, 61.

36. Ladd and Ladd, *Bachelder Papers*, 1:173.

37. Campbell, "Baptism of Fire: The Ninth Massachusetts Battery at Gettysburg, July 2, 1863," 71.

38. Bigelow, *The Peach Orchard*, 18; "Bigelow's Battery," Brake Collection, USAMHI; Ladd and Ladd, *Bachelder Papers*, 1:174.

39. Bigelow, *The Peach Orchard*, 17–18; Naisawald, *Grape and Canister*, 309.

40. Quoted in McNeily, *Barksdale's Mississippi Brigade at Gettysburg*, 248.

41. John Bigelow to John Bachelder, n.d., in Bachelder Papers, NHHS.

42. *OR*, vol. 27, pt. 1, 886.

43. Baker, *History of the Ninth Massachusetts Battery*, 61.

44. Ibid. Long after the war, Bigelow was able to secure a Medal of Honor for Charles Reed, the gunner who had rescued him on July 2, 1863 ("How the Battle Was Won," *Minneapolis Journal*, August 31, 1895).

45. *OR*, vol. 27, pt. 1, 882, 897; Woods, "Defending Watson's Battery," 46; Ladd and Ladd, *Bachelder Papers*, 1:168; Maine Gettysburg Commission, *Maine at Gettysburg*, 327. McGilvery had commanded the 6th Maine before he ascended to a reserve brigade command (Maine Gettysburg Commission, *Maine at Gettysburg*, 326).

46. *OR*, vol. 27, pt. 1, 882.

47. Ladd and Ladd, *Bachelder Papers*, 1:169.

48. Appleton, *History of the Fifth Massachusetts Battery*, 631.

49. Naisawald, *Grape and Canister*, 313.

50. *OR*, vol. 27, pt. 1, 660. The valor of Watson's battery has been recently

debated. Harry Pfanz wrote that the battery's "stay [in the Peach Orchard] was a short one that added no glory to its reputation." James Woods, however, carefully examined the battery's actions and found that several eyewitnesses had mistaken Malbone F. Watson's battery for James Thompson's battery, which apparently did not perform nobly on the battlefield. Pfanz used an adjective from the battery's official report to help frame his opinions. Lt. Charles C. MacConnell wrote that the battery's actions were "unexceptionable" or irreproachable. Pfanz apparently misread the word as "unexceptional" or ordinary (Woods, "Defending Watson's Battery, 41–42; Pfanz, *Gettysburg—The Second Day*, 317, 347).

51. McNeily, Barksdale's Mississippi Brigade at Gettysburg, 249.

52. *OR*, vol. 27, pt. 1, 474, *New York at Gettysburg*, vol. 2, 906; Beyer and Keydel, *Deeds of Honor*, 240–41. According to David Martin, Lt. Samuel Peeples of the battery "took a musket and led the charge himself, driving the enemy from the guns, and retaking everything that was lost" (*OR*, vol. 27, pt. 1, 660). The guns were not serviceable, as the gunners had thought to take the rammers and friction primers with them (Woods, "Defending Watson's Battery," 42).

53. *OR*, vol. 27, pt. 1, 883. One of Rorty's men wrote long after the battle that the battery was removed from the rest of the 2nd Corps line because of "Captain Rorty's great and hurried desire to distinguish himself" (D. W. Linsday, "War Services of Battery B, First N. Y. Lt. Artillery").

54. Linsday, "War Services of Battery B, First N. Y. Lt. Artillery."

55. *OR*, vol. 27, pt. 1, 883.

56. Ibid., 897.

57. Ibid. The role of Dow's battery was outrageously glorified after the war. While it did help repulse Barksdale's charge, in actuality, it was the timely arrival of the Federal infantry that sealed the Mississippians' fate. The title of Dow's article in the *New York Times* (June 29, 1913) says it all: "How One Brave Battery Saved the Federal Left." The article stated that the role of Dow's battery was as important as John Bigelow's and just as heroic. This viewpoint was echoed to a lesser extent in Maine Gettysburg Commission, *Maine at Gettysburg*, 327–28.

58. Benjamin Humphreys to John Bachelder, May 1, 1876, in Bachelder Papers, NHHS; John Bigelow to W. R. Warner, August 22, 1913, 9th Massachusetts folder, GNMP; Pfanz, *Gettysburg—The Second Day*, 405–6; Maine Gettysburg Commission, *Maine at Gettysburg*, 328.

59. *OR*, vol. 27, pt. 1, 591, 883; Charles A. Phillips to John Bachelder, March 7, 1866, Bachelder Papers, NHHS.

60. *OR*, vol. 27, pt. 1, 883.

61. Scott, "The Artillery at Gettysburg."

62. *OR*, vol. 27, pt. 1, 591, 886, 898.

63. Wise, *The Long Army of Lee*, 656.

64. Murray, *E. P. Alexander and the Artillery Action in the Peach Orchard*, 76, 88.

65. Ibid., 84–86.

66. Alexander, "Artillery Fighting at Gettysburg," 3:359–60.

67. Murray, *E. P. Alexander and the Artillery Action in the Peach Orchard*, 85.

68. Imhof, *Gettysburg—Day Two: A Study in Maps*, 130; Shultz, "Gulian V. Weir's 5th U.S. Artillery, Battery C," 83.

69. Wilcox, "Letter from General C. M. Wilcox, 26 March 1877," 111–17.

70. *OR*, vol. 27, pt. 1, 880. Although Weir had been with the battery since its formation, this was the first battle that he fought as its commander. The battery was also a bit rusty, as it played but minor roles in the battles of Fredericksburg and Chancellorsville (Shultz, "Gulian V. Weir's 5th U.S. Artillery, Battery C," 77).

71. Shultz, "Gulian V. Weir's 5th U.S. Artillery, Battery C," 84.

72. Gulian V. Weir to father, July 5, 1863, copy in 5th U.S. Artillery, Battery C folder, GNMP.

73. *OR*, vol. 27, pt. 1, 880. The closest 3rd Corps troops, Carr's brigade, were actually to his left, not right.

74. *OR*, vol. 27, pt. 2, 631.

75. L. B. Johnson, "A Limited Review of What One Man Saw of the Battle of Gettysburg," 5th Florida folder, GNMP.

76. Francis Heath to John Bachelder, October 12, 1889, Bachelder Papers, NHHS.

77. Ibid. It appears that Hancock mistook Seeley's battery for Weir's (Shultz, "Gulian V. Weir's 5th U.S. Artillery, Battery C," 85).

78. Shultz, "Gulian V. Weir's 5th U.S. Artillery, Battery C," 85.

79. *OR*, vol. 27, pt. 1, 880.

80. Ibid., 873; Shultz, "Gulian V. Weir's 5th U.S. Artillery, Battery C," 86.

81. *OR*, vol. 27, pt. 1, 880.

82. Ibid.

83. Gulian V. Weir took his own life in 1886. Many believed that he was tortured by the temporary loss of his three guns and by his absence when they were recaptured. According to a contemporary, he "took a rifle and put a bullet through his troubled heart" (Pfanz, *Gettysburg—The Second Day*, 378). A recent article, however, argued that Weir was dying of melanoma and decided to end his suffering (Shultz, "Gulian V. Weir's 5th U.S. Artillery, Battery C," 95).

84. *OR*, vol. 27, pt. 1, 416.

85. Ladd and Ladd, *The Bachelder Papers*, 3:1355.

86. Naisawald, *Grape and Canister*, 208–9.

87. Rhodes, *The History of Battery B, First Rhode Island Light Artillery*, 200–201.

88. Coco, *From Ball's Bluff to Gettysburg . . . And Beyond*, 197–98.

89. Rhodes, *The History of Battery B, First Rhode Island Light Artillery*, 201; Grandchamp, "Brown's Company B, 1st Rhode Island Artillery at the Battle of Gettysburg,"

88; Ladd and Ladd, *Bachelder Papers*, 3:1403; Gottfried, *Stopping Pickett: The History of the Philadelphia Brigade*, 158–59.

90. Rhodes, *The History of Battery B, First Rhode Island Light Artillery*, 201.

91. Ibid., 202; Grandchamp, "Brown's Company B, 1st Rhode Island Artillery at the Battle of Gettysburg," 88.

92. Foote, "Marching in Clover."

93. Rhodes, *The History of Battery B, First Rhode Island Light Artillery*, 203–4; Grandchamp, "Brown's Company B, 1st Rhode Island Artillery at the Battle of Gettysburg," 88.

94. Rhodes, *The History of Battery B, First Rhode Island Light Artillery*, 203.

95. Reichardt, *Diary of Battery A, First Rhode Island*, 95.

96. Fuger, "Cushing's Battery at Gettysburg," 406–7.

97. Gottfried, "Wright's Charge on July 2, 1863," 77; Gottfried, *Stopping Pickett: The History of the Philadelphia Brigade*, 161–62.

98. Rhodes, *The History of Battery B, First Rhode Island Light Artillery*, 203–4.

99. Shultz, "Gulian V. Weir's 5th U.S. Artillery, Battery C," 88.

100. *OR*, vol. 27, pt. 1, 501.

101. *OR*, vol. 27, pt. 1, 662.

102. *New York at Gettysburg*, 3:1139.

103. Pfanz, *Gettysburg—The Second Day*, 239.

104. *OR*, vol. 27, pt. 1, 660; Parker, *Henry Wilson's Regiment: History of the Twenty-Second Massachusetts Infantry*, 313.

105. Parker, *Henry Wilson's Regiment: History of the Twenty-Second Massachusetts Infantry*, 313.

106. *OR*, vol. 27, pt. 1, 662.

107. Ibid.

108. Minnigh, *History of Company K, 1st Penn'a Reserves*, 26; *Pennsylvania at Gettysburg*, 1:278. According to Minnigh, the German officer sought out the brigade the following day and exclaimed, "The Pennsylvania Reserves saved mine pattery, by—. I gets you fellers all drunk mit beer."

109. Powell, *The Fifth Army Corps*, 535.

110. Scott, "On Little Round Top."

111. *OR*, vol. 27, pt. 1, 235; Barnett, "The Severest and Bloodiest Artillery Fight I Ever Saw," 78.

112. *OR*, vol. 27, pt. 1, 695.

Chapter 7: *July 2: Cemetery Hill and Culp's Hill*

1. Pfanz, *Gettysburg—Culp's Hill and Cemetery Hill*, 168.

2. *OR*, vol. 27, pt. 2, 543. In a classic case of showmanship, characteristic of

many who fought at Gettysburg, Col. Charles S. Wainwright described the situation as favoring the Confederates: "They were on higher ground, and having plenty of room were able to place their guns some thirty yards apart, while ours were not over twelve . . . the limbers stood absolutely crowded together" (Nevins, *A Diary of Battle*, 242–43).

3. *OR*, vol. 27, pt. 1, 358.

4. Memoir of John William Hatton, Accession Number 9243, DLC, Library of Congress. James Stewart's smoothbore Napoleons were not used in this fight.

5. Moore, *The Story of a Cannoneer Under Stonewall Jackson*, 197.

6. Stewart, "Battery B Fourth United States Artillery at Gettysburg," 374–75.

7. Ibid.

8. Osborn, *The Eleventh Corps Artillery at Gettysburg*, 66.

9. Howard, "Campaign and Battle of Gettysburg," 63.

10. *Pennsylvania at Gettysburg*, 2:901.

11. "Account of Cooper's Actions at Gettysburg," Cooper's Battery folder, GNMP; *Pennsylvania at Gettysburg*, 2:901.

12. "Account of Cooper's Actions at Gettysburg," Cooper's Battery folder, GNMP; Nevins, *A Diary of Battle*, 243.

13. Nevins, *A Diary of Battle*, 244; Callihan, "Captain Michael Wiedrich's Company I, First New York Light Artillery at Gettysburg," 86.

14. Marvel, *The First New Hampshire Battery*, 46.

15. Nevins, *A Diary of Battle*, 243.

16. Kiefer, *History of the 153d Pennsylvania*, 86.

17. Osborn, *Trials and Triumphs: The Record of the Fifty-fifth Ohio Volunteer Infantry*, 100–101.

18. Underwood, *The Three Years' Service of the Thirty-Third Massachusetts Infantry Regiment*, 123.

19. Steuben Coon, "Letter," 60th New York folder, GNMP.

20. Goldsborough, *The Maryland Line in the Confederate States Army*, 146.

21. Pfanz, *Gettysburg—Culp's Hill and Cemetery Hill*, 180; *OR*, vol. 27, pt. 2, 518.

22. William J. Seymour, "Memoirs," William L. Clements Library, University of Michigan.

23. *OR*, vol. 27, pt. 1, 870; Brady, *Hurrah for the Artillery*, 252.

24. *OR*, vol. 27, pt. 1, 870.

25. *OR*, vol. 27, pt. 2, 543–44.

26. Bohannon, *The Giles, Alleghany and Jackson Artillery*, 36.

27. *OR*, vol. 27, pt. 1, 870.

28. Stiles, *Four Years Under Marse Robert*, 217–18.

29. Scott, "The Artillery at Gettysburg"; Goldsborough, *The Maryland Line in the Confederate Army*, 146–47.

30. *OR*, vol. 27, pt. 2, 544.

31. Ibid.

32. *OR*, vol. 27, pt. 2, 544; Seymour, *The Civil War Memoirs of Captain William Seymour*, 74; Moore, *The Charlottesville, Lee Lynchburg and Johnson's Bedford Artillery*, 96.

33. *OR*, vol. 27, pt. 2, 456.

34. Nevins, *A Diary of Battle*, 243. Historian Jay Jorgensen suggested that two of Charles Raine's and two of William F. Dement's guns were left on the hill. It is unclear where he received this information (Jorgensen, "The Confederate Artillery at Gettysburg," 33). It appears that part of Latimer's arm was blown off during the battle. An additional portion was probably amputated afterward. Gangrene, however, set in, and he died on August 1, 1863 (Wise, *The Long Arm of Lee*, 653).

35. *OR*, vol. 27, pt. 1, 363.

36. *OR*, vol. 27, pt. 2, 544; Busey and Martin, *Regimental Strengths and Losses at Gettysburg*, 286; *OR*, vol. 27, pt. 1, 365, 773; Sgt. David Nichol to father, July 9, 1863, Battery E, 1st Pennsylvania folder, GNMP.

37. Buell, *The Cannoneer*, 91–92; Naisawald, *Grape and Canister*, 319.

38. Wise, *The Long Arm of Lee*, 652–53; *OR*, vol. 27, pt. 2, 604. Wise believed that three batteries from Capt. Willis J. Dance's battalion did open fire on Cemetery Hill from their positions on Seminary Ridge. The first-person accounts, however, do not support this suggestion, and the range would have been extreme.

39. Archer, *The Hour Was One of Horror*, 32–35.

40. Capt. R. J. Hancock, Letter, John Daniel Papers, University of Virginia.

41. R. Stark Jackson, Letter, Louisiana State University Library.

42. William Seymour, Journal, William L. Clements Library, University of Michigan.

43. Kiefer, *History of the 153d Pennsylvania*, 219–20.

44. *OR*, vol. 27, pt. 1, 358; Pfanz, *Gettysburg—Culp's Hill and Cemetery Hill*, 252; Tomasak, "An Encounter with Battery Hell," 33. As Wainwright was positioning Ricketts's guns, he said, "This is the key to our position on Cemetery Hill and must be held, and in case you are charged here, you will not limber up and leave under any circumstances, but fight your battery as long as you can" (R. Bruce Ricketts to John P. Nicholson, September 10, 1893, GNMP).

45. William Seymour, Journal, William L. Clements Library, University of Michigan.

46. Kiefer, *History of the 153d Pennsylvania*, 152.

47. Tomasak, "An Encounter with Battery Hell," 36–37.

48. Andrew Harris to John Bachelder, March 14, 1881, Bachelder Papers, NHHS.

49. Maine Gettysburg Commission, *Maine at Gettysburg*, 94–95.

50. Edward N. Whittier, "The Left Attack (Ewell's) at Gettysburg," 3:86–87.

51. *OR*, vol. 27, pt. 1, 363.

52. *OR*, vol. 27, pt. 2, 280.

53. Capt. R. J. Hancock, Letter, John Daniel Papers, University of Virginia.

54. Ryder, *Reminiscences of Three Years' Service in the Civil War by a Cape Cod Boy*, 35.

55. Hewett, *Supplement to the Official Records of the Union and Confederate Armies*, 5:218.

56. Maine Gettysburg Commission, *Maine at Gettysburg*, 95.

57. *OR*, vol. 27, pt. 1, 361.

58. Buell, *The Cannoneer*, 83.

59. *OR*, vol. 27, pt. 1, 361.

60. Joseph Todd, Diary, Battery F & G, 1st Pennsylvania folder, GNMP.

61. Nevins, *A Diary of Battle*, 245.

62. R. Bruce Ricketts to John Bachelder, March 2, 1866, Bachelder Papers, NHHS.

63. Nevins, *A Diary of Battle*, 245.

64. Jones, *Cemetery Hill*, 88–89.

65. Nevins, *A Diary of Battle*, 245–46.

66. Pfanz, *Gettysburg—The Second Day*, 268–70; Clark, *North Carolina Regiments*, 1:313–14; Lt. C. B. Brockway to D. Conaughy, March 5, 1864, Ricketts's Battery folder, GNMP.

67. *OR*, vol. 27, pt. 2, 280.

68. R. Bruce Ricketts to John Bachelder, December 3, 1883, Bachelder Papers, NHHS.

69. Moore, "Charge of the Louisianians."

70. Ladd and Ladd, *The Bachelder Papers*, 1:238.

71. Turner, "Ricketts' Role Remembered at Gettysburg Anniversary."

72. *OR*, vol. 27, pt. 1, 358.

73. Goldsborough, *The Maryland Line in the Confederate States Army*, 116.

74. *OR*, vol. 27, pt. 2, 536, 538, 539.

75. Hamblen, *Connecticut Yankees at Gettysburg*, 72–73

76. Coy, "Letter."

77. *OR*, vol. 27, pt. 1, 761, 775; *OR*, vol. 27, pt. 2, 447, 504; Williams, *From the Cannon's Mouth*, 230.

78. *OR*, vol. 27, pt. 1, 761.

79. Pfanz, *Gettysburg—Culp's Hill and Cemetery Hill*, 176.

80. *OR*, vol. 27, pt. 1, 870.

81. Pfanz, *Gettysburg—Culp's Hill and Cemetery Hill*, 285.

82. *OR*, vol. 27, pt. 2, 836; *OR*, vol. 27, pt. 1, 870.

83. Storrs, *The Twentieth Connecticut*, 92–93.

84. J. R. Lynn, "At Gettysburg."

85. Toombs, *New Jersey Troops in the Gettysburg Campaign*, 274–75; Brown, *The Twenty-Seventh Indiana Volunteer Infantry in the War of the Rebellion, 1861 to 1865*, 387.

86. Armstrong, *Twenty-fifth Virginia Infantry and Ninth Virginia Infantry*, 64.

87. Benjamin Jones, "Memoirs," CWTI Collection, USAMHI.

88. Storrs, *The Twentieth Connecticut*, 92–93; *OR*, vol. 27, pt. 1, 784.

89. *OR*, vol. 27, pt. 1, 801.

90. Ibid., 785; Cruikshank, "Memoirs."

91. *OR*, vol. 27, pt. 1, 871.

Chapter 8: *July 3: Up to 1:00 P.M.*

1. Wise, *The Long Arm of Lee*, 658–59. According to a story published in the *Pittsburgh Commercial Gazette* after the war, George G. Meade ordered the artillerymen on Cemetery Hill to replenish their ammunition chests with unexploded Confederate shells that were everywhere in abundance. Meade was even seen assisting in this effort (Time-Life Books, *Arms and Equipment of the Confederacy*, 283).

2. Chilton, *Unveiling and Dedication of the Monument to Hood's Texas Brigade*, 340; Adelman, "The Fight for and Location of the 4th New York Independent Battery at Gettysburg," 63.

3. Alexander, "Artillery Fighting at Gettysburg," 361. The Washington Artillery was a venerable unit, having been formed in 1838, and had proved its worth on many Civil War battlefields. It remained in reserve when it arrived in Gettysburg on July 2 (Wise, *The Long Arm of Lee*, 93).

4. Alexander, *Military Memoirs of a Confederate*, 416.

5. Rollins, "Lee's Artillery Preparation for Pickett's Charge," 44.

6. Ibid., 45, 50.

7. *OR*, vol. 27, pt. 2, 320.

8. Hess, *Pickett's Charge*, 23.

9. Stewart, *Pickett's Charge*, 118.

10. Alexander, *Military Memoirs of a Confederate*, 418–19.

11. Priest, *Into the Fight*, 30; Wood, *Reminiscences of Big I*, 44; *OR*, vol. 27, pt. 2, 434.

12. *OR*, vol. 27, pt. 2, 375, 388; Elmore, "The Grand Cannonade: A Confederate Perspective," 109; Jorgensen, "Confederate Artillery at Gettysburg," 31, 34; Priest, *Into the Fight*, 187–89.

13. Priest, *Into the Fight*, 50.

14. *OR*, vol. 27, pt. 2, 619.

15. Alexander, *Fighting for the Confederacy*, 245.

16. Alexander, "Artillery Fighting at Gettysburg," 361.

17. Alexander, *Fighting for the Confederacy*, 246.

18. Alexander, *Military Memoirs of a Confederate*, 418.

19. Hunt, "The Third Day at Gettysburg," 3:371–72.

20. Alexander, "Artillery Fighting at Gettysburg," 362.

21. E. Porter Alexander to John Bachelder, May 3, 1876, Bachelder Papers, NHHS.

22. Hess, *Pickett's Charge*, 30–31.

23. According to John Priest, there were actually eight howitzers that came from the following commands: one each from George Ward's and James W. Wyatt's batteries and two each from James V. Brooke's, Joseph Graham's, and Charles R. Grandy's batteries (Priest, *Into the Fight*, 24, 183).

24. Sifakis, *Who Was Who in the Civil War*, 511.

25. Graham, "An Awful Affair," 47.

26. *OR*, vol. 27, pt. 2, 673. These seven guns were kept near the northwest corner of Spangler's Woods (Christ, *The Struggle for the Bliss Farm at Gettysburg, July 2nd and 3rd, 1863*, 64).

27. Cockrell, *Gunner with Stonewall: Reminiscences of William Thomas Poague*, 74.

28. *OR*, vol. 27, pt. 2, 605.

29. *OR*, vol. 27, pt. 2, 603, 604; Elmore, "The Grand Cannonade: A Confederate Perspective," 102.

30. Alexander, *Military Memoirs of a Confederate*, 418–19.

31. Rollins, "Lee's Artillery Preparation for Pickett's Charge," 46–47.

32. *OR*, vol. 27, pt. 2, 352.

33. Rollins, "Lee's Artillery Preparation for Pickett's Charge," 47–48.

34. Cole, *Civil War Artillery at Gettysburg*, 205, 206, 213, 280.

35. OR, vol. 27, pt. 1, 238; Lewis, *History of Battery E, First Rhode Island Light Artillery*, 209.

36. Naisawald, *Grape and Canister*, 324–25.

37. *OR*, vol. 27, pt. 1, 883; Shultz, *Double Canister at Ten Yards*, 6–7, 20; Hess, *Pickett's Charge*, 114. Milton's two Napoleons were deployed just north of Abraham Bryan's barn, about seventy-five yards south of George A. Woodruff's. John G. Turnbull's two guns were posted in the southern portion of Bryan's orchard (Shultz, *Double Canister at Ten Yards*, 20).

38. Murray, "Cowan's, Cushing's, and Rorty's Batteries in Action During the Pickett-Pettigrew-Trimble Charge," 39, 41; Hunt, "The Third Day at Gettysburg," 3:371; Hankin, *History of the First Regiment of Artillery*, 170. An Irish immigrant, Capt. James M. Rorty had only been in America since 1857. He was captured at First Bull Run and later made a daring escape from a Richmond prison (Murray, "Cowan's, Cushing's, and Rorty's Batteries in Action During the Pickett-Pettigrew-Trimble Charge," 41).

39. Osborn, *The Eleventh Corps Artillery at Gettysburg*, 30. Frederick M. Edgell's battery had moved from the left of the first line to a reserve position in a cornfield during the evening of July 2; it was then moved into the second line in the cemetery when the barrage began (Marvel, *First New Hampshire Battery*, 47, 49).

40. Osborn, *The Eleventh Corps Artillery at Gettysburg*, 70.

41. Naisawald, *Grape and Canister*, 329; Shultz, "Double Canister at Ten Yards," 19.

42. Henry Hunt to John Bachelder, January 6, 1866, Bachelder Papers, NHHS.

43. *OR*, vol. 27, pt. 1, 238.

44. *OR*, vol. 27, pt. 2, 434.

45. Alexander, *Fighting for the Confederacy*, 245.

46. *OR*, vol. 27, pt. 1, 591.

47. Fuger, "Cushing's Battery at Gettysburg," 407.

48. Brown, "By Hand to the Front," 13.

49. Ibid., 10.

50. *OR*, vol. 27, pt. 2, 673–74.

51. Rhodes, *History of Battery B, First Rhode Island Light Artillery*, 208.

52. Fuger, "Cushing's Battery at Gettysburg," 407.

53. Smith, "Bloody Angle." Smith incorrectly believed that only two caissons exploded, and that they did so on July 2.

54. Naisawald, *Grape and Canister*, 331.

55. Fuger, "Cushing's Battery at Gettysburg," 407.

56. Christ, *Struggle for the Bliss Farm at Gettysburg, July 2nd and 3rd, 1863*, 65.

57. Alexander, "Artillery Fighting at Gettysburg," 362.

58. Carmichael, "Every Map of the Field Cries Out About It," 278.

59. Alexander, *Military Memoirs of a Confederate*, 420.

60. Alexander to Bachelder, May 3, 1876.

61. *OR*, vol. 27, pt. 2, 388, 434.

62. Ibid., 434.

63. Ibid., 375–76, 379.

64. Ibid., 434; Scott, "The Artillery at Gettysburg."

65. Rhodes, *History of Battery B, First Rhode Island Light Artillery*, 208.

66. Aldrich, *The History of Battery A, First Regiment Rhode Island Light Artillery in the War to Preserve the Union*, 211.

67. Priest, *Into the Fight*, 11; John Cheves Haskell Account, John Warwick Daniel Papers, Duke University.

68. Alexander, *Fighting for the Confederacy*, 245.

69. Alexander, "Artillery Fighting at Gettysburg," 362.

70. Ibid., 362.

71. Alexander, "Artillery Fighting at Gettysburg," 363.

72. Alexander, *Military Memoirs of a Confederate*, 421–22.

73. Ibid., 422.

74. Shultz and Rollins, "A Combined and Concentrated Fire," 52.

75. Alexander, *Military Memoirs of a Confederate*, 420; Alexander to Bachelder, May 3, 1876.

76. Alexander, *Military Memoirs of a Confederate*, 420; Alexander, *Fighting for the Confederacy*, 258.

77. Alexander to Bachelder, May 3, 1876.

Chapter 9: *July 3: The Great Cannonade*

1. Owen, *In Camp and Battle with the Washington Artillery*, 248.

2. Alexander, "The Great Charge and Artillery Fighting at Gettysburg," 362. L. Van Loan Naisawald estimated that the 140 or so Confederate cannon fired at a rate of sixty to seventy-five shots a minute (Naisawald, *Grape and Canister*, 332).

3. Brown, "By Hand to the Front," 13.

4. *OR*, vol. 27, pt. 1, 239.

5. Schultz, *Double Canister at Ten Yards*, 3–4.

6. *OR*, vol. 27, pt. 2, 388.

7. Aldrich, *The History of Battery A, First Regiment Rhode Island Light Artillery in the War to Preserve the Union*, 211.

8. Gibbon, *Personal Recollections of the Civil War*, 178.

9. Stewart, *Pickett's Charge*, 131.

10. Hunt, "The Third Day at Gettysburg," 371–72.

11. *OR*, vol. 27, pt. 1, 750.

12. Account of Brig. Gen. Henry Hunt, January 20, 1873, Bachelder Papers, NHHS. Hunt apparently never informed Hancock of his plan. Whether such a discussion would have altered Hancock's opinions can only be speculated (Kross, "I Do Not Believe That Pickett's Division Would Have Reached Our Line," 293–94).

13. Hunt to Bachelder, January 6, 1866.

14. *OR*, vol. 27, pt. 1, 480. According to Hunt, "Capt. Hazard informed him [Hancock] of my orders and begged him not to insist upon his own, but to this he would not listen and compelled a rapid rely to the enemy" (Account of Brig. Gen. Henry Hunt, January 20, 1873, Bachelder Papers, NHHS).

15. Rhodes, *History of Battery B, First Rhode Island Light Artillery*, 209; *OR*, vol. 27, pt. 1, 239; Alexander, *Military Memoirs of a Confederate*, 422; Grandchamp, "Brown's Company B, 1st Rhode Island Artillery at the Battle of Gettysburg," 88; Barnett, "Union Artillery on July 3," 222. A fight between Hunt and Hancock erupted after the war. Hancock explained his actions in taking command of the artillery assigned to his corps: "It is thought to be common sense and much safer, that those

commanders, who fight the troops in time of war, and are responsible for success or failure of the operations, should have the same control . . . over other arms . . . rather than to have them subject to the command of officers, who would not be responsible in the event of a loss, while the Chief of Artillery of the army would have had no responsibility in that event" (Hancock letter quoted in Henry Hunt to John Bachelder, February 1882, Bachelder Papers, NHHS).

16. Account of Brig. Gen. Henry Hunt, January 20, 1873, Bachelder Papers, NHHS.

17. *OR*, vol. 27, pt. 1, 884; Longacre, *The Man Behind the Guns*, 174.

18. *OR*, vol. 27, pt. 1, 885, 888, 889; Longacre, *The Man Behind the Guns*, 174.

19. Ladd and Ladd, *Bachelder Papers*, 3, 1798

20. *Pennsylvania at Gettysburg*, 2:911.

21. *OR*, vol. 27, pt. 1, 883.

22. *Pennsylvania at Gettysburg*, 2:911.

23. Applegate, *History of the Fifth Massachusetts Battery*, 652.

24. *OR*, vol. 27, pt. 1, 884.

25. Ibid.

26. *OR*, vol. 27, pt. 1, 885.

27. Applegate, *History of the Fifth Massachusetts Battery*, 652.

28. Henry Hunt to John Bachelder, January 6, 1866, Bachelder Papers, NHHS.

29. Henry Hunt to John Bachelder, January 20, 1863, Bachelder Papers, NHHS.

30. Galloway, "Gettysburg: The Battle and the Retreat," 388.

31. Koleszar, *Ashland, Bedford, and Taylor Virginia Light Artillery*, 24.

32. Moore, *The Richmond Fayette, Hampden, Thomas, and Blount's Lynchburg Artillery*, 79.

33. Marye, "The First Gun at Gettysburg," 1231.

34. Carmichael, *The Purcell, Crenshaw and Letcher Artillery*, 188.

35. Ibid., 189.

36. McDermott, *A Brief History of the 69th Regiment Pennsylvania Veteran Volunteers*, 30; Survivors of the Seventy-Second Regiment of Pennsylvania Volunteers, Plaintiffs, vs. Gettysburg Battlefield Memorial Association . . . (Supreme Court of Pennsylvania, Middle District, May Term, 1891, Nos. 20 and 30), 266.

37. Gibbon, *Personal Recollections of the Civil War*, 147.

38. Smith, "Bloody Angle."

39. Downey, *The Guns at Gettysburg*, 125. After fighting in the battle of First Bull Run, Cushing served through the winter of 1862 as a 2nd Corps staff officer. He returned to the artillery in command of his own battery early in 1863 and fought with distinction at Chancellorsville (Sifakis, *Who Was Who in the Civil War*, 161).

40. Smith, "Bloody Angle"; Brown, *Cushing of Gettysburg: The Story of a Union Artillery Commander*, 7, 18.

41. Smith, "Bloody Angle."

42. Gibbon, *Personal Recollections of the Civil War*, 148.

43. Rhodes, *The History of Battery B, First Rhode Island Light Artillery*, 209.

44. Ibid., 210.

45. Waitt, *History of the Nineteenth Regiment, Massachusetts Volunteer Infantry*, 237.

46. Frederick Oesterle, Memoirs, CWTI Collection, USAHMI.

47. Bandy and Freeland, *The Gettysburg Papers*, 1:432–33.

48. Shultz, *Double Canister at Ten Yards*, 24.

49. Linsday, "War Services of Battery B, First N.Y. Lt. Artillery." The battery had been in the artillery reserve until July 1, when it was transferred to the 2nd Corps. At that time, Hancock made the change in command. The men were still smarting over the slight to Sheldon (Shultz, *Double Canister at Ten Yards*, 11–12).

50. E. Corbin letter, Gregory A. Coco Collection, GNMP.

51. Jackson, "The Battle of Gettysburg," 1:179.

52. Purcell, "The Nineteenth Massachusetts at Gettysburg," 282–83.

53. Waitt, *History of the Nineteenth Regiment, Massachusetts Volunteer Infantry*, 235–36; OR, vol. 27, pt. 1, 443.

54. Waitt, *History of the Nineteenth Regiment, Massachusetts Volunteer Infantry*, 236.

55. Stewart, *Pickett's Charge*, 149.

56. Scott, "The Artillery at Gettysburg"; Osborn, *The Eleventh Corps Artillery at Gettysburg*, 31.

57. Benjamin W. Thompson, "Personal Narrative of Experiences in the Civil War, 1861–1865," Civil War Times Illustrated Collection, USAMHI; *OR*, vol. 27, pt. 1, 473; George Yost Letter, 126th New York folder, GNMP; *New York at Gettysburg*, 2:800, 801, 889.

58. Haines, *History of the Men of Company F*, 41; Washburn, *A Complete Military History and Record of the 108th Regiment N. Y. Vols. from 1862 to 1894*, 50; Diary of Francis Wafer, 108th New York file, GNMP.

59. Priest, *Into the Fight*, 55.

60. Naisawald, *Grape and Canister*, 335.

61. Cerbin, "Pettit's Batt. At Gettysburg."

62. Smith, "Bloody Angle."

63. Ibid.

64. Rollins, "Lee's Artillery Prepares for Pickett's Charge," 42; Cole, *Civil War Artillery at Gettysburg*, 243–44.

65. Henry Hunt to W. T. Sherman, February 1882, Bachelder Papers, NHHS.

66. Wilkeson, "A Tragedy at Gettysburg," 387.

67. Hunt, "The Third Day at Gettysburg," 373.

68. *OR*, vol. 27, pt. 1, 879.

69. Gulian V. Weir to Winfield Hancock, December 7, 1885, Bachelder Papers, NHHS.

70. OR, vol. 27, pt. 1, 239; Shultz, *Double Canister at Ten Yards*, 25.

71. Osborn, *The Eleventh Corps Artillery at Gettysburg*, 32.

72. Jorgensen, "Confederate Artillery at Gettysburg," 36; *OR*, vol. 27, pt. 2, 603.

73. *OR*, vol. 27, pt. 2, 603.

74. White, *Contributions to a History of the Richmond Howitzer Battalion*, 207–8.

75. *OR*, vol. 27, pt. 2, 603.

76. *Marietta (GA) Register*, July 17, 1863.

77. Osborn, *The Eleventh Corps Artillery at Gettysburg*, 34–35.

78. *OR*, vol. 27, pt. 1, 871.

79. *OR*, vol. 27, pt. 2, 456.

80. Ibid., 604.

81. Osborn, *The Eleventh Corps Artillery at Gettysburg*, 35–36.

82. Ibid., 37.

83. Elmore, "The Grand Cannonade: A Confederate Perspective," 102.

84. White, *Contributions to a History of the Richmond Howitzer Battalion*, 207–8.

85. Osborn, *The Eleventh Corps Artillery at Gettysburg*, 37–38; Collier and Collier, "Sgt. John W. Chase, Company A, 1st Massachusetts Light Artillery," 72. There was much movement of the batteries under Osborn's command. Frederick M. Edgell's battery, which was initially in a cornfield in the rear of the cemetery, was moved up to take George W. Norton's position. Capt. William H. McCartney's 1st Massachusetts Light Artillery Battery A (six Napoleons) briefly occupied Edgell's old position before pulling out. Two of Elijah D. Taft's cannon that were firing to the west were replaced by the three in the cemetery after one of them burst (Osborn, *The Eleventh Corps Artillery at Gettysburg*, 45–47; *OR*, vol. 27, pt. 1, 891).

86. *OR*, vol. 27, pt. 1, 689.

87. Browne, "Battery H, 1st Ohio Light Artillery: Controversy in the Cemetery," 125–26.

88. Johnston, *Four Years a Soldier*, 249, 253.

89. Ibid., 254; Gunn, *Twenty-fourth Virginia*, 45; *OR*, vol. 27, pt. 2, 650–51.

90. Hilary Herbert to Edward Alexander, August 8, 1869, McLaws Papers, Southern Historical Collection, University of North Carolina.

91. Clark, *A Glance Backward*, 39.

92. Turney, "The First Tennessee at Gettysburg," 535.

93. Birkett Fry to John Bachelder, December 27, 1877, Bachelder Papers, NHHS.

94. Benedict, *Vermont in the Civil War*, 12–13

95. Wheelock Veazey to John Bachelder, December 1863, Bachelder Papers, NHHS.

96. *OR*, vol. 27, pt. 1, 449.

97. Gibbon, *Personal Recollections of the Civil War*, 178.

98. Carmichael, *Lee's Young Artillerist*, 103.

99. Fuger, "Cushing's Battery at Gettysburg," 408.

100. *OR*, vol. 27, pt. 1, 690.

101. Murray, *"Hurrah for the Ould Flag!"* 68–74; New York Monuments Commission for the Battlefields of Gettysburg, *In Memoriam, Alexander Steward Webb*, 65, hereafter cited as *In Memoriam*.

102. Cowan, "When Cowan's Battery Withstood Pickett's Splendid Charge."

103. Ibid. The twenty-two-year-old Cowan had immigrated from Scotland with his parents prior to the war and was in college at its outbreak. Initially an infantry private, Cowan joined the artillery and rose rapidly in command (Shultz, *Double Canister at Ten Yards*, 18).

104. Shultz, *Double Canister at Ten Yards*, 36; *OR*, vol. 27, pt. 1, 896.

105. *OR*, vol. 27, pt. 1, 896.

106. Naisawald, *Grape and Canister*, 340; Hess, *Pickett's Charge*, 145.

107. Alexander, *Military Memoirs of a Confederate*, 418–19. According to the Confederate 3rd Corps artillery officers' official reports, all but Lt. Col. John Garnett's battalion participated in the cannonade (*OR*, vol. 27, pt. 2, 635, 653, 673–74, 675, 678).

108. Alexander, *Fighting for the Confederacy*, 251.

109. Wise, *The Long Arm of Lee*, 668.

110. Osborn, "The Artillery at Gettysburg."

111. Stewart, *Pickett's Charge*, 146.

112. Alexander, *Military Memoirs of a Confederate*, 423.

113. Ibid.

114. Ibid. Earl Hess believes that Alexander was actually referring to the 2nd Corps artillery in the center of the Federal line (Hess, *Pickett's Charge*, 160).

115. Alexander, "The Great Charge and Artillery Fighting at Gettysburg," 364.

116. Hunt to Bachelder, January 6, 1866.

117. Kross, "I Do Not Believe That Pickett's Division Would Have Reached Our Line," 297–298; *OR*, vol. 27, pt. 1, 239.

118. Hunt, "The Third Day at Gettysburg," 374. It was an aide to Hancock who informed Hunt that Meade's aides were carrying orders for him to cease all artillery fire. The impetus for Meade's orders appears to be Brig. Gen. Gouverneur Warren, who carefully watched the unfolding events atop Little Round Top. Maj. Thomas W. Osborn claimed that he was responsible for suggesting the idea of gradually reducing the intensity of the firing to Hunt (Osborn, *The Eleventh Corps Artillery at*

Gettysburg, 39; Ladd, *The Bachelder Papers*, 1:430–31; Cooksey, "Forcing the Issue: Brig. Gen. Henry Hunt at Gettysburg on July 3, 1863," 80).

119. Alexander, *Fighting for the Confederacy*, 261.

120. *OR*, vol. 27, pt. 2, 352.

121. Stewart, *Pickett's Charge*, 154.

122. Rollins, "Lee's Artillery Preparation for Pickett's Charge," 51.

123. Elmore, "The Grand Cannonade: A Confederate Perspective," 100, 107, 110, 111; Hess, *Pickett's Charge*, 162, 164.

124. Galloway, "Gettysburg," 388.

125. Hess, *Pickett's Charge*, 162; Hewett, *Supplement to the Official Records of the Union and Confederate Armies*, 5:310.

126. Hunt, "The Third Day at Gettysburg," 374.

127. McIntosh, "A Review of the Gettysburg Campaign by One Who Participated," 136–37.

128. *OR*, vol. 27, pt. 2, 352.

129. Rollins, "Lee's Artillery Prepares for Pickett's Charge," 42; Cole, *Civil War Artillery at Gettysburg*, 243–44; Carmichael, "Every Map of the Field Cries Out About It," 273–74.

130. Cole, *Civil War Artillery at Gettysburg*, 283.

131. Hunt, "The Third Day at Gettysburg," 373–74.

132. Carmichael, "Every Map of the Field Cries Out About It," 279.

133. Lee, *General Lee*, 293.

134. Carmichael, "Every Map of the Field Cries Out About It," 275, 281.

Chapter 10: *July 3: The Pickett-Pettigrew-Trimble Charge*

1. Harrison and Busey, *Nothing but Glory*, 42; "Terrific Fight of Third Day," *Scranton (PA) Truth*, July 3, 1913; Eppa Hunton, Letter, University of Virginia.

2. Smith, "Bloody Angle."

3. Hunt to Bachelder, January 6, 1866.

4. Account of Brig. Gen. Henry Hunt, January 20, 1873, Bachelder Papers, NHHS; Walker, "General Hancock and the Artillery at Gettysburg," 3:385–87.

5. Walker, "The Charge of Pickett's Division," 222.

6. Shotwell, "Virginia and North Carolina in the Battle of Gettysburg," 91–92.

7. John H. Lewis, Memoirs, Brake Collection, USAMHI.

8. William B. Robertson, Account, Daniel's Papers, University of Virginia.

9. Fields, *Twenty-Eighth Virginia Infantry*, 26.

10. "Terrific Fight of Third Day," *Scranton (PA) Truth*, July 3, 1913.

11. *OR*, vol. 27, pt. 2, 376; Alexander, *Fighting for the Confederacy*, 262.

12. Reid, "Hugh Garden's Battery: Recollections of Gettysburg."

13. Rittenhouse, "The Battle of Gettysburg as Seen from Little Round Top," 43. In reviewing eighteen cases where solid shot was used against infantry during the battle, Thomas Elmore found that in fourteen cases only one man was struck, and two were struck in the remaining four cases (Elmore, "The Grand Cannonade: A Confederate Perspective," 102).

14. Rittenhouse, "The Battle of Gettysburg as Seen from Little Round Top," 43.

15. *OR*, vol. 27, pt. 1, 883.

16. Sgt. Joseph Todd, Diary, GNMP.

17. Shultz, *Double Canister at Ten Yards*, 13, 45; *OR*, vol. 27, pt. 1, 364; *Pennsylvania at Gettysburg*, 2:902.

18. *OR*, vol. 27, pt. 1, 884.

19. Ibid., 888.

20. Applegate, *History of the Fifth Massachusetts Battery*, 653.

21. *OR*, vol. 27, pt. 1, 896, 899. Capt. Jabez Daniel's 9th Michigan Artillery (six ordnance rifles), assigned to Capt. James M. Robertson's 1st Brigade of the cavalry corps horse artillery, wheeled into position during the bombardment.

22. *OR*, vol. 27, pt. 1, 239.

23. Ibid., 480.

24. Ladd and Ladd, *Bachelder Papers*, 1:431–32.

25. *In Memoriam*, 65.

26. Andrew Cowan, "Undated Remarks," Alexander Stewart Collection, Yale University Library; Cowan, "When Cowan's Battery Withstood Pickett's Charge." Cowan's battery was the only 6th Corps unit to see action at Gettysburg. William H. McCartney's and William A. Harn's batteries were sent to Cemetery Hill, and Leonard Martin's and John H. Butler's arrived at Zeigler's Grove at the end of the Pickett-Pettigrew-Trimble Charge (Scott, "The Artillery at Gettysburg").

27. Cowan, "When Cowan's Battery Withstood Pickett's Charge."

28. Andrew Cowan to John Bachelder, August 26, 1866, Bachelder Papers, NHHS. Cowan wrote to Colonel Bachelder after the war that T. Frederick Brown's battery was "almost annihilated by the heavy cannonade." The excitement for Brown's battery was not over yet. One of the four pieces took a wrong turn, and because several of the horses were killed or wounded, the gun had to be abandoned. When the gunners returned after the battle, they found that some enterprising soldiers, probably from another battery, had taken the gun barrel and all of the other usable equipment, leaving but the carcass behind (Rhodes, "The Gettysburg Gun," 21, 23, 27). Bachelder apparently suggested to Cowan that a painting of his battery in action would be on the cover of his new map. Cowan graciously declined, writing, "Brown was there before me and was not withdrawn until he had done all that men and guns could do[;] he was entitled to be shown in front and my battery directly behind, as nearly in the same place as possible" (Cowan to Bachelder, November 24, 1885, Bachelder Papers, NHHS).

29. *In Memoriam*, 65.

30. Cowan, "When Cowan's Battery Withstood Pickett's Charge." Cowan apparently asked Cushing to watch the gun on the north side of the Copse of Trees so he could remain with his other five guns to the south of it (Scott, "The Artillery at Gettysburg").

31. *OR*, vol. 27, pt. 1, 753.

32. Fuger, "Cushing's Battery at Gettysburg," 408; Murray, "Cowan's, Cushing's, and Rorty's Batteries in Action During the Pickett-Pettigrew-Trimble Charge," 48.

33. Smith, "Bloody Angle."

34. Ibid.

35. Anthony McDermott to John Bachelder, June 2, 1886, Bachelder Papers, NHHS.

36. Lewis, "Memoirs"; Crocker, *Gettysburg: Pickett's Charge and Other War Addresses*, 43; Gregory, *Thirty-Eight Virginia Infantry*, 40; Clement, *The History of Pittsylvania County, Virginia*, 249.

37. Lewis, "Memoirs"; Crocker, *Gettysburg: Pickett's Charge and Other War Addresses*, 43; Gregory, *Thirty-Eight Virginia Infantry*, 40; Clement, *The History of Pittsylvania County, Virginia*, 249.

38. *OR*, vol. 27, pt. 1, 239.

39. Ibid., 480.

40. Shultz, *Double Canister at Ten Yards*, 51; *New York at Gettysburg*, 3:1183–84.

41. *OR*, vol. 27, pt. 1, 690.

42. Cowan, "When Cowan's Battery Withstood Pickett's Charge."

43. Ibid.; Hewett, *Supplement to the Official Records of the Union and Confederate Armies*, 5:213–14.

44. Fuger, "Cushing's Battery at Gettysburg," 408. Cushing's second wound was exceedingly painful as it probably removed his testicles. The blood oozed out of the wound and onto his hand as he held it while continuing to direct the actions of his gunners (Ladd and Ladd, *Bachelder Papers*, 3:1978).

45. Brown, "By Hand to the Front," 14; Fuger, "Cushing's Battery at Gettysburg," 408; Brown, *Cushing of Gettysburg*, 249. Lt. Anthony McDermott of the 69th Pennsylvania recalled that Cushing was at the wall with the men, encouraging them. He heard his last words as, "That's excellent, keep that range" (McDermott to Bachelder, June 2, 1886).

46. Landregan, "Battery A, 4th U.S.—Its Savage Work on Pickett's Column."

47. Newberry, "Cushing's Battery"; Murray, "Cowan's, Cushing's, and Rorty's Batteries in Action During the Pickett-Pettigrew-Trimble Charge," 50.

48. Gottfried, *Stopping Pickett*, 175–76; Arrington, *The Medal of Honor at Gettysburg*, 32.

49. *In Memoriam*, 67.

50. Andrew Cowan to John Bachelder, December 2, 1885.

51. *In Memoriam*, 67.

52. Cowan, "When Cowan's Battery Withstood Pickett's Charge"; *In Memoriam*, 67. After the war, Cowan tried in vain to learn the identity of the young officer so that he could return the sword to his family. He finally gave it to the Pickett's Division veteran group with the hope that they could find its rightful place.

53. Cowan to Bachelder, August 26, 1966.

54. *New York at Gettysburg*, 3:1184; Shultz, *Double Canister at Ten Yards*, 57.

55. Jim Decker to sister, July 6, 1863, Rorty's Battery folder, GNMP.

56. Cerbin, "Pettit's Battery at Gettysburg."

57. Osborn, *The Eleventh Corps Artillery at Gettysburg*, 40.

58. Ibid., 42.

59. Cole, *Civil War Artillery at Gettysburg*, 186.

60. Osborn, *The Eleventh Corps Artillery at Gettysburg*, 79–80.

61. Winschel, "The Gettysburg Diary of Lieutenant William Peel," 105.

62. *OR*, vol. 27, pt. 2, 651.

63. Winschel, "The Gettysburg Diary of Lieutenant William Peel," 105.

64. Clark, *North Carolina Regiments*, 5:125; *OR*, vol. 27, pt. 2, 651.

65. Haines, *History of the Men of Company F*, 42.

66. *OR*, vol. 27, pt. 1, 467.

67. Page, *History of the Fourteenth Regiment, Connecticut Volunteer Infantry*, 151.

68. Simmons, *A Regimental History of the One Hundred and Twenty-fifth New York State Volunteers*, 135–36.

69. Osborn, *The Eleventh Corps Artillery at Gettysburg*, 42.

70. *OR*, vol. 27, pt. 1, 893.

71. Osborn, "The Artillery at Gettysburg."

72. Hewett, *Supplement to the Official Records of the Union and Confederate Armies*, 5:415.

73. Franklin Sawyer, *A Military History of the 8th Regiment, Ohio Volunteer Infantry* (Cleveland: Fairbanks & Co., Printers, 1881) 132.

74. Gottfried, "To Fail Twice: Brockenbrough's Brigade at Gettysburg," 72, 74; *OR*, vol. 27, pt. 1, 893.

75. O'Sullivan, *Fifty-fifth Virginia Infantry*, 55; Musselman, *Forty-seventh Virginia Infantry*, 53; Hewett, *Supplement to the Official Records of the Union and Confederate Armies*, 5:415.

76. Hewett, *Supplement to the Official Records of the Union and Confederate Armies*, 5:410.

77. Aldrich, *The History of Battery A, First Regiment Rhode Island Light Artillery in the War to Preserve the Union*, 213–14.

78. Page, *History of the Fourteenth Regiment, Connecticut Volunteer Infantry*, 152.

79. Tully McCrea, Letter, March 30, 1904, Brake Collection, USAMHI.

80. Ibid.

81. Charles Morgan, report, Bachelder Papers, NHHS.

82. Ladd and Ladd, *Bachelder Papers*, 3:1977; *OR*, vol. 27, pt. 1, 480; Haskin, *History of the First Regiment of Artillery*, 169–70; Tully McCrea, Letter, March 30, 1904. Woodruff apparently died in great anguish as he was concerned that being shot in the back was an indicator of his lack of valor. He apparently asked those with him to safeguard his reputation (Ladd and Ladd, *Bachelder Papers*, 3:1977).

83. Shultz, *Double Canister at Ten Yards*, 52.

84. Gulian V. Weir, "Recollections of July 3rd," GNMP.

85. Ibid.

86. Reichardt, *Diary of Battery A, First Rhode Island*, 96.

87. Trinque, "Arnold's Battery and the 26th North Carolina," 66; Shultz, "Gulian V. Weir's 5th U.S. Artillery, Battery C," 93; Aldrich, *The History of Battery A, First Regiment Rhode Island Light Artillery in the War to Preserve the Union*, 217.

88. Weir, "Recollections of July 3rd."

89. *OR*, vol. 27, pt. 1, 880. According to David Shultz, many of the Confederates that Weir's battery mowed down were actually trying to surrender (Shultz, "Gulian V. Weir's 5th U.S. Artillery, Battery C," 94).

90. Homer Baldwin to father, July 7, 1863, GNMP.

91. Weir, "Recollections of July 3rd.

92. Ibid.

93. Osborn, "The Artillery at Gettysburg."

94. E. Porter Alexander, "The Great Charge and Artillery Fighting at Gettysburg," 3:366.

95. *New York at Gettysburg*, 3:1329.

96. Ames, *History of Battery G*, 79.

97. Shultz, *Double Canister at Ten Yards*, 59.

98. *OR*, vol. 27, pt. 2, 620.

99. Herbert, "A Short History of the 8th Alabama Regiment."

100. "James Wentworth Wrote Diary of Civil War Days," *Perry News*, n.d., 5th Florida File, GNMP.

101. *OR*, vol. 27, pt. 1, 364.

102. Raymond Reid Papers, St. Augustine (FL) Historical Society.

103. *OR*, vol. 27, pt. 2, 620.

104. *OR*, vol. 27, pt. 1, 364.

105. Ibid., 898; Callihan, "A Cool, Clear Headed Old Sailor," 50.

106. *OR*, vol. 27, pt. 2, 320.

107. Ladd and Ladd, *Bachelder Papers*, 3:490; Alexander, *Fighting for the Confederacy*, 262; Glenn Dedmondt, *Southern Bronze*, 97. There is some confusion about

the number of guns participating in the advance. Alexander noted that there were about twelve. Priest (*Into the Fight*, 96–97) believed that fourteen guns were involved. The calculations included two guns from each of the following batteries: Edward S. McCarthy's, Henry A. Battles's section of the 4th Company of the Washington Artillery, and C. H. C. Brown's section of the 1st Company of the Washington Artillery, and four guns each from C. W. Motes's and George V. Moody's batteries. The actual number may have been closer to nineteen, for Fairfax Downey noted that five additional guns came from Hugh R. Garden's and James Reilly's batteries (Mathis W. Henry's Battalion) (*The Guns at Gettysburg*, 141). Using Alexander's estimate that he advanced two out of every five guns, Paul Cooksey speculated that as many as thirty may have advanced (Cooksey, "Forcing the Issue: Brig. Gen. Henry Hunt at Gettysburg on July 3, 1863," 85).

108. Priest, *Into the Fight*, 97.

109. Rice, "Repelling Lee's Last Blow at Gettysburg," 3:389.

110. Alexander, *Military Memoirs of a Confederate*, 424. These guns included two Napoleons from Edward S. McCarthy's battery, two howitzers and two 10-pounder Parrotts from C. W. Motes's battery (Priest, *Into the Fight*, 96–97).

111. Benedict, *Army Life in Virginia*, 179–80.

112. Alexander, *Military Memoirs of a Confederate*, 429.

113. Carmichael, "Every Map of the Field Cries Out About It," 282.

114. *OR*, vol. 27, pt. 2, 321.

115. Ibid., 389.

116. Ibid., 885.

117. Scott, "On Little Round Top."

118. *OR*, vol. 27, pt. 2, 435; Hess, *Pickett's Charge*, 181.

119. Reid, "Hugh Garden's Battery"; Gottfried, "The Story of Henry's Artillery Battalion at Gettysburg," 37.

120. Reid, "Hugh Garden's Battery"; Dedmondt, *Southern Bronze*, 98–99. William McQueen survived his painful thigh wound and returned to the army. He was wounded later near Petersburg in October 1864. He returned home to South Carolina to mend, but approaching Federal troops caused him to take up arms again on Sunday, April 9, 1865, commanding a scratch force of militia and an old cannon. During the ensuing fighting, a cannonball crashed into McQueen, killing him instantly (Clemmer, *Valor in Gray*, 310).

121. *OR*, vol. 27, pt. 1, 690.

122. *OR*, vol. 27, pt. 2, 376, 381.

123. Poague, *Gunner with Stonewall*, 73.

124. Ibid., 76.

125. Hess, *Pickett's Charge*, 181–82.

126. Alexander, *Fighting for the Confederacy*, 264.

127. *OR*, vol. 27, pt. 1, 240.

128. Hunt, "The Third Day at Gettysburg," 373.

129. *OR*, vol. 27, pt. 2, 321.

130. Alexander, *Fighting for the Confederacy*, 280.

131. Longstreet, *From Manassas to Appomattox: Memoirs of the Civil War in America*, 388.

132. Carmichael, "Every Map of the Field Cries Out About It," 277–78.

Chapter 11: *July 3: Cavalry Actions and Aftermath*

1. *OR*, vol. 27, pt. 2, 267; Sears, *Gettysburg*, 259–60.

2. Trout, *Galloping Thunder*, 292–93; Ladd and Ladd, *Bachelder Papers*, 2:1252; 3:1442.

3. Gottfried, *Maps of Gettysburg*, 271; Shultz and Rollins, "The Most Accurate Fire Ever Witnessed," 74–75.

4. Longacre, *The Cavalry at Gettysburg*, 228–29.

5. Trout, *Galloping Thunder*, 294; Sears, *Gettysburg*, 461.

6. Trout, *Galloping Thunder*, 294–95; Hewett, *Supplement to the Official Records of the Union and Confederate Armies*, 5:285.

7. Shultz and Rollins, "The Most Accurate Fire Ever Witnessed," 76; *OR*, vol. 27, pt. 1, 957.

8. Wittenberg, *Gettysburg's Forgotten Cavalry Actions*, 13–14, 21; Shultz and Rollins, "The Most Accurate Fire Ever Witnessed," 71; Gottfried, *The Maps of Gettysburg*, 273.

9. Shultz and Rollins, "The Most Accurate Fire Ever Witnessed," 77.

10. Ide, "The First Vermont Cavalry in the Gettysburg Campaign," 16.

11. Wittenberg, *Gettysburg's Forgotten Cavalry Actions*, 20–22; Shultz and Rollins, "The Most Accurate Fire Ever Witnessed," 77; *OR*, vol. 27, pt. 1, 993.

12. Shultz and Rollins, "The Most Accurate Fire Ever Witnessed," 77; Wittenberg, *Gettysburg's Forgotten Cavalry Actions*, 20–22; Oates, *The War Between the Union and Confederacy*, 236; Gottfried, *The Maps of Gettysburg*, 275; Thomas McCarthy, "Battle of Gettysburg July 1st 2 & 3d 1863," Brake Collection, USAMHI.

13. Wittenberg, *Gettysburg's Forgotten Cavalry Actions*, 23, 25.

14. Ibid., 27, 34–37.

15. Alexander, *Military Memoirs of a Confederate*, 426.

16. Hunt, "The Third Day at Gettysburg," 375.

17. Evans, *Confederate Military History: South Carolina*, 595–96.

18. Elmore, "The Grand Cannonade: A Confederate Perspective," 107.

19. Scott, "The Artillery at Gettysburg"; Naisawald, *Grape and Canister*, 351; *OR*, vol. 27, pt. 1, 241.

20. *OR*, vol. 27, pt. 1, 243. Pendleton reported that his artillery lost fourteen caissons as well (*OR*, vol. 27, pt. 2, 355–56).

21. Lt. Col. Hilary P. Jones reported that he temporarily lost three guns because of incorrect ammunition, not two, as Gregory reported (*OR*, vol. 27, pt. 2, 495).

22. *OR*, vol. 27, pt. 1, 362, 757.

23. *OR*, vol. 27, pt. 2, 611.

24. Henry Hunt to Mary Hunt, July 4, 1863, Henry Jackson Hunt Papers, Library of Congress; Callihan, "A Cool, Clear Headed Old Sailor: Freeman McGilvery at Gettysburg," 51.

25. Sifakis, *Who Was Who in the Civil War*, 299, 656, 682.

26. Ibid., 480–81.

27. Ibid., 666.

28. Pfanz, *Gettysburg—The Second Day*, 482.

29. Callihan, "A Cool, Clear Headed Old Sailor: Freeman McGilvery at Gettysburg," 51.

30. Sifakis, *Who Was Who in the Civil War*, 6, 207, 498–99, 496, 511, 686; Wise, *The Long Arm of Lee*, 720; Krick, *Lee's Colonels*, 248.

31. Krick, *Lee's Colonels*, 171; Sifakis, *Who Was Who in the Civil War*, 6.

32. Krick, *Lee's Colonels*, 349; Sifakis, *Who Was Who in the Civil War*, 110, 183, 417.

33. Sifakis, *Who Was Who in the Civil War*, 237–38.

34. Gottfried, "The Story of Henry's Artillery Battalion at Gettysburg," 40; Krick, *Lee's Colonels*, 162, 333.

35. Warner, *Generals in Gray*, 69–70; Sifakis, *Who Was Who in the Civil War*, 80, 175.

BIBLIOGRAPHY

"A Letter from the Army." *Savannah (GA) Republican*, July 22, 1863.

Adams, A. J. "The Fight at the Peach Orchard." *National Tribune*, April 23, 1885.

Adelman, Garry E. "The Fight for and Location of the 4th New York Independent Battery at Gettysburg." *Gettysburg Magazine* 26 (January 2002): 53–68.

———, and Timothy H. Smith. *Devil's Den: A History and Guide*. Gettysburg: Thomas Publications, 1997.

Aldrich, Thomas M. *The History of Battery A, First Regiment Rhode Island Light Artillery in the War to Preserve the Union*. Providence, RI: Snow & Farnham, 1904.

Alexander, Edward P. "Artillery Fighting at Gettysburg." In *Battles and Leaders of the Civil War*, edited by Robert U. Johnson and Clarence C. Buel. 4 vols. New York: T. Yoseloff, 1958–59, 3:357–68.

———. "Confederate Military Service." *Southern Historical Society Papers* 11 (1886): 98–113.

———. *Fighting for the Confederacy*. Chapel Hill: University of North Carolina Press, 1989.

———. *Military Memoirs of a Confederate*. New York: Scribners, 1907.

Ames, Nelson. *History of Battery G, First Regiment, New York Light Artillery*. Marshalltown, IA: Marshall Printing Company, 1900.

Applegate, John S. *Reminiscences and Letters of George Arrowsmith of New Jersey*. Red Bank, NJ: John H. Cook, 1893.

Appleton, Nathan. *History of the Fifth Massachusetts Battery*. Boston: Luther Cowles, 1902.

Archer, John M. *The Hour Was One of Horror*. Gettysburg: Thomas Publications, 1997.

Armstrong, Richard L. *Twenty-Fifth Virginia Infantry and Ninth Virginia Infantry*. Lynchburg, VA: H. E. Howard, 1990.

Arrington, B. T. *The Medal of Honor at Gettysburg*. Gettysburg: Thomas Publications, 1996.

Bachelder, John B. "Movements of the 151st Penn. Vols. of Biddle's Brigade." In *Bachelder Papers*, edited by David L. Ladd and Audrey J. Ladd. Dayton, OH: Morningside Press, 1994–95.

Baker, Levi. *History of the Ninth Massachusetts Battery*. South Framingham, MA: Lakeview Press, 1888.

Bandy, Ken, and Florence Freeland. *The Gettysburg Papers*. Dayton, OH: Morningside Bookshop, 1978–.

Barnett, Bert. "The Batteries Fired with Very Decided Effect." In *The Gettysburg Campaign and First Day of Battle: Papers of the Tenth Gettysburg National Military Park Seminar*. Gettysburg: Gettysburg National Military Park, 2006.

———. "If Ever Men Stayed by Their Guns." In *Leadership in the Campaign and Battle of Gettysburg: Papers of the Ninth Annual Gettysburg National Military Park Seminar*. Gettysburg: Gettysburg National Military Park, 2002.

———. "The Severest and Bloodiest Artillery Fight I Ever Saw." In *The Army of Northern Virginia in the Gettysburg Campaign: Papers of the Seventh Gettysburg National Military Park Seminar*. Gettysburg: Gettysburg National Military Park, 1999.

———. "Union Artillery on July 3." In *Mr. Lincoln's Army: Papers of the Sixth Gettysburg National Military Park Seminar*. Gettysburg: Gettysburg National Military Park, 1998.

Barziza, Decimus. *The Adventures of a Prisoner of War*. Austin, TX: n.p., 1964.

Bates, Samuel P. *History of Pennsylvania Volunteers, 1861–1865*. 5 vols. Wilmington, NC: Broadfoot, 1993.

Benedict, George G. *Army Life in Virginia*. Burlington, VT: Free Press Association, 1895.

———. *Vermont in the Civil War*. Burlington, VT: Free Press Association, 1886–88.

Bennett, Brian A. *Sons of Old Monroe: A Regimental History of Patrick O'Rorke's 140th New York Volunteers*. Dayton, OH: Morningside, 1992.

Beyer, W. F., and O. F. Keydel. *Deeds of Honor*. Stamford, CT: Longmeadow Press, 1994.

Bigelow, John. *The Peach Orchard*. Minneapolis, MN: Kimball-Storer, 1910.

Blake, Henry N. *Three Years in the Army of the Potomac*. Boston: Lee and Shepard, 1865.

Bloodgood, John D. *Personal Reminiscences of the War*. New York: Hunt and Eaton, 1893.

Bohannon, Keith S. *The Giles, Alleghany and Jackson Artillery*. Lynchburg, VA: H. E. Howard, 1990.

Bowen, Edward R. "Collis' Zouaves: The 114th Pennsylvania at Gettysburg." *Philadelphia Weekly Times*, June 22, 1887.

Bradley, Thomas. "At Gettysburg." *National Tribune*, February 4, 1886.

Brady, James P. *Hurrah for the Artillery*. Gettysburg: Thomas Publications, 1992.

Brown, B. "Some Recollections of Gettysburg," *Confederate Veteran Magazine* 31 (1931): 53.

Brown, Edmund. *The Twenty-Seventh Indiana Volunteer Infantry in the War of the Rebellion, 1861 to 1865*. Monticello, IN: n.p., 1899.

Brown, Henri L. *History of the 3d Regiment, Excelsior Brigade*. Jamestown, NY: Journal Print Company, 1902.

Brown, Kent Masterson. "By Hand to the Front." *Virginia Country's Civil War* 4, no. 1, 4–15.

———. *Cushing of Gettysburg: The Story of a Union Artillery Commander*. Lexington: University of Kentucky Press, 1993.

Brown, Varina D. *A Colonel at Gettysburg and Spotsylvania*. Columbia, SC: State Company, 1931.

Browne, Edward C., Jr. "Battery H, 1st Ohio Light Artillery: Controversy in the Cemetery." *Gettysburg Magazine* 26 (January 2002): 115–28.

Buell, Augustus C. *The Cannoneer*. Washington, DC: National Tribune, 1890.

Busey, John W., and David G. Martin. *Regimental Strengths and Losses at Gettysburg*. Hightstown, NJ: Longstreet House, 1986.

Caldwell, J. *A Brigade of South Carolinians*. Marietta, GA: Continental Book Co., 1951.

Calef, John H. "Gettysburg Notes: The Opening Gun." *Journal of Military Service Institution* 40 (1907): 40–57.

Callihan, David L. "Captain Michael Wiedrich's Company I, First New York Light Artillery at Gettysburg." *Gettysburg Magazine* 33 (July 2005): 82–88.

———. "A Cool, Clear Headed Old Sailor: Freeman McGilvery at Gettysburg." *Gettysburg Magazine* 31 (July 2004): 46–51.

Campbell, Eric. "Baptism of Fire: The Ninth Massachusetts Battery at Gettysburg, July 2, 1863." *Gettysburg Magazine* 5 (July 1991): 47–78.

Carmichael, Peter S. "Every Map of the Field Cries Out About It." In *Three Days at Gettysburg*, edited by Gary W. Gallagher. Kent, OH: Kent State University Press, 1999.

———. *Lee's Young Artillerist*. Charlottesville, VA: University of Virginia, 1995.

———. *The Purcell, Crenshaw and Letcher Artillery*. Lynchburg, VA: H. E. Howard, 1990.

Carrington, James. "First Day on the Left at Gettysburg." *Southern Historical Society Papers* 37 (1909): 326–37.

———. "First Day on the Left at Gettysburg." *Richmond Times Dispatch*, February 19, 1905.

Cerbin, F. "Pettit's Batt. at Gettysburg." *National Tribune*, February 3, 1910.

Chamberlin, Thomas. *History of the One Hundred and Fiftieth Regiment, Pennsylvania Volunteers*. Philadelphia: Lippincott, 1895.

Chilton, Frank B. ed. *Unveiling and Dedication of the Monument to Hood's Texas Brigade*. Houston: F. B. Chilton, 1911.

Christ, Elwood. *The Struggle for the Bliss Farm at Gettysburg, July 2nd and 3rd, 1863*. Baltimore: Butternut and Blue Press, 1993.

Clark, George. *A Glance Backward*. Houston: Rein and Sons, 1914.

Clark, Walter, ed. *Histories of the Several Regiments and Battalions from North Carolina in the Great War, 1861–'65*. 5 vols. Raleigh: E. M. Uzzell, 1901.

Clement, Maude Carter. *The History of Pittsylvania County, Virginia*. Lynchburg, VA: J. P. Bell Co., 1929.

Clemmer, Gregg S. *Valor in Gray*. Staunton, VA: Hearthside, 1998.

Cockrell, Monroe F., ed. *Gunner with Stonewall: Reminiscences of William Thomas Poague*. Jackson, TN: McCowat-Mercer, 1957.

Coco, Gregory A. *A Concise Guide to the Artillery at Gettysburg*. Gettysburg: Thomas Publications, 1998.

———, ed. *From Ball's Bluff to Gettysburg . . . and Beyond*. Gettysburg: Thomas Publications, 1994.

Coddington, Edwin B. *The Gettysburg Campaign*. Dayton, OH: Morningside, 1979.

Cole, Philip M. *Civil War Artillery at Gettysburg: Organization, Equipment, Ammunition and Operations*. Cambridge, MA: DaCapo Press, 2002.

Collier, Calvin L. *"They'll Do to Tie To!"—The Story of the Third Regiment, Arkansas Infantry, C.S.A.* Little Rock: J. D. Warren, 1959.

Collier, John S., and Bonnie B. Collier. "Sgt. John W. Chase, Company A, 1st Massachusetts Light Artillery." *Gettysburg Magazine* 32 (January 2005): 71–73.

Colston, F. M. "Gettysburg as I Saw It." *Confederate Veteran Magazine* 5 (1897): 551–53.

Cook, John D. S. "Personal Reminiscences of Gettysburg." *Kansas MOLLUS*, 321–41.

Cooksey, Paul Clark. "Forcing the Issue: Brig. Gen. Henry Hunt at Gettysburg on July 3, 1863." *Gettysburg Magazine* 30 (January 2004): 77–88.

Cowan, Andrew. "When Cowan's Battery Withstood Pickett's Splendid Charge." *New York Herald*, July 2, 1911.

Coxe, John. "The Battle of Gettysburg." *Confederate Veteran* 21 (1913): 433–36.

Crocker, James F. *Gettysburg: Pickett's Charge and Other War Addresses*. Portsmouth, VA: W. A. Fiske, 1906.

Daniel, John W. "First Day on the Left at Gettysburg." *Richmond Times Dispatch*, February 19, 1905.

Dedmondt, Glenn. *Southern Bronze: Capt. Garden's (S.C.) Artillery Company During the War Between the States*. Columbia, SC: Palmetto Bookworks, 1993.

Dew, Charles B. *Ironmaker to the Confederacy: Joseph R. Anderson and the Tredegar Iron Works*. Wilmington, NC: Broadfoot, 1987.

Dodge, Theodore. "Left Wounded on the Field." *Putnam's Monthly Magazine* 4 (1869): 317–26.

Dow, Edwin. "How One Brave Battery Saved the Federal Left." *New York Times*, June 29, 1913.

Downey, Fairfax. *The Guns at Gettysburg*. New York: David Mackay Co., 1958.

Dyer, John P. *The Gallant Hood*. Indianapolis: Bobbs-Merrill, 1950.

Elmore, Thomas L. "The Grand Cannonade: A Confederate Perspective." *Gettysburg Magazine* 19 (July 1998): 100–111.

Evans, Clement A. *Confederate Military History: South Carolina*. Secaucus, NJ: Blue and Gray Press, n.d.

Farley, Porter. "Bloody Round Top." *National Tribune*, May 3, 1883.

Felton, Silas. "The Iron Brigade Battery." In *Giants in Their Tall Black Hats: Essays on the Iron Brigade*, edited by Alan T. Nolan and Sharon Eggleston. Bloomington: Indiana University Press, 1998.

———. "Pursuing the Elusive 'Cannoneer.'" *Gettysburg Magazine* 9 (July 1993): 33–40.

Fields, Frank E. *Twenty-Eighth Virginia Infantry*. Lynchburg, VA: H. E. Howard, 1985.

Fleet, C. R. "First Appearance of the Confederate States Flag with White Field." *Southern Historical Society Papers* 32 (1904): 240.

Foote, Frank. "Marching in Clover." *Philadelphia Weekly Times*, October 8, 1881.

Frassanito, William A. *Gettysburg: A Journey in Time*. New York: Scribner, 1975.

Freeman, Douglas S. *Lee's Lieutenants: A Study in Command*. 3 vols. New York: Scribner, 1944.

Fuger, Frederick. "Cushing's Battery at Gettysburg." *Journal of the Military Service Institution of the United States* 41 (1907): 405–10.

Galloway, Felix Richard. "Gettysburg: The Battle and the Retreat." *Confederate Veteran* 21 (1913): 388–89.

Garnett, John J. *Gettysburg*. New York: J. M. Hill, 1888.

Gibbon, John. *The Artillerists' Manual, Compiled from Various Sources, and Adapted to the Service of the United States*. 1860. Reprint, Dayton, OH: Morningside, 1991.

———. *Personal Recollections of the Civil War*. 1928. Reprint, Dayton, OH: Morningside, 1978.

Goldsborough, W. W. *The Maryland Line in the Confederate States Army*. Baltimore: Butternut and Blue, 1983.

Goolsby, J. C. "Crenshaw's Battery, Pegram's Battalion, Confederate States Artillery." *Southern Historical Society Papers* 28 (1900): 336–76.

Gottfried, Bradley M. *Brigades of Gettysburg*. New York: DaCapo, 2002.

———. "To Fail Twice: Brockenbrough's Brigade at Gettysburg." *Gettysburg Magazine* 23 (July 2000): 66–75.

———. *The Maps of Gettysburg*. New York: Savas Beatie, 2007.

———. *Roads to Gettysburg: Lee's Invasion of the North*. Shippensburg, PA: White Mane, 2001.

———. *Stopping Pickett: The History of the Philadelphia Brigade*. Shippensburg, PA: White Mane, 1999.

———. "The Story of Henry's Artillery Battalion at Gettysburg, " *Gettysburg Magazine* 34 (January 2006): 30–40.

———. "Wright's Charge on July 2, 1863." *Gettysburg Magazine* 17 (July 1997): 70–82.

Graham, Joseph. "An Awful Affair." *Civil War Times Illustrated*, April 1984, 47.

Grandchamp, Robert. "Brown's Company B, 1st Rhode Island Artillery at the Battle of Gettysburg." *Gettysburg Magazine* 36 (January 2007): 86–94.

Gregory, G. H. *Thirty-Eight Virginia Infantry*. Lynchburg, VA: H. E. Howard, 1988.

Gunn, Ralph W. *Twenty-Fourth Virginia*. Lynchburg, VA: H. E. Howard, 1987.

Haines, William P. *History of the Men of Company F*. Camden, NJ: Magrath, 1897.

Hamblen, Charles P. *Connecticut Yankees at Gettysburg*. Kent, OH: Kent State University Press, 1993.

Hankins, Samuel W. *Simple Story of a Soldier*. Nashville, TN: n.d.

Hanifen, Michael. *History of Battery B, First New Jersey Artillery*. Ottawa, IL: Republican-Times Printers, 1905.

Haskell, John Cheves. *The Haskell Memoirs*. Edited by Gilbert E. Govan and James W. Livingood. New York: Putnam, 1960.

Haskin, William Lawrence. *History of the First Regiment of Artillery*. Portland, ME: Thurston and Company, 1879.

Haynes, Martin A. *History of the Second New Hampshire: Its Camps, Marches, and Battles*. Manchester, NH: n.p., 1865.

Hazen, Samuel R. "Fighting the Good Fight: The 140th New York and Its Work on Little Round Top." *National Tribune*, September 13, 1894.

Herdegen, Lance J., and William J. Beaudot. *In the Bloody Railroad Cut of Gettysburg*. Dayton, OH: Morningside, 1990.

Hess, Earl J. *Pickett's Charge*. Chapel Hill: University of North Carolina Press, 2001.

Hewett, Janet B., et al., eds. *Supplement to the Official Records of the Union and Confederate Armies*. 100 vols. Wilmington, NC: Broadfoot, 1994–98.

"How the Battle Was Won." *Minneapolis Journal*, August 31, 1895.

Howard, Oliver O. "Campaign and Battle of Gettysburg." *Atlantic Monthly* 38 (July 1876): 48–71.

Hubler, Simon. "Just a Plain Unvarnished Story." *New York Times*, June 29, 1913.

Hunt, Henry. "The Second Day at Gettysburg." In *Battles and Leaders of the Civil War*, edited by Robert U. Johnson and Clarence C. Buel. 4 vols. New York: T. Yoseloff, 1958–59, 3:290–313.

———. "The Third Day at Gettysburg." In *Battles and Leaders of the Civil War*, edited by Robert U. Johnson and Clarence C. Buel. 4 vols. New York: T. Yoseloff, 1958–59, 3:269–385.

Ide, Horace K. "The First Vermont Cavalry in the Gettysburg Campaign." *Gettysburg Magazine* 14 (January 1996): 16.

Imhof, John D. *Gettysburg—Day Two: A Study in Maps*. Baltimore: Butternut and Blue, 1999.

Jackson, H. W. "The Battle of Gettysburg." *Illinois MOLLUS*, 1:145–84.

Johnston, David E. *Four Years a Soldier*. Princeton, WV: n.p., 1887.

Jones, Terry L. *Cemetery Hill*. New York: DaCapo Press, 2003.

Jorgensen, Jay. "Clark's Battery B, 1st New Jersey Artillery on July 2, 1863." *Gettysburg Magazine* 31 (July 2004): 39–45.

———. "Confederate Artillery at Gettysburg." *Gettysburg Magazine* 24 (January 2001): 19–37.

Kepf, Kenneth M. "Dilger's Battery at Gettysburg." *Gettysburg Magazine* 4 (January 1991): 49–64.

Kershaw, Joseph. "Kershaw's Brigade at Gettysburg." In *Battles and Leaders of the Civil War*, edited by Robert U. Johnson and Clarence C. Buel. 4 vols. New York: T. Yoseloff, 1958–59, 3:331–38.

Kiefer, W. R. *History of the One Hundred and Fifty-Third Regiment Pennsylvania Volunteers Infantry*. Easton, PA: Chemical Publishing Co., 1909.

Klein, Maury. *Edward Porter Alexander*. Athens: University of Georgia Press, 1971.

Koleszar, Marilyn B. *Ashland, Bedford, and Taylor Virginia Light Artillery*. Lynchburg, VA: H. E. Howard, 1994.

Krick, Robert. "Three Confederate Disasters on Oak Ridge." In *Three Days at Gettysburg*, edited by Gary W. Gallagher. Kent, OH: Kent State University Press, 1999.

Krick, Robert K. *The Fredericksburg Artillery*. Lynchburg, VA: H. E. Howard, 1986.

———. *Lee's Colonels: A Biographical Register of the Field Officers of the Army of Northern Virginia*. Dayton, OH: Morningside, 1984.

———. *Parker's Virginia Battery, C.S.A.* Berryville: Virginia Book Co., 1975.

Kross, Gary M. "I Do Not Believe That Pickett's Division Would Have Reached Our Line." In *Three Days at Gettysburg*, edited by Gary W. Gallagher. Kent, OH: Kent State University, 1999.

Ladd, David L., and Audrey J. Ladd. *Bachelder Papers*. Dayton, OH: Morningside Press, 1994–95.

Laboda, Lawrence R. *From Selma to Appomattox*. Shippensburg, PA: White Mane, 1994.

Landregan, Rody. "Battery A, 4th U.S.: Its Savage Work on Pickett's Column." *National Tribune*, September 2, 1909.

Lee, Alfred E. "Reminiscences of Gettysburg Battle." *Lippincott's Magazine of Popular Literature and Science*, July 1883, 54–55.

Lee, Fitzhugh. *General Lee*. Wilmington, NC: Broadfoot, 1989.

Leftwich, George J. "The Carreer of a Veteran." *Aberdeen (MS) Examiner*, August 22, 1913.

Lewis, George. *History of Battery E, First Rhode Island Light Artillery*. Providence, RI: Snow and Farnham, 1892.

Linsday, D. W. "War Services of Battery B, First N. Y. Lt. Artillery." *Baldwinsville (NY) Gazette and Farmers' Journal*, September 3, 1896.

Lokey, J. W. "Wounded at Gettysburg." *Confederate Veteran* 22 (1914): 400.

Longacre, Edward G. *The Cavalry at Gettysburg*. Lincoln, NE: Bison Books, 1986.

———. *The Man Behind the Guns: A Biography of Henry Jack Hunt, Chief of Artillery, Army of the Potomac*. South Brunswick, NJ: A. S. Barnes, 1977.

Longstreet, James. *From Manassas to Appomattox: Memoirs of the Civil War in America*. Secaucus, NJ: Blue and Grey Press, 1984.

Loring, William E. "Gettysburg." *National Tribune*, July 9, 1885.

Lynn, J. R. "At Gettysburg." *National Tribune*, October 7, 1897.

McDermott, Anthony W. *A Brief History of the 69th Regiment Pennsylvania Veteran Volunteers*. Ann Arbor, MI: University Microfilm, 1968.

McIntosh, David Gregg. "The Gettysburg Campaign." Virginia Historical Society.

———. "A Review of the Gettysburg Campaign by One Who Participated." *Southern Historical Society Papers* 37 (1909): 74–143.

McKelvey, Blake. "George Breck's Civil War Letters from Reynolds' Battery." *Rochester (NY) Historical Society Publications* 22 (1944): 91–149.

McLaws, Lafayette. "Gettysburg." *Southern Historical Society Papers* 7 (1879): 72–73.

McLean, James L., Jr. *Cutler's Brigade at Gettysburg*. Baltimore: Butternut and Blue, 1987.

McNeily, J. S. "Barksdale's Mississippi Brigade at Gettysburg." *Publications of the Mississippi Historical Society* 14 (1914): 231–65.

Maine Gettysburg Commission. *Maine at Gettysburg: Report of Maine Commissioners Prepared by the Executive Committee*. Portland, ME: Lakeside Press, 1898.

Marbaker, Thomas B. *History of the Eleventh New Jersey Volunteers*. Highstown, NJ: Longstreet House, 1990.

Marcus, Edward. *A New Canaan Private in the Civil War*. New Canaan, CT: New Canaan Historical Society, 1984.

Martin, A. P. "Little Round Top." *Gettysburg Compiler*, October 24, 1899.

Martin, David G. *Confederate Monuments at Gettysburg*. Conshohocken, PA: Combined Books, 1995.

———. *Gettysburg, July 1*. Conshohocken, PA: Combined Books, 1996.

Marvel, William. *The First New Hampshire Battery*. South Conway, NH: Lost Cemetery Press, 1985.

Marye, John L. "The First Gun at Gettysburg." *American Historical Register* 2 (1895): 1225–32.

Matthews, Richard E. *The 149th Pennsylvania Volunteer Infantry Unit in the Civil War*. Jefferson, NC: McFarland, 1994.

Minnigh, Henry N. *History of Company K, 1st Penn'a Reserves*. Duncansville, PA: "Home Print" Publisher, 1891.

Moore, Edward A. *The Story of a Cannoneer Under Stonewall Jackson*. Freeport, NY: Books for Library Press, 1971.

Moore, L. E. C. "Charge of the Louisianians." *National Tribune*, August 5, 1909.

Moore, Robert H. *The Charlottesville, Lee Lynchburg and Johnson's Bedford Artillery*. Lynchburg, VA: H. E. Howard, 1990.

———. *The Danville, Eight Star New Market and Dixie Artillery*. Lynchburg, VA: H. E. Howard, 1989.

———. *The Richmond Fayette, Hampden, Thomas, and Blount's Lynchburg Artillery*. Lynchburg, VA: H. E. Howard, 1991.

Moran, Frank E. "About Gettysburg." *National Tribune*, November 6, 1890.

———. "A Fire Zouave, Memoirs of a Member of the 73rd New York." *National Tribune*, November 6, 13, 1890.

———. "A New View of Gettysburg." *Philadelphia Weekly Times*, April 22, 1882.

Murray, R. L. "The Artillery Duel in the Peach Orchard July 2, 1863," *Gettysburg Magazine* 36 (January 2007): 72.

———. *Artillery Tactics of the Civil War*. Wolcott, NY: Benedum Books, 1998.

———. "Cowan's, Cushing's, and Rorty's Batteries in Action During the Pickett-Pettigrew-Trimble Charge." *Gettysburg Magazine* 35 (July 2006): 39–53.

———. *E. P. Alexander and the Artillery Action in the Peach Orchard*. Wolcott, NY: Benedum Books, 2000.

———. *"Hurrah for the Ould Flag!"* Wolcott, NY: Benedum Books, 1998.

Musselman, Homer D. *Forty-Seventh Virginia Infantry*. Lynchburg, VA: H.E. Howard, 1991.

Naisawald, L. VanLoan. *Grape and Canister: The Story of the Field Artillery in the Army of the Potomac*. New York: Oxford University Press, 1960.

Nash, Eugene A. *A History of the 44th New York Infantry*. Chicago: Donnelly, 1911.

Nevins, Allen, ed. *A Diary of Battle: The Personal Journal of Colonel Charles S. Wainwright*. New York: Harcourt, Brace, and World, 1962.

Newberry, M. F. "Cushing's Battery." *National Tribune*, March 23, 1911.

Newton, George W. *Silent Sentinels*. New York: Savas Beatie, 2005.

New York Monuments Commission. *Final Report of the Battlefield of Gettysburg*. 3 vols. Albany: J. B. Lyon, 1902.

New York Monuments Commission for the Battlefields of Gettysburg. *In Memoriam, Alexander Steward Webb*. Albany: J. B. Lyon, 1916.

Nicholas, Richard Ludlum, and Joseph Servis. *Powhatan, Salem and Courtney Henrico Artillery*. Lynchburg, VA: H.E. Howard, 1997.

Oates, William C. *The War Between the Union and Confederacy and Its Lost Opportunities*. New York: Neale Publishing Company, 1905.

Osborn, Hartwell. *Trials and Triumphs*. Chicago: A. C. McClurg & Co., 1904.

Osborn, Thomas W. "The Artillery at Gettysburg." *Philadelphia Weekly Times*, May 31, 1879.

———. *The Eleventh Corps Artillery at Gettysburg: The Papers of Major Thomas Ward Osborn, Chief of Artillery*. Edited by Herb S. Crumb. Hamilton, NY: Edmonston Publishing, 1991.

O'Sullivan, Richard. *Fifty-fifth Virginia Infantry*. Lynchburg, VA: H. E. Howard, 1989.

Owen, William M. *In Camp and Battle with the Washington Artillery*. Boston: Ticknor and Company, 1885.

Page, Charles D. *History of the Fourteenth Regiment, Connecticut Volunteer Infantry*. Meriden, CT: Horton Printing Company, 1906.

Parker, John L. *Henry Wilson's Regiment: History of the Twenty-Second Massachusetts Infantry*. Boston: Rand Avery Co., 1887.

Pennsylvania Gettysburg Battle-field Commission. *Pennsylvania at Gettysburg: Ceremonies at the Dedication of the Monuments Erected by the Commonwealth of Pennsylvania to Major General George G. Meade, Major General Winfield S. Hancock, Major General John F. Reynolds and to Mark the Positions of the Pennsylvania Commands Engaged in the Battle*. 2 vols. Harrisburg, PA: W. S. Ray, 1904.

Perrin, Abner. "A Little More Light on Gettysburg." *Mississippi Valley Historical Review* 24 (1938): 519–25.

Pfanz, Harry W. *Gettysburg—Culp's Hill and Cemetery Hill*. Chapel Hill: University of North Carolina Press, 1993.

———. *Gettysburg—The First Day*. Chapel Hill: University of North Carolina Press, 2001.

———. *Gettysburg—The Second Day*. Chapel Hill: University of North Carolina Press, 1987.

Piatek, Frank J. "Col. Charles Wainwright's Account of Cooper's Company B, 1st Pennsylvania Light Artillery on East Cemetery Hill: A Case of Mistaken Identity?" *Gettysburg Magazine* 36 (January 2007): 95–102.

Poague, William T. *Gunner with Stonewall*. Jackson, TN: McCowat-Mercer Press, 1957.

Polley, J. B. *Hood's Texas Brigade: Its Marches, Its Battles, Its Achievements*. New York: Neale, 1910.

Powell, William H. *The Fifth Army Corps*. London: Putnam's, 1896.

Priest, John Michael. *Into the Fight*. Shippensburg, PA: White Mane, 1998.

Purcell, Hugh Devereux. "The Nineteenth Massachusetts at Gettysburg." *Essex Institute Historical Collection* 99 (1963): 277–88.

Purifoy, John. "The Artillery at Gettysburg." *Confederate Veteran* 32 (1924): 424–27.

Reichardt, Theodore. *Diary of Battery A, First Rhode Island*. Providence, RI: N. B. Williams, 1865.

Reid, James M. "Hugh Garden's Battery: Recollections of Gettysburg." *Sumter Herald*, August 29, 1902.

Rhodes, John H. *The History of Battery B, First Rhode Island Light Artillery*. Providence, RI: Snow & Farnham, 1894.

Rice, Edmund. "Repelling Lee's Last Blow at Gettysburg." In *Battles and Leaders of the Civil War*, edited by Robert U. Johnson and Clarence C. Buel. 4 vols. New York: T. Yoseloff, 1958–59, 3:387–90.

Rittenhouse, Benjamin F. "The Battle of Gettysburg as Seen from Little Round Top." *District of Columbia MOLLUS*, 1 (1887): 35–47.

Rollins, Richard. "Lee's Artillery Preparation for Pickett's Charge." *North and South* 2 (September 1999): 41–55.

Ryder, John J. *Reminiscences of Three Years' Service in the Civil War by a Cape Cod Boy*. New Bedford, MA: Reynolds Printing, 1928.

Sawyer, Franklin. *A Military History of the 8th Regiment, Ohio Volunteer Inf'y: Its Battles, Marches and Army Movements*. Cleveland, OH: Fairbanks & Co., 1881.

Scott, James K. P. "The Artillery at Gettysburg." Gettysburg National Military Park.

Scott, Thomas. "On Little Round Top." *National Tribune*, August 2, 1894.

Sears, Stephen. *Gettysburg*. New York: Houghton Mifflin, 2003.

Seymour, William J. *The Civil War Memoirs of Captain William Seymour*. Edited by Terry L. Jones. Baton Rouge: Louisiana State University Press, 1991.

Shotwell, Randolph A. "Virginia and North Carolina in the Battle of Gettysburg." *Our Living and Our Dead* 4 (1876): 80–97.

Shultz, David. *Double Canister at Ten Yards*. Redondo Beach, CA: Rank and File Publications, 1995.

———. "Gulian V. Weir's 5th U.S. Artillery, Battery C." *Gettysburg Magazine* 18 (January 1998): 77–95.

———, and Richard Rollins. "A Combined and Concentrated Fire." *North and South* 2 (March 1999): 39–63.

———, and Richard Rollins. "The Most Accurate Fire Ever Witnessed: Federal Horse Artillery in the Pennsylvania Campaign." *Gettysburg Magazine* 33 (July 2005): 44–81.

Shumate, W. T. "With Kershaw at Gettysburg." *Philadelphia Weekly Times*, May 6, 1882.

Sifakis, Stewart. *Who Was Who in the Civil War*. New York: Facts on File, 1988.

Simmons, Ezra D. *A Regimental History of the One Hundred and Twenty-fifth New York State Volunteers*. New York: Judson Printing Company, 1888.

Small, Abner. *The Road to Richmond: The Civil War Memoirs of Major Abner Small of the Sixteenth Maine Volunteers*. Edited by Harold Adams Small. Berkeley and Los Angeles: University of California Press, 1939.

Smith, Abram P. *History of the Seventy-Sixth Regiment, New York Volunteers*. Cortland, NY: Truair, Smith and Miles, 1867.

Smith, Charles N. "Account." *New York Times*, July 3, 1913.

Smith, Christopher. "The Bloody Angle." *Buffalo (NY) Evening News*, May 29, 1894.

Smith, Gerald J. *One of the Most Daring of Men*. Murfreesboro, TN: Southern Heritage Press, 1997.

Smith, James E. *A Famous Battery and Its Campaigns*. Washington, DC: Lowdermilk and Company, 1892.

Stewart, George R. *Pickett's Charge: A Microhistory of the Final Attack at Gettysburg*. Boston: Houghton Mifflin, 1959.

Stewart, James. "Battery B Fourth United States Artillery at Gettysburg." In *The Gettysburg Papers*, edited by Ken Bandy and Florence Freeland. Dayton, OH; Morningside, 1978–, 1:364–77.

Stiles, Robert. *Four Years Under Marse Robert*. New York: Neale Publishing Company, 1903.

Storrs, John. *The Twentieth Connecticut*. Ansonia, CT: Naugatuck Valley Sentinel, 1886.

Survivors of the Seventy-second Regiment of Pennsylvania Volunteers, Planintiffs. vs. Gettysburg Battlefield Memorial Association . . . Supreme Court of Pennsylvania, Middle District, May Term, 1891, nos. 20 and 30.

Taylor, Eugene G. *Gouveneur Kemple Warren: The Life and Letters of an American Soldier*. Boston: Houghton-Mifflin, 1932.

"Terrific Fight of Third Day." *Scranton (PA) Truth*, July 3, 1913.

Tervis, C. V. *History of the Fighting Fourteenth*. Brooklyn, NY: Brooklyn Eagle Press, 1911.

Thomas, Dean S. *Cannons: An Introduction to Civil War Artillery*. Arendtsville, PA: Thomas Publications, 1985.

Time-Life Books, eds. *Arms and Equipment of the Confederacy*. Alexandria, VA: Time-Life Books, 1996.

Tomasak, Pete. "An Encounter with Battery Hell." *Gettysburg Magazine* 12 (January 1995): 30–41.

Toombs, Samuel. *New Jersey Troops in the Gettysburg Campaign*. Orange, NJ: Evening Mail Publishing House, 1888.

Trinque, Bruce A. "Arnold's Battery and the 26th North Carolina." *Gettysburg Magazine* 12 (January 1995): 61–67.

Trout, Robert J. *Galloping Thunder*. Mechanicsburg, PA: Stackpole, 2002.

Tucker, A. W. "Orange Blossoms: Services of the 124th New York at Gettysburg." *National Tribune*, January 21, 1886.

Turner, George A. "Ricketts' Role Remembered at Gettysburg Anniversary." *Enterprise*, July 3, 1992.

Turney, J. B. "The First Tennessee at Gettysburg." *Confederate Veteran* 8 (1900): 535–37.

Underwood, Adin B. *The Three Years' Service of the Thirty-Third Massachusetts Infantry Regiment*. Boston: A. Williams, 1881.

U.S. War Department. *The War of the Rebellion: A Compilation of the Official Records of the Union and Confederate Armies*. 128 vols. Washington, DC: U.S. Government Printing Office, 1880–1901.

Waitt, Ernest L. *History of the Nineteenth Regiment, Massachusetts Volunteer Infantry*. Salem, MA: Salem Press, 1906.

Walker, Francis A. "General Hancock and the Artillery at Gettysburg." In *Battles and Leaders of the Civil War*, edited by Robert U. Johnson and Clarence C. Buel. 4 vols. New York: T. Yoseloff, 1958–59, 3:385–87.

Walker, James H. "The Charge of Pickett's Division." *Blue and Gray* 1 (1893): 222–23.

Ward, Fanny B. *Sketch of Battery "I" First Ohio Artillery*. n.p., n.d.

Warner, Ezra J. *Generals in Gray: Lives of the Confederate Commanders*. Baton Rouge: Louisiana State University Press, 1959.

Washburn, George H. *A Complete Military History and Record of the 108th Regiment N. Y. Vols. from 1862 to 1894*. Rochester, NY: E. R. Andrews, 1894.

Wert, Jeffry D. *Gettysburg, Day Three*. New York: Simon and Schuster, 2001.

Weygant, Charles H. *History of the One Hundred and Twenty-Fourth Regiment*. Newburgh, NY: Journal Printing House, 1877.

Wheeler, Richard. *Witness to Gettysburg*. New York: Harper & Row, 1987.

Wheeler, William. *Letters of William Wheeler of the Class of 1855, Yale College*. Cambridge, MA: Houghton, 1875.

White, William S. *Contributions to a History of the Richmond Howitzer Battalion*. Richmond: C. McCarthy, 1883–86.

Whittier, Edward N. "The Left Attack (Ewell's) at Gettysburg." *Massachusetts MOLLUS*, 3:86–87.

Wilcox, Cadmus M. "Letter from General C. M. Wilcox, 26 March 1877." In "Causes of the Confederate Defeat at Gettysburg." *Southern Historical Society Papers* 4 (1877): 111–17.

Wilkeson, Samuel. "A Tragedy at Gettysburg." *Confederate Veteran* 21 (1913): 387–88.

Williams, Alpheus S. *From the Cannon's Mouth*. Detroit: University of Detroit Press, 1959.

Winschel, Terrence J. "The Gettysburg Diary of Lieutenant William Peel." *Gettysburg Magazine* 9 (July 1993): 98–108.

Winslow, George. "Ón Little Round Top." *National Tribune*, July 26, 1879.
Wise, Jennings C. *The Long Arm of Lee: The History of the Army of Northern Virginia*. New York: Oxford University Press, 1959.
Wittenberg, Eric J. *Gettysburg's Forgotten Cavalry Actions*. Gettysburg: Thomas Publications, 1998.
Wood, W. Nathaniel. *Reminiscences of Big I*. Jackson, TN: McCowat-Mercer Press, 1956.
Woods, James A. "Defending Watson's Battery," *Gettysburg Magazine* 9 (July 1993): 41–47.

INDEX